NUCLEAR RECEPTORS IN DRUG METABOLISM

NUCLEAR RECEPTORS IN DRUG METABOLISM

Edited by

WEN XIE
University of Pittsburgh

WILEY

A John Wiley & Sons, Inc., Publication

Library of Congress Cataloging-in-Publication Data:

Nuclear receptors in drug metabolism / [edited by] Wen Xie.
   p. ; cm.
  Includes bibliographical references and index.
  ISBN 978-0-470-08679-7 (cloth)
  1. Drugs–Metabolism. 2. Nuclear receptors (Biochemistry) 3. Genetic regulation.
I. Xie, Wen, 1967–
  [DNLM: 1. Receptors, Cytoplasmic and Nuclear–metabolism. 2. Drug Delivery
Systems–methods. 3. Pharmaceutical Preparations–metabolism. 4. Receptors,
Steroid–metabolism. QV 38 N964 2008]
  RM301.55.N83 2008
  615′.7–dc22

                              2008029786

Printed in the United States of America

10 9 8 7 6 5 4 3 2 1

CONTENTS

PREFACE

I am happy to present you with this special-topic book *Nuclear Receptors in Drug Metabolism.*

It has been widely appreciated that drug-induced changes in the expression and/or activity of drug-metabolizing enzymes and transporters can affect the absorption and elimination of drugs, thereby altering the therapeutic or toxicological responses to a drug. The molecular mechanisms by which drugs regulate the expression of drug-metabolizing enzymes and transporters have been elusive despite the cloning in 1991 of aryl hydrocarbon receptor (AhR), the first "xenobiotic receptor" and a PAS domain transcriptional factor.

A major breakthrough was achieved in 1998 when two groups, led by Steven Kliewer (then at Glaxo Wellcome) and Ronald Evans (at the Salk Institute), first reported the cloning and characterization of the xenobiotic nuclear receptor pregnane X receptor (PXR). The xenobiotic receptor identity for the constitutive androstane receptor (CAR), a receptor cloned by David Moore's lab, was subsequently revealed. This effort was spearheaded by the laboratories of Masahiko Negishi at the National Institute of Environmental Health Sciences and David Moore at the Baylor College of Medicine. Since 1998, combinations of molecular biology, mouse genetics, structural biology, and drug metabolism studies have led to the conclusion that PXR and CAR can function as master regulators of the xenobiotic responses by regulating phase I and phase II enzymes as well as drug transporters. It has also become evident that the nuclear receptor mediated xenobiotic regulation represents a complex regulatory network. The complexity is manifested by the observations that multiple receptors are involved in the regulation; each receptor is capable of regulating multiple xenobiotic targets; there is extensive cross talk between receptors; and receptors exhibit a distinctive, yet overlapping, spectrum of ligands. Finally, the receptor-mediated regulation of enzymes and transporters only impacts drug metabolism, but can also influence many pathophysiological conditions by affecting the homeostasis of endogenous substances. This book is intended to offer a comprehensive review of the historical discoveries and current development of nuclear hormone receptors as transcriptional regulators of drug-metabolizing enzymes and transporters. The implications of this regulation in drug metabolism, drug development, clinical drug use, therapeutic potentials, and pharmacogenetics are particularly emphasized.

I want to thank all the contributing authors, who are experts in the forefront of this emerging and exciting field of research. Special thanks is extended to Mr Jonathan T. Rose, Editor at John Wiley & Sons, Inc., who has been inspirational in proposing the book topic and extremely helpful in all stages of the development of this book.

Wen Xie, MD, PhD
Associate Professor
Center for Pharmacogenetics
University of Pittsburgh

ABBREVIATIONS

2-AAF	2-Acetylaminofluorene
ABC	ATP binding cassette
ACC	Acetyl CoA carboxylase
ACTH	Adrenocorticotropic hormone
ADME	Absorption, distribution, metabolism, and excretion
AF-2	Activation function 2 domain
AhR	Aryl hydrocarbon receptor
AKR	Aldo-ketoreductase
Alb	Albumin
ALT	Alanine aminotransferase
AO/XO	Aldehyde oxidase/xanthine oxidase (molybdenum hydroxylases)
APAP	Acetaminophen
ARE	Antioxidant response element
ASBT	Apical sodium-dependent bile acid transporter
ASC-2	Activating signal cointegrator-2
ATP	Adenosine triphosphate
AUC	Area under curve
AZT	Zidovudine
BBB	Blood–brain barrier
BCRP	Breast cancer resistance protein
BDL	Bile duct ligation
BHA	Butylated hydroxyanisole
tBHQ	*tert*-Butyl-hydroquinone
BR	Bilirubin
BSEP	Bile salt export pump
CAR	Constitutive androstane receptor

CARM-1   Coactivator-associated arginine methyltransferase 1
CAT   Catalase
CBZ   Carbamazepine
CCRP   Cytoplasmic CAR retention protein
CDCA   Chenodeoxycholic acid
CGD   Cholesterol gallstone disease
ChIP   Chromatin immunoprecipitation
CITCO   6-(4-Chlorophenyl)imidazo[2,1-b]thiazole-5-carbaldehyde
    O-(3,4-dichlorobenzyl)oxime
CN   Crigler–Najar (syndrome)
COMT   Catechol-O-methyltransferase
CPT1   Carnitine palmitoyltransferase 1
CREB   cAMP response element binding protein
CSI   Cholesterol saturation index
CTX   Cerebrotendinous xanthomatosis
CYP   Cytochrome P450 (CYP450)
Cyp7a1   Cholesterol 7α-hydroxylase
Cyp7b1   Oxysterol 7α-hydroxylase
D3T   3H-1,2-dithiol-3-thione
DAS   Diallyl sulfide
DBD   DNA binding domain
DDI   Drug–drug interaction
DEX   Dexamethasone
DHEA   Dehydroepiandrosterone
DME   Drug-metabolizing enzyme
DMPK   Drug metabolism and pharmacokinetics
DMSO   Dimethyl sulfoxide
Dox   Doxycycline
DRE   Dioxin response element
DTg   Double transgenic
ECI   Enoyl-CoA isomerase
EH   Epoxide hydrolase
EM   Extensive metabolizer
EMSA   Electrophoretic mobility shift assay
EPHX   Epoxide hydrolase
EpRE   Electrophilic response element
ER   Estrogen receptor
EST   Estrogen sulfotransferase
FABP   Fatty acid binding protein
FAE   Long chain free fatty acid elongase
FAS   Fatty acid synthase
FFA   Free fatty acid
FMO   Flavin mono-oxygenase
FoxO1   Forkhead transcription factor O1
FXR   Farnesoid X receptor

G6Pase	Glucose-6-phosphatase
GR	Glucocorticoid receptor
GRIP-1	Glucocorticoid receptor-interacting protein 1
GSH	Glutathione
GST	Glutathione S-transferase
3β-Hsd	3β-Hydroxysteroid dehydrogenase
HAH	Halogenated aromatic hydrocarbon
HAT	Histone acetyltransferase
HDAC	Histone deacetylase
HDL	High density lipoprotein
HETE	Hydroxyeicosatetraenoic acid
HMGCS2	3-Hydroxy-3-methylglutaryl Co-A synthase 2
HNE	4-Hydroxynonenal
HNF	Hepatocyte nuclear factor
HPA	Hypothalamus–pituitary–adrenal
IND	Investigative new drug
IR	Inverted repeat
LBD	Ligand binding domain
LCA	Lithocholic acid
LDL	Low density lipoprotein
LDLR	Low density lipoprotein receptor
LRH-1	Liver-related homologue-1
LXR	Liver X receptor
LXRE	LXR responsive element
MAO	Monoamine oxidase
MBI	Metabolism-based inactivation
MCD	Methionine and choline deficient
MDR	Multidrug resistance protein
MPTP	1-Methyl-4-phenyl-1,2,3,6-tetrahydropyridine
MODY	Maturity onset diabetes of the young
MRP	Multidrug resistance related protein
MT	Methyltransferase
Mtp	Microsomal triglyceride transfer protein
NADPH	Nicotinamide adenine dinucleotide phosphate
NASH	Nonalcoholic steatohepatitis
NAT	N-acetyltransferase
NCE	New chemical entity
NCoR	Nuclear receptor corepressor
NF-κB	Nuclear factor κB
NLS	Nuclear localization signal
Npc1l1	Niemann-Pick C1 like 1
NQO1	NAD(P)H:quinone oxidoreductase 1
NR	Nuclear receptor
Nrf2	Nuclear factor erythroid 2-related factor 2
Ntcp	Na^+-taurocholate cotransport proteins

OAT	Organic anion transporter
OATP	Organic anion transporter polypeptide
OCT	Organic cation transporter
OCTN	Organic cation transporters novel
Oct-1	Octamer transcription factor 1
Ostα	Organic solute transporter α
Ostβ	Organic solute transporter β
PAH	Polycyclic aromatic hydrocarbon
PAH	Planar aromatic hydrocarbon
PAPS	3′-Phosphoadenosine 5′-phosphosulfate
PAPSS2	3′-Phosphoadenosine 5′-phosphosulfate synthetase 2
PB	Phenobarbital
PBC	Primary biliary cirrhosis
PBP	PPAR binding protein
PBMC	Peripheral blood mononuclear cells
PBREM	Phenobarbital-responsive enhancer module
PCN	Pregnenolone-16α-carbonitrile
PEPCK	Phosphoenoylpyruvate carboxylase
PEPT	Peptide (co)transporter
P-gp	P-glycoprotein
PHZ	Phenylhydrazine
PM	Poor metabolizer
PMA	Phorbol 12-myristate
PMRT1/5	Protein arginine methyltransferases
PPA2	Protein phosphatase A2
PPAR	Peroxisome proliferator-activated receptor
PRIP	PPAR-interacting protein
PRMTs	Protein arginine methyltransferases
PXR	Pregnane X receptor
PK	Pharmacokinetics
PD	Pharmacodynamics
RAC3	Receptor-associated coactivator 3
RAR	Retinoid acid receptor
RID	Receptor interacting domain
RIF	Rifampicin
ROR	Retinoid-related orphan receptor
RORE	ROR responsive element
RS domain	Serine/arginine domain
RXR	Retinoid X receptor
SAM	S-adenosyl-L-methionine
SCD-1	Stearoyl CoA desaturase-1
SHP	Small heterodimer partner
SJW	St. John's wort
SLC	Solute carrier family
SOD	Superoxide dismutase

SXR	Steroid and xenobiotic receptor
SMRT	Silencing mediator of retinoic acid and thyroid hormone receptor
SNP	Single nucleotide polymorphism
SRC-1	Steroid receptor coactivator 1
SREBP	Sterol regulatory element binding protein
SULT	Sulfotransferase
SV 40	Simian virus 40
tBHQ	*tert*-Butyl-hydroquinone
TCPOBOP	1,4-Bis[2-(3,5 dichloropyridyloxy)] benzene
TCDD	2,3,7,8-Tetrachlorodibenzo-p-dioxin
TEA	Tetraethylammonium
Tg	Transgenic
THR	Thyroid hormone receptor
TI	Therapeutic index
TMT	Thiol methyltransferase
TPMT	Thiopurine methyltransferase
TRAP 220	Thyroid hormone receptor associated protein 220
TSH	Thyroid-stimulating hormone
TSO	*trans*-Stilbene oxide
tTA	Tetracycline-responsive transcriptional activator
UDCA	Ursodeoxycholic acid
UDP	Uridine diphosphate
UDPGA	UDP-glucuronic acid
UGT	UDP-glucuronosyltransferase
VD_{ss}	Volume of distribution
VDR	Vitamin D receptor
VLDL	Very low density lipoprotein
VP	Viral protein
WT	Wild type
XME	Xenobiotic-metabolizing enzyme
XO	Xanthine oxidase
XRE	Xenobiotic response elements
XRS	Xenobiochemical response signal

CONTRIBUTORS

Alfin D. N. Vaz, Department of Pharmacokinetics, Dynamics and Metabolism, Pfizer Global Research and Development, Pfizer Inc., Groton, CT, USA

Chandra Prakash, Department of Pharmacokinetics, Dynamics and Metabolism, Pfizer Global Research and Development, Pfizer Inc., Groton, CT, USA

Erin G. Schuetz, Department of Pharmaceutical Sciences, St. Jude Children's Research Hospital, Memphis, TN, USA

Gernot Zollner, Division of Gastroenterology and Hepatology, Medical University of Graz, Graz, Austria

H. Eric Xu, Laboratory of Structural Sciences, Van Andel Research Institute, Grand Rapids, MI, USA

Haibiao Gong, Department of Biotechnology, LI-COR Biosciences, Lincoln, NE, USA

Jatinder Lamba, Department of Pharmaceutical Sciences, St. Jude Children's Research Hospital, Memphis, TN, USA

John Y. L. Chiang, Department of Microbiology, Immunology and Biochemistry, Northeastern Ohio Universities College of Medicine, Rootstown, OH, USA

Junichiro Sonoda, Gene Expression Laboratory, Howard Hughes Medical Institute, The Salk Institute for Biological Studies, La Jolla, CA, USA

Martin Wagner, Division of Gastroenterology and Hepatology, Medical University of Graz, Graz, Austria

Michael Trauner, Division of Gastroenterology and Hepatology, Medical University of Graz, Graz, Austria

Olivier Barbier, Faculty of Pharmacy, Laval University and Research Center of the "Centre Hospitalier de l'Université Laval—CRCHUL", Québec, Canada

Oliver Burk, Department of Pharmacogenomics, Dr Margarete Fischer-Bosch Institute of Clinical Pharmacology, Stuttgart, Germany

Richard B. Kim, Division of Clinical Pharmacology, Department of Medicine and Department of Physiology and Pharmacology, Schulich School of Medicine and Dentistry, The University of Western Ontario, Ontario, Canada

Ronald M. Evans, Gene Expression Laboratory, Howard Hughes Medical Institute, The Salk Institute for Biological Studies, La Jolla, CA, USA

Rommel G. Tirona, Division of Clinical Pharmacology, Department of Medicine and Department of Physiology and Pharmacology, Schulich School of Medicine and Dentistry, The University of Western Ontario, Ontario, Canada

Taira Wada, Center for Pharmacogenetics, Department of Pharmaceutical Sciences, University of Pittsburgh, Pittsburgh, PA, USA

Tao Li, Gene Expression Laboratory, Howard Hughes Medical Institute, The Salk Institute for Biological Studies, La Jolla, CA, USA

Wen Xie, Center for Pharmacogenetics, Department of Pharmaceutical Sciences, University of Pittsburgh, Pittsburgh, PA, USA

X. Edward Zhou, Laboratory of Structural Sciences, Van Andel Research Institute, Grand Rapids, MI, USA

1

DRUG METABOLISM: SIGNIFICANCE AND CHALLENGES

CHANDRA PRAKASH AND ALFIN D. N. VAZ

Department of Pharmacokinetics, Dynamics and Metabolism, Pfizer Global Research and Development, Pfizer Inc., Groton, CT, USA

1.1 INTRODUCTION

Searching for new drugs is a very time-consuming and expensive endeavor, taking approximately 10–12 years and on the order of $900 million to bring a new drug to market [1, 2]. It has been estimated that for every 5000 new chemical entities (NCEs) evaluated in a discovery program, only 1 is approved for market [3]. Even after a drug is marketed, there is the possibility of some undesired side effects, which were not seen in earlier clinical trials. In these cases, the drug is either withdrawn from the market or acquires a warning label (black box). Therefore, efforts are being made to reduce attrition of drug candidates during the various stages of their development to bring safer compounds to market. The major reasons for the failure of the NCEs are lack of *in vivo* efficacy, serious undesired side effects, and unfavorable drug metabolism and pharmacokinetics (DMPK). Therefore, in addition to potency and selectivity, drug candidates are selected on the basis of DMPK properties, such as desired clearance, oral bioavailability, low potential of drug–drug interactions, and acceptable metabolism/toxicology profiles in preclinical species [4, 5]. In support of this need, and as a consequence of increased knowledge within the drug metabolism discipline, new approaches have been developed that include extensive *in vitro* methods using human and animal hepatic cellular and subcellular systems, recombinant human drug-metabolizing enzymes, transgenic animals and cell lines stably expressing human transporters, increased automation for higher throughput screens, sensitive analytical

Nuclear Receptors in Drug Metabolism   Edited by Wen Xie
Copyright © 2009 John Wiley & Sons, Inc.

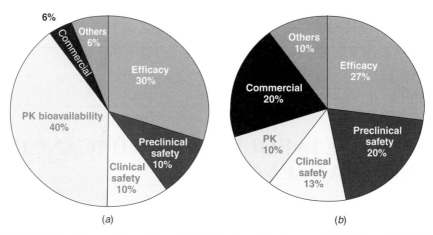

FIGURE 1.1   Reasons for attrition of NCEs in drug development: (*a*) 1990 and (*b*) 2000. (Modified from Ref. 6)

technologies, and *in silico* computational models to assess drug metabolism aspects of the NCEs. Recent data suggest that these approaches have reduced the attrition due to DMPK issues from 40% of total attrition in 1990 to 10% in 2000 (Figure 1.1) [6].

The predictive power of *in vitro* studies using animal and human hepatocellular and subcellular fractions and/or recombinant enzymes [7–9], *in silico* models [10–12], and *in vivo* studies using a range of experimental animal models [13, 14] has advanced considerably due to an ever increasing understanding of the relationships between *in vitro* and *in vivo* drug metabolism and disposition. These *in vitro* studies include

1. absorption/transport studies in Caco-2 cells or cell lines over expressing various transporters;
2. metabolic stability and metabolite formation in liver microsomes, S-9, hepatocytes or recombinant cytochrome P450 enzymes;
3. cytochrome P450 inhibition and induction;
4. plasma protein binding;
5. reactive metabolite (glutathione adduct formation).

The *in vivo* studies include

1. pharmacokinetic studies in laboratory animals via various routes of adminis-tration (oral, intravenous, subcutaneous, etc.);
2. tissue distribution (e.g., brain penetration);
3. metabolite identification and clearance pathways in animals and humans using radiotracers;
4. PK/PD relationship;
5. specific studies using genetically engineered mouse models.

Among ADME (absorption, distribution, metabolism, and excretion) properties, metabolism of an NCE by the host system can be one of the most important determinants of its pharmacokinetic disposition. It is the biochemical process by which compounds are converted to more hydrophilic (water-soluble) entities, which not only enhance their elimination from the body but also lead to compounds that are generally pharmacologically inactive and relatively nontoxic. However, metabolic transformation of an NCE at times can lead to the formation of metabolites with pharmacological activity [15] and/or toxicity [16–18]. Additionally, metabolism can be the main cause of poor bioavailability and drug–drug interactions via inhibition or induction of drug-metabolizing enzymes [19, 20]. Therefore, determination of metabolic rate and biotransformation pathways of an NCE, in animals and humans, and evaluation of pharmacological and toxicological consequences of its metabolites are very critical to pharmaceutical development [21]. At the lead optimization stage, information on the metabolic fate of the NCEs can direct medicinal chemists to synthesize metabolically more stable analogs by blocking sites of metabolism, and potentially creating NCEs with superior pharmacology and safety profiles. Knowledge of the major human metabolites of an NCE early in its development is useful to enable the judicious selection of animal species used for safety evaluation, to ensure that the selected animal species are exposed to all major metabolites formed in humans [22, 23]. Subsequently, major circulatory metabolites in humans can be synthesized for the evaluation of their pharmacological activity.

The metabolism of drugs has traditionally been classified into two reaction classes, phase I and phase II. Phase I reactions include hydroxylation, dealkylation, deamination, N- or S-oxidation, reduction, and hydrolysis. These reactions introduce or unmask a functional group (e.g., –OH, –COOH, –NH_2, or –SH) within a molecule to enhance its hydrophilicity. Phase II or conjugation biotransformations include glucuronidation, sulfation, methylation, acetylation, and amino acid (glycine, glutamic acid, and taurine) and glutathione (GSH) conjugation. The cofactors of these reactions react with functional groups that are either present on the NCE or are introduced during phase I biotransformation. Most phase II biotransformation reactions result in a large increase in drug hydrophilicity, thus greatly promoting the excretion of foreign chemicals via urine and/or bile, and, with few exceptions, are generally pharmacologically inactive. While phase I and phase II reactions are thought of as acting sequentially in the biotransformation of drugs, NCEs, and xenobiotics, these reactions occur independently, and often phase II enzymes become the primary metabolic route.

1.2   PHASE I DRUG-METABOLIZING ENZYMES

The phase I reactions are mediated primarily by liver enzymes such as cytochrome P450 (CYP450), FAD-containing mono-oxygenase (FMO), monoamine oxidase (MAO), molybdenum hydroxylase (aldehyde oxidase/xanthine oxidase; AO/XO), aldo-ketoreductase (AKR), epoxide hydrolase (EH), and esterase.

1.2.1   Cytochrome P450

CYP450 is a superfamily of hemoproteins, responsible for the oxidative metabolism as well as metabolic activation of the vast majority of xenobiotics (drugs, dietary components, and pollutants) and endogenous substrates (e.g., steroids, cholesterol, and bile acids). The CYP450 system possesses three known types of activities. CYP450 enzymes, acting as mono-oxygenases, activate molecular oxygen with electrons from NADPH via NADPH-CYP450 reductase, and insert one atom of molecular oxygen into the substrate while reducing the other atom of oxygen to water (equation 1.1). As a result, the xenobiotics can undergo hydroxylations, epoxidations, N-, S-, or O-dealkylations, deaminations, N- or S-oxidations, and oxidative dehalogenations:

$$RH + O_2 + NADPH + H^+ \rightarrow ROH + H_2O + NAD(P)^+ \qquad (1.1)$$

The second activity commonly referred to as the oxidase activity of CYP450 involves electron transfer from reduced CYP450 to molecular oxygen with the formation of superoxide anion radical and H_2O_2 (equations 1.2a and 1.2b):

$$NADPH + O_2 \rightarrow O_2^- + NAD(P)^+ \qquad (1.2a)$$

$$2NADPH + 2H^+ + O_2 \rightarrow H_2O_2 + NAD(P)^+ \qquad (1.2b)$$

The third activity of P450 system, known as reductase activity, involves direct electron transfer to reducible substrates such as quinones and proceeds readily under anaerobic conditions.

CYP450 superfamily members are grouped into subfamilies on the basis of amino acid sequence homology. The drug-metabolizing CYP450 enzymes are confined to subfamilies 1, 2, 3, and 4. These subfamilies are further divided into isoforms. There are approximately 57 human CYP450 isoforms exhibiting major differences with respect to their catalytic specificity and patterns of tissue expression [24, 25]. Only a small group of these isoforms is involved in xenobiotic transformations. In the human liver, there are at least 18 distinct CYP450 isozymes while only 10 isoforms from families 1, 2, and 3 (CYP1A2, CYP2A6, CYP2B6, CYP2C8, CYP2C9,

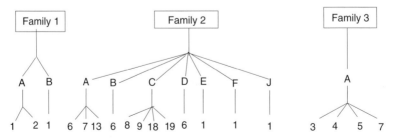

FIGURE 1.2   Human cytochrome P450 genes in the drug-metabolizing CYP1, CYP2, and CYP3 families (Modified from Ref. 24).

CYP2C19, CYP2D6, CYP2E1, CYP2F1, and CYP3A4) are responsible for the hepatic metabolism of most of the marketed drugs (Figure 1.2) [24].

CYP1A subfamily. CYP1A subfamily consists of two members CYP1A1 and 1A2. CYP1A1 is present predominantly in extrahepatic tissue such as lung, small intestine, placenta and kidney, and at a very low level in the liver. On the other hand, CYP1A2 is mainly confined to the liver and expressed at a very low level in extrahepatic tissue. Both CYP1A1 and 1A2 play an important role in the metabolic activation of polycyclic aromatic hydrocarbons (PAHs), aromatic amines, and heterocyclic amines [26]. Despite the high similarity in their primary structures, CYP1A1 and 1A2 have different substrate specificities. For example, dibenzo[a]pyrene, a potent carcinogen, is oxidized almost exclusively by CYP1A1 in humans to highly mutagenic diol-epoxides [27] while acetanilide is metabolized primarily by CYP1A2. In human liver, CYP1A2 accounts for ∼15% of the total CYP content and is involved in the metabolism of ∼4% of marketed drug including acetaminophen, phenacetin, tacrine, ropinirole, riluzole, theophylline, and caffeine [25, 28].

CYP1B subfamily. CYP1B1 is expressed at a low level in heart, brain, placenta, lung, liver, and kidney [29], but it is present at much higher levels in tumor cells compared with the surrounding normal tissue [30]. CYP1B1 catalyzes 2- and 4-hydroxylation of 17β-estradiol [31]. It also participates in the metabolic activation of a number of procarcinogens, including PAHs, PAH-dihydrodiols, and aromatic amines [32, 33]. CYP1B1 is also involved in the metabolism of some clinically relevant anticancer agents used in the treatment of hormone-mediated cancer.

CYP2A subfamily. CYP2A subfamily includes CYP2A6, 2A7, and 2A13 in humans (Figure 1.2). CYP2A6 is expressed in the liver and accounts for ∼4% of total hepatic CYP450, whereas 2A7 and 2A13 appear to be expressed at even lower levels. CYP2A6 is the principal and perhaps the sole catalyst for human liver microsomal coumarin 7-hydroxylation. CYP2A6 is involved in the oxidation of only 3% of drugs such as nicotine, cyclophosphamide, ifosfamide, and fadrozole [25, 34, 35]. In addition, CYP2A6 also plays an important role in the activation of several procarcinogens and promutagens especially the nitrosamines [36]. CYP2A6 might also be involved in the metabolic activation of 3-methylindole in human pulmonary microsomes [37]. CYP2A13 is predominantly expressed in the human respiratory tract and significantly involved in the activation of aflatoxin B_1 [38].

CYP2B subfamily. CYP2B subfamily includes CYP2B6 and 2B7 isoforms. CYP2B6 is expressed in the liver and in some extrahepatic tissues, whereas CYP2B7 mRNA expression was detected in lung tissue. CYP2B6 accounts for ∼1% of total hepatic CYP content. CYP2B6 is involved in the metabolism of only 4% of the marketed drugs, such as the anticancer drugs cyclophosphamide and tamoxifen [39], the anesthetics ketamine and propofol [40], and procarcinogens, such as the environmental contaminants aflatoxin B1 and dibenzanthracene [41]. Significant interindividual differences from 25- to 250-fold have been reported in hepatic CYP2B6 expression [42]. Recent studies have reported that females express significantly higher amounts of CYP2B6 than males [43].

CYP2C subfamily. To date, four members of the human CYP2C subfamily (CYP2C8, 2C9, 2C18, and 2C19) have been identified and all together account for ~25% of the total CYP in human liver [25]. CYP2C8 and 2C9 are the major isoforms, accounting for 35% and 60%, respectively, of total human CYP2C, whereas CYP2C18 (4%) and CYP2C19 (1%) are the minor expressed CYP2C isoforms [44]. In addition to liver, CYP2C9 mRNA is also detected in the kidney, testes, adrenal gland, prostate, ovary, and duodenum, and 2C19 is detected in duodenum. CYP2C subfamily is involved in the metabolism of ~25% of drugs on the market. CYP2C8 is involved in the metabolism of retinol and retinoic acid, arachidonic acid, benzo[a]pyrene, and anticancer drug paclitaxel. CYP2C9 plays a critical role in the metabolism of a number of clinically significant drugs, including tolbutamide, phenytoin, S-warfarin, ibuprofen, diclofenac, piroxicam, tenoxicam, mefenamic acid, losartan, glipizide, and torasemide. CYP2C19 metabolizes (S)-mephenytoin, omeprazole, imipramine diazepam, some barbiturates, and proguanil.

CYP2D subfamily. In humans, only one isoform, CYP2D6, is expressed in various tissues including the liver, kidney, placenta, brain, breast, lungs, and small intestine. It is expressed at a low level in human liver, accounting for only ~3% of total CYP protein. In the small intestine, CYP2D6 is expressed in the duodenum and jejunum and not in the ileum and colon. CYP2D6 is responsible for the metabolism of numerous therapeutically used drugs [25, 45]. Approximately 7–10% of the Caucasian population shows an inherited deficiency in this enzyme due to the presence of one or several mutant alleles at the CYP2D6 gene locus [46]. These subjects are characterized by the poor metabolizer (PM) phenotype. Compared with normal or extensive metabolizers (EMs), PM subjects demonstrate markedly greater AUC values for parent drugs that are metabolized by CYP2D6, and therefore require lower doses to achieve therapeutic effects [47].

CYP2E subfamily. CYP2E1 is the only gene of this subfamily. It is expressed in many tissues, such as the nose, lung, and the liver. CYP2E1 accounts for ~7% of total CYP in the human liver and is involved in the metabolism of only 3% of the drugs, such as acetaminophen, caffeine, and chlorzoxazone, the latter being considered a marker of CYP2E1 activity [48]. CYP2E1 has a ubiquitous role in the metabolism and activation of an array of solvent carcinogens (which also induce its expression), such as N-nitrosamines, benzene, styrene, carbon tetrachloride, acrylonitrile, and ethylene glycol. CYP2E1 is the most active CYP enzyme in forming reactive oxygen intermediates, such as superoxide radical, causing tissue injury [49].

CYP2F subfamily. Only one functional gene has been found in the human CYP2F subfamily. The expression of CYP2F1 is highly tissue selective with highest expression observed in the lung and little or no hepatic expression [50]. Substrates for CYP2F1 include ethoxycoumarin, propoxycoumarin, and pentoxyresorufin, but not ethoxyresorufin. Recombinant CYP2F1 is capable of activating two prototypical pneumotoxicant, 3-methylindole and naphthalene. CYP2F1 metabolizes naphthalene to its highly toxic intermediate, naphthalene-1,2-epoxide, and 3-methyl indole to its dehydrogenated pneumotoxic metabolite 3-methyleneindolenine [51, 52].

CYP2J subfamily. CYP2J2 is the only gene of this subfamily. It is known to be expressed in many extrahepatic tissues and may play a role in the oxidative bioactivation of arachidonic acid to form epoxyeicosatrienoic acids, which modulate bronchial smooth muscle tone and airway transepithelial ion transport [53]. CYP2J2 is also active toward other compounds such as linoleic acid and testosterone. Recently, it has been reported that CYP2J2 is involved in the intestinal first-pass metabolism of an antihistamine drug, astemizole [54].

CYP2S subfamily. In humans, CYP2S1 appears to be the sole member of a new subfamily, CYP2S. The CYP2S1 gene is located at the proximal end on chromosome 19q13.2 CYP2 gene cluster [55]. Of interest, CYP2S1 is closely related to CYP2F1 (47–49% identity) and is induced by dioxin in a human lung epithelial cell line, suggesting the possibility that CYP2S1 may participate in the metabolism of toxic and carcinogenic compounds [39]. Recent studies using heterologously expressed CYP2S1 in yeast have shown that CYP2S1 is able to metabolize naphthalene [56]. Therefore, it is speculated that CYP2S1 might also play a role in naphthalene-induced lung toxicity [57].

CYP3A subfamily. The CYP3A subfamily of CYP450 in humans is composed of several enzymes and accounts for ∼28% of total hepatic P450 content. The human CYP3A family is clinically very important because it has been shown to catalyze the metabolism of an amazingly large number of structurally diverse xenobiotics and endobiotics. It is estimated that CYP3A forms participate in the metabolism of 35–50% of all marketed drugs [25, 58]. The human CYP3A subfamily includes CYP3A4, CYP3A5, CYP3A7 [59], and CYP3A43 [60]. CYP3A4 is the major human liver CYP3A enzyme, whereas CYP3A5 is present in only ∼20% of human liver. CYP3A4 and CYP3A5 are also expressed in the stomach, lungs, small intestine, and renal tissue. Most of the CYP3A4 substrates are also metabolized by CYP3A5 [61]. CYP3A7 and CYP3A43 isozymes seem to play only a minor role in the metabolism of drugs. In fact, CYP3A7 is only present in fetal liver, whereas CYP3A43, which is expressed in liver, appears to be very restricted, both in terms of its activity and expression [62]. The highest concentration of transcript expression of CYP3A43 is in the prostate, whereas hepatic mRNA concentration is only 0.2–5% that of CYP3A4 [62]. Some examples of drugs metabolized by CYP3A are terfenadine, the benzodiazepines midazolam and triazolam, quinidine, lidocaine, carbamazepine, nifedipine, tacrolimus, dapsone, erythromycin, and dextromethorphan [63]. In addition to drugs, CYP3A is involved in the oxidation of a variety of endogenous substrates, such as steroids, bile acids, and retinoic acid [64].

1.2.2 Flavin Mono-Oxygenase (FMO)

FMOs are NADPH-dependent and oxygen-dependent microsomal flavoenzymes, which oxygenate a number of drugs and xenobiotics that contain a "soft-nucleophile" heteroatom such as nitrogen, sulfur, and phosphorus. Unlike CYPs, which generally use sequential one-electron transfer, the FMOs are two-electron oxygenating enzymes for N-oxidation. The microsomal FMO enzyme family is comprised of five isozymes (designated FMO1 to FMO5), whose expression is tissue specific. FMO1 is predominantly expressed in human kidney and FMO2 in lung and kidney [65]. FMO3 is the prominent isozyme in adult human liver, FMO4 is more broadly distributed in liver, kidney, small intestine and lung, and FMO5 is expressed in human liver, lung, small intestine, and kidney [66]. Of all the FMOs isozymes, FMO3 has a wide substrate specificity, including the physiologically and plant-derived tertiary amine, trimethylamine, tyramine, and nicotine; commonly used drugs including cimetidine, ranitidine, clozapine [65, 66], methimazole, itopride, ketoconazole, tamoxifen, and sulindac sulfide; and agrichemicals, such as organophosphates and carbamates [65, 66].

1.2.3 Monoamine Oxidase

MAOs are mitochondrial flavoproteins containing one covalently bound FAD cofactor. Two isozymes, termed as MAO-A and MAO-B, are known for the MAO enzyme family. They catalyze the oxidative deamination of structurally diverse amines including neurotransmitters dopamine, norepinephrine, serotonin, tyramine, and 2-phenylethylamine, and some drugs and xenobiotics that contain cyclic and acyclic alkylamine functional groups [67, 68]. The MAO reaction cycle involves two half reactions, as shown in equations 1.3a and 1.3c:

$$RCH_2NH_2 + FAD \rightarrow RCH = NH_2 + FADH_2 \qquad (1.3a)$$

$$RCH = NH_2 + H_2O \rightarrow RCH = O + NH_3 \qquad (1.3b)$$

$$FADH_2 + O_2 + 2H^+ \rightarrow FAD + H_2O_2 \qquad (1.3c)$$

A two-electron oxidation results in the imine and reduced protein-bound FAD (equation 1.3a). The imine is then nonenzymatically hydrolyzed to the carbonyl compound (equation 1.3b). In the second half reaction, the reduced FAD ($FADH_2$) is reoxidized by molecular oxygen producing hydrogen peroxide (equation 1.3c).

MAOs are expressed in most mammalian tissues that complicate pharmacokinetic predictions when MAOs are involved in metabolism. Inhibitors of MAOs are used in psychiatry for the treatment of depressive disorders and in neurology for the treatment of Parkinson's disease. MAO-A and MAO-B play a critical role in the bioactivation of 1-methyl-4-phenyl-1,2,3,6-tetrahydropyridine (MPTP) to a toxic metabolite that induces Parkinson-like effects [68].

1.2.4   Molybdenum Hydroxylases

Molybdenum hydroxylases (i.e., AO and XO) are flavoproteins that contain in addition to a FAD, a pterine cofactor coordinated to a molybdenum atom, and an iron sulfur center for their catalytic activity. They catalyze the two-electron oxidation of substrates with transfer to molecular oxygen to produce H_2O_2, and insert an atom of oxygen from water into a wide range of N-heterocycles and aldehydes via two-electron redox reaction as shown in equation 1.4:

$$RH + H_2O \rightarrow ROH + 2e^- + 2H^+ \tag{1.4}$$

AO and XO are cytosolic enzymes and are closely related. However, they differ in their substrate/inhibitor specificities. AO is involved in the metabolism of several clinically significant drugs such as famciclovir, zaleplon, zonisamide, and ziprasidone [69–72]. XO has a narrower substrate specificity than AO and is mainly active toward purines and pyrimidines. XO plays a role in the oxidation of several chemotherapeutic agents and has been implicated in the bioactivation of mitomycin B [73].

1.2.5   Epoxide Hydrolase

EH converts potentially reactive epoxides to *trans*-dihydrodiols. Originally, it was thought to be localized solely in the endoplasmic reticulum; subsequent studies demonstrated there are distinct microsomal (EH1) and cytosolic (EH2) forms of the enzyme. The EH1 gene is polymorphic and the expressed enzyme has two metabolic functions, detoxification and bioactivation, depending on the particular substrate. EH1 hydrolyzes epoxides, derived from aromatics and alkenes by CYP450 enzymes, to the corresponding dihydrodiols through the *trans* addition of water. Dihydrodiols from PAHs can be substrates for further transformation by CYP450 enzymes to dihydrodiol epoxides such as (+)-*anti*–benzo[*a*]pyrene (B[*a*]P)-7,8-diol-9,10 epoxide, the most mutagenic and carcinogenic metabolite of B[*a*]P [74].

1.2.6   Esterase/Amidase

These enzymes are widely distributed in various tissue types and belong to the neutral, acidic, or metalloproteinase classes. While their roles have been recognized when drug structures contain ester or amide functions, very little characterization of the specific enzymes involved have been reported. Commonly, the ester function is introduced into drug structures as a means of masking carboxylate functions to increase absorption by increasing lipophilicity as with the antiviral drugs such as valganciclovir and oseltamivir that are converted *in vivo* to ganciclovir and oseltamivir carboxylate by intestinal and hepatic esterases [75, 76]. Amide functions are commonly used in linking drug substructures. These bonds are generally not susceptible to amidase activity since typically they are not natural amino acid derived. However, when such functions do present themselves in the NCEs, the potential for hydrolysis as a mechanism for clearance should be considered and stability in whole blood and hepatic subcellular matrices should be determined.

1.3   PHASE II CONJUGATIVE ENZYMES

Phase II reactions are catalyzed by conjugative enzymes, such as UDP-glucuronosyltransferase (UGT), sulfotransferase (SULT), glutathione *S*-transferase (GST), *N*-acetyltransferase (NAT) and methyltransferase (*N*-methyl-, thiomethyl-, and thiopurinemethyl-). Glutathione conjugates are further metabolized to cysteine and *N*-acetyl cysteine adducts. Most phase II reactions result in a compound's concomitant increase in hydrophilicity and decrease in volume of distribution (VD_{ss}), which together greatly facilitate its excretion from the body.

1.3.1   Uridine Diphosphate Glucuronosyltransferase (UGT)

UGT family of enzymes catalyzes the transfer of glucuronic acid from UDP glucuronic acid to available substrates to form the water-soluble glucuronide conjugates, suitable for excretion. Thus, glucuronidation is a major detoxification pathway of endo- and xenobiotics in man and other animals. Beside human liver, the gut and kidneys are two important sites of glucuronidation. The UGT family of enzymes is subdivided into two subfamilies, UGT1 (1A1, 1A3, 1A4, 1A5 1A6, 1A7, 1A8, 1A9, and 1A10) and UGT2 (2A1, 2B4, 2B7, 2B10, 2B11, 2B15, 2B17, and 2B28), on the basis of sequence homology [77]. All classes of drugs containing a wide range of acceptor groups including phenols, alcohols, aliphatic and aromatic amines, thiols, acidic carbon atoms, and carboxylic acids are substrates for UGTs and this pathway has been estimated to account for ~35% of all drugs metabolized by phase II drug-metabolizing enzymes [78]. Human UGTs can also form N-linked glucuronides of several tertiary amine drugs [79]. Some acyl glucuronides are reactive intermediates that bind covalently to macromolecules causing potential toxicity. Several drugs that contain a carboxylic acid such as diclofenac, ketoprofen, suprofen, and tolmetin are glucuronidated to form a reactive acyl glucuronide that could be associated with some of the observed toxicity of these drugs [80].

1.3.2   Sulfotransferase

SULT family of enzymes catalyzes the transfer of sulfite (SO_3^-) from 3'-phosphoadenosine 5'-phosphosulfate (PAPS) to a hydroxyl or an amino group on an acceptor molecule to form the water-soluble sulfonate or sulfamate conjugates and thereby aid in their excretion via the kidneys or bile. SULTs are capable of sulfonating hydroxyl group of a wide range of substrates including phenols, primary and secondary alcohols, *N*-hydroxy arylamines, and *N*-hydroxy heterocyclic amines. Amino groups of arylamines such as 2-naphthylamine are also sulfonated by SULTs. SULTs have a wide tissue distribution and play an important role in the detoxification, metabolism, and bioactivation of numerous xenobiotics, many dietary and environmental mutagens, drugs, neurotransmitters, and hormones [81]. In humans, three SULT families, SULT1, SULT2, and SULT4, have been identified that contain at least 13 distinct members. SULT1 and SULT2 families are the largest and are responsible for sulfonating the greatest number of endogenous and xenobiotic compounds.

Studies using recombinant enzymes demonstrated that many promutagens are activated with high selectivity by an individual SULT form.

1.3.3 Glutathione S-Transferase

The GST family of enzymes catalyzes the nucleophilic attack of the tripeptide (γ-glu-cys-gly) (GSH) on a wide variety of soft electrophiles such as epoxides and quinones, formed during phase I oxidation of xenobiotics, generally resulting in their elimination and detoxification. There are two GST superfamilies: (1) the membrane-bound GST isozymes and leukotriene C_4 synthetase and (2) the cytosolic-soluble enzymes, each of which displays different intracellular distribution and distinct catalytic as well as noncatalytic binding properties. Thirteen different human GST subunits, GSTA1 through GSTA4, GSTM1 through GSTM5, GSTP1, GSTT1 and GSTT2, and GSTZ1, have been identified belonging to seven distinct classes: alpha (α), mu (μ), omega (ω), pi (π), sigma (σ), theta (θ), and zeta (ξ) [82]. GSTs appear to be ubiquitously distributed in human tissues. Some examples of clinically significant drugs that form glutathione conjugates include acetaminophen, sulfonamides, irinotecan, carbamazepine, rotonavir, clozapine, procainmide, hydralazine, cyclosporine A, diclofenac, estrogens, and tamoxifen [83].

1.3.4 N-Acetyltransferase

NAT family of enzymes catalyzes the acetyl-CoA-dependent acetylation of aryl amines and arylhydrazines and N-hydroxyarylamine including many drugs and carcinogens. In most cases, this reaction is generally considered to result in the detoxification of potentially toxic exogenous compounds. However, NATs are also involved in bioactivation reactions via O-acetylation of N-hydroxyarylamines to unstable acetoxy esters that decompose to highly reactive mutagens that form adducts with cellular macromolecules. Humans express two distinct isozymes, designated NAT1 and NAT2 [84]. Clinically relevant substrates of NAT include isoniazid, procainamide, aminoglutethimide, sulphamethoxazole, 5-aminosalicylic acid, hydralazine, phenelzine, and dapsone. Also, NATs play an important role in the metabolism of industrial and environmental carcinogens, including 2-naphthylamine, benzidine, 2-aminofluorene, and 4-aminobiphenyl, as well as of potentially carcinogenic heterocyclic amines found in well-cooked red meat and cigarette smoke [85].

1.3.5 Methyl Transferase (MT)

MT family of enzymes catalyzes the O-, S-, or N-methylation of drugs, hormones, and neurotransmitters, and utilizes S-adenosyl-L-methionine (SAM) as a methyl donor. Catechol-O-methyltransferase (COMT) is the most extensively investigated drug-metabolizing MT. It plays an important role in the biotransformation of both endogenous and exogenous catechols. COMT has a rather broad substrate specificity for structures that contain catechol moieties and is often involved in the methylation of drugs that have been metabolized by the CYP system to generate catechol functions [86, 87].

S-Methylation is also an important pathway in the biotransformation of many sulfur-containing drugs. At least two separate enzymes, thiol methyltransferase (TMT) and thiopurine methyltransferase (TPMT) are known to catalyze S-methylation in humans [87]. TMT, a membrane-bound enzyme, catalyzes the S-methylation of captopril, D-penicillamine, and other aliphatic sulfhydryl compounds such as 2-mercaptoethanol. On the other hand TPMT, a cytosolic enzyme, catalyzes the S-methylation of aromatic and heterocyclic sulfhydryl compounds including 6-mercaptopurine and other thiopurines. Recently, S-methyltransferase has been shown to play a critical role in the metabolism of the antipsychotic drug, ziprasidone, in humans [72, 88]. Both TPMT and TMT have been shown to be genetically polymorphic in humans.

Histamine-N-methyltransferase is a hepatic enzyme, although its inhibition by various drugs has been reported [87, 89], its ability to methylate drugs has not been established. Several protein arginine methyltransferases (PRMTs) are known and their physiological role in producing mono- and dimethylated arginine is known [90]. However, their role in N-methylating drug molecules has yet to be identified. Thus, except for COMT and S-methyltransferases the other MTs including DNA methylating enzymes are not known to play a role in drug metabolism.

1.4   DRUG EFFLUX TRANSPORTERS

Carrier-mediated transport of drugs and their metabolites has recently been recognized as an important issue in pharmaceutical research. There is a wealth of information that suggests that transporters are responsible both for the uptake and efflux of drugs and other xenobiotics, and may be key determinants of the disposition of a drug [91, 92]. Transporter proteins are divided into two categories: (1) the adenosine triphosphate (ATP) binding cassette (ABC) transporter superfamily and (2) the solute carrier (SLC) family of proteins.

1.4.1   ABCB1 (P-Glycoprotein, MDR1)

The first member of ABC efflux transporter family to be discovered was P-glycoprotein (P-gp; gene ABCB1 or MDR1). The importance of P-gp or MDR1 was first recognized in the multidrug resistance of tumor cells in response to chemotherapy treatment [93]. However, constitutive expression of P-gp in many normal tissues such as liver, kidney, intestine, blood–brain barrier, and lymphocytes demonstrates its role in modulation of the absorption, distribution, metabolism, excretion, and toxicology behaviors of some drugs and NCEs in development [94]. Experiments with *mdr1a* knock-out mice revealed that P-gp not only limits the CNS entry of drugs, but it also reduces the oral absorption of drugs by extruding them from enterocytes back into the intestinal lumen [95]. Therefore, the inhibition and induction of P-gp transporter can lead to significant drug–drug interaction and, most importantly, to drug treatment resistance. Talinolol, fexofenadine, and digoxin are the most commonly used probe substrates for assessing the modulation of P-gp.

1.4.2   ABCC1 (Multidrug Resistance Related Protein1, MRP1)

ABCC1 or MRP1 is a 190-kDa transport protein that was originally cloned from a doxorubicin-selected lung cancer cell line. MRP1 has now been found to be expressed in a range of solid and hematological tumors, and has been demonstrated to transport a wide array of structurally diverse anticancer drugs such as oxorubicin, vincristine, and methotrexate [96]. In addition, MRP1 also transports many glutathione, glucuronide, and sulfate conjugated organic anions, such as leukotriene C_4, 17β-estradiol-glucuronide, and estrone 3-sulfate, respectively. MRP1 is also found in normal tissues throughout the human body and plays a significant role in tissue defense from toxic agents. Thus the expression levels and activity of MRP1 are important considerations in drug development and chemical toxicity.

1.4.3   ABCC2 (Multidrug Resistance Related Protein2, MRP2)

ABCC2 or MRP2 is located on the bile canalicular membrane and is involved in the biliary excretion of glucuronide, glutathione, and sulfate conjugates of lipophilic compounds (i.e., drugs) as well as endogenous compounds such as hormone and bilirubin conjugates. In addition, MRP2 can also transport uncharged compounds in cotransport with glutathione and thus can modulate the disposition of many drugs. MRP2 is predominantly expressed at the hepatocyte canalicular membrane [97]. It is also expressed in other tissues such as the kidney, as well as the intestinal entrocytes [97, 98] and apical membrane of brain capillary endothelial cells [99]. In humans, a genetic deficiency of MRP2 results in a disease known as Dubin–Johnson syndrome [100]. In addition, altered MRP2 function can change the ADME properties of many clinically important drugs including cancer chemotherapeutics (irinotecan, methotrexate, and vinblastine), antibiotics (ampicillin and rifampin), antihyperlipidemics, and angiotensin-converting enzyme inhibitors.

1.4.4   ABCG2 (Breast Cancer Resistance Protein, BCRP)

ABCG2 or BCRP, a 655 amino acid peptide with an ability to extrude a wide variety of chemical compounds from the cells, was first cloned from mitoxantrone and anthracycline-resistant breast and colon cancer cell lines [101, 102]. BCRP is a half-transporter efflux pump with an intracellular N and C terminus and an intracellular ATP binding site, followed by six putative transmembrane segments. It is expressed in liver, small intestine, placenta, kidney, mammary gland, and endothelial cells at the blood–brain barrier, and is believed to be involved in protecting tissues from xenobiotic accumulation and resulting toxicity [103]. It is capable of transporting sulfate and glucuronide conjugated organic anions, at least *in vitro*. It affects intestinal absorption/bioavailability, organ distribution, hepatic/renal elimination, and plasma clearance of substrate drugs such as topotechan [104]. Downregulation of BCRP was suggested to contribute to cellular adaptation to folate deficiency and homeostasis [105].

1.5 DRUG UPTAKE TRANSPORTERS

The drug uptake (SLC) family of transporter is the largest superfamily of transporters. This family includes 31 transporters from organic anion transporter polypeptides (OATPs), organic anion transporters (OATs), organic cation transporters (OCTs), peptide cotransporters (PEPTs), and sodium–bile acid cotransporter classes. Only OATP, OAT, OCT, and PEPT are primarily involved with the transport of drugs/xenobiotics.

1.5.1 Organic Anion Transporter Polypeptides

The OATP transport proteins function as organic solute exchangers with broad substrate specificity that includes organic anions, cations, and neutral compounds [106]. The OATP transporters are expressed in organs such as the intestine, liver, and blood–brain barrier and considered to be the key determinant in the cellular uptake of many endogenous and exogenous chemicals, including drugs in clinical use. To date, eight different OATPs (OATP1A2, OATP1B1, OATP2B1, OATP3A1, OATP4A1, OATP1C1, OATP1B3, and OATP 2A1) have been cloned from human tissues [107].

1.5.2 Organic Cation Transporter

OCTs are responsible for reabsorption and excretion of a wide variety of organic cations, such as quinidine, cimetidine, procainamide, vecuronium, and cardiac glycosides as well as endogenous substances including, dopamine, epinenephrine, norepinephrine, creatinine, and choline. The OCT transporters are expressed in human organs such as the intestine, liver, proximal tubules of kidney, brain, arota, skeletal muscle, prostate, adrenal gland, salivary gland, and fetal lung. To date, three different OCTs (OCT1, OCT2, and OCT3) have been identified in humans [108, 109].

1.5.3 Organic Anion Transporter

OAT family plays a critical role in the renal excretion and detoxification of a wide variety of compounds including drugs, toxins, hormones, and neurotransmitter metabolites. OATs are primarily expressed in the kidney. In addition, OAT expression has been detected in liver, placenta, brain, and choroid plexus. At least five different OATs (OAT1, OAT2, OAT3, OAT4, and Urat1) have been detected in human tissues [109].

1.5.4 Peptide Transporter

The PEPTs are hydrogen ion-dependent transporters that transport small peptides and proteinlike compounds such as cephalosporins, penicillin, enalapril, and captopril. Humans express two distinct PEPTs designated PEPT1 and PEPT2. PEPT1 is involved with the oral absorption of drugs from small intestine while PEPT2 contributes to the proximal tubular reabsorption of drugs [108].

1.6   CHALLENGES IN DRUG METABOLISM

1.6.1   Prediction of Metabolism and Pharmacokinetics in Humans

1.6.1.1 **In silico** *Computational Tools.* In recent years, several *in silico* computational methods have become available for prediction of metabolism and pharmacokinetics of the NCEs and are increasingly becoming a part of the drug discovery process to select candidates with desirable ADME properties [110–112]. De Graaf et al. [110] have described how *in silico* computational approaches can be used to understand, rationalize, and predict the activity and substrate selectivity of CYPs. Mechanism-based quantum chemical calculations on substrates and the enzyme, pharmacophore modeling of ligands, and protein homology modeling in combination with automated docking and molecular dynamics simulations have been used to rationalize and predict ligand binding and formation of metabolites. Several of these models, especially for CYP2D6 [113] and CYP2C9 [114], have been shown to have a reasonably good predictive value concerning qualitative metabolism and substrate inhibitor selectivity. Programs such as Meteor and Metasite use database-based evaluations or de novo computational methods based on bond energies and active site docking to predict metabolic transformations and hot spots. While these methods assist in predicting metabolic outcomes, they are still limited and neither software is capable of predicting rates of metabolism at the sites that are predicted. These tools cannot be used to guide new chemical synthesis that incorporates metabolic concerns without experimental verification.

1.6.1.2 **In Vitro–In Vivo** *Extrapolation.* The ability to predict human pharmacokinetic from *in vitro* and *in vivo* models with a reasonable degree of accuracy (<2.0-fold variation) is an ongoing challenge with many different approaches yielding varying degrees of success. Interspecies allometric scaling of pharmacokinetic parameters is the earliest of the approaches and is still a tool used in predicting human pharmacokinetic behavior of drugs. The method is based on empirically observed relationships between physiological processes and body weights of mammals [115]. Several enhancements have been introduced to the original method such as maximum life-span potential, brain weight, body surface area, rule of exponents, and protein binding [115–117]. The allometric methods assume that clearance mechanisms are similar across species and do not consider interspecies specificity differences in drug-metabolizing enzymes and metabolic pathways. A recent analysis of 103 compounds for allometric scaling incorporating the enhanced methods where data were available for rat, dog, and monkey showed that the success rate for human projection from preclinical models was suboptimal (18–53%) with or without the enhanced methods [118]. A major limitation of the allometric methods is the lack of incorporation of metabolic differences in preclinical species and humans.

In vitro methods using liver slices, hepatocytes, and liver subcellular fractions, such as microsomes, have also been developed, and are used to varying extents in clearance prediction. Methods based on half-life of NCEs in metabolically active liver microsomes are used to project hepatic extraction from which clearance predictions are calculated [119]. The limitation of this approach is that prediction is limited to those NCEs for which hepatic metabolism by oxidative enzymes of the endoplasmic reticulum dominate the overall clearance mechanism. When drug-metabolizing enzymes other than CYPs or FMOs dominate the metabolic clearance, this approach falls short. Hepatocytes contain the entire cellular metabolic machinery and would be considered the better choice for hepatic clearance prediction. However, unlike human liver microsomes, which can be prepared from frozen liver tissue and effectively stored at $-80°C$, fresh human hepatocytes are not readily accessible, and the activities of some drug-metabolizing enzymes have been shown to decline with age of hepatocyte cultures, thus limiting their use. Cryopreserved human hepatocytes are used in a limited capacity to predict hepatic clearance and in evaluating overall metabolism of NCEs. When allometric scaling was combined with *in vitro* metabolism-based scaling for 10 highly metabolized drugs, a significant improvement was noted for the predicted human clearance compared to either method alone [120]. A recent retrospective analysis of pharmacokinetic data for a large set of diverse drugs suggests that human pharmacokinetic parameters can be predicted from rat pharmacokinetic data within a threefold variation by a fixed exponent approach [121]. This method allows for early binning of NCEs into low-, medium-, or high-clearance categories at a very early stage in drug discovery.

In the past two decades, success in predicting human pharmacokinetics from preclinical and *in vitro* techniques has dropped the attrition rate for NCEs due to failure to achieve the pharmacokinetic thresholds necessary for drug action in humans from $\sim40\%$ to less than 10% [6]. Thus, while success has been achieved in this area, clearly much progress is still necessary in preclinical models to achieve greater success in eliminating attrition due to pharmacokinetics. A more direct approach using drug microdosing in humans to predict pharmacokinetic behavior early in the drug development process is under investigation. In this approach, NCEs are administered at doses significantly below therapeutic levels (>100-fold lower than projected therapeutic dose or 100 µg) in order to get an early readout on their pharmacokinetic behavior [122, 123]. A microdosing study with five marketed drugs (warfarin ZK253 (schering), midazolam, erythromycin, and diazepam), known to be problematic in extrapolation to humans from either *in vitro* or animal models, has shown good concordance in the pharmacokinetic parameters obtained under micro- and therapeutic-dosing conditions for all except erythromycin [124]. Because of the extrmely low dose, this method does not require the same extent of animal safety testing as is required by a traditional investigative new drug (IND) application to the Food and Drug Administration [124–126]. Microdosing has become possible because of newer analytical tools such as liquid chromatography coupled high sensitivity soft ionization mass spectrometry and more recently at significantly higher sensitivity with accelerator mass spectrometry [127].

1.6.1.3 Species Specificity and Limitation of Animal Models. The primary goals in the use of preclinical studies in animals, typically rodent (rat and mouse), dog, monkey, and rabbit, are to extrapolate pharmacokinetic parameters and establish clearance mechanisms applicable to human drug administration, determine toxicity limiting dosages, and evaluate long-term carcinogenic and teratogenic potential of the NCEs. Drug-metabolizing enzymes in mammals have common ancestral roots. However, the substrate and reaction rate specificities of NCEs are known to vary considerably among species [128, 129]. These variations in activity and specificity are particularly important for pharmacokinetic parameters when drug clearance is metabolically driven, and also when preclinical species are considered for the toxicological model. For toxicological studies, identification of major metabolites from *in vitro* and *in vivo* studies in rodent, dog, and monkey compared to those from human *in vitro* reactions with hepatocytes and microsomes are particularly useful to identify the preclinical species with optimal coverage of exposure to potential circulating metabolites in humans. Interspecies specificity of some drug-metabolizing enzymes can differ significantly. For example, when AO is involved in metabolic clearance, species selection for pharmacokinetics and toxicology testing is critical. AO activity is high in monkey and human, moderate in rodent, and not detected in dog liver. Complicating the situation is that significant species differences exist in both specificity and activity of this enzyme. Thus, in addition to identifying metabolites, the reaction phenotype for the major metabolites of an NCE at early stages in the drug development process can be critical in selection of the species for pharmacokinetic and toxicological testing [130].

1.6.2   Drug–Drug Interactions

Drug–drug interactions (DDIs) have received considerable attention in the pharmaceutical industry because in recent years several prominent drugs have been withdrawn from the market in the US and Europe due to serious adverse events as a result of significant DDIs [131]. DDIs are caused when the clearance of one drug is influenced by a coadministered drug [131]. Therefore, the consequences of these DDIs can range from loss of therapeutic efficacy to the introduction of potential lethal toxic effects. Although DDIs can occur during absorption, distribution, and elimination phases following initial drug administration, metabolism seems to be the predominant mechanism for such interactions. Since most marketed drugs are eliminated from the body at least in part by metabolism, inhibition and/or induction of DMEs by one drug could have a significant effect on the disposition of another drug. The NCEs can either be perpetrators or victims of such interactions, and early evaluation of NCEs for their potential to function in either capacity is critical to the development of new drug candidates and involves the identification of drug-metabolizing enzymes inhibited by NCEs (in particular CYPs) as well as the metabolic pathways by which they are cleared [130]. High specificity substrates for *in vitro/in vivo* use in drug interaction assays are known for six major isoforms of CYP enzymes involved in drug metabolism.

1.6.2.1 Biotransformation Reaction Phenotyping. As described earlier, a majority of drugs are cleared by metabolic transformation, generally first by oxidation to more polar metabolites, by hepatic enzymes. Knowledge of the specific enzyme(s) especially CYPs, involved in the metabolism of and inhibition by the NCE, particularly of the major pathways, is critical for evaluation of the potential liabilities that an NCE may have with respect to DDI [130]. Metabolic flux of an NCE through several CYP-dependent pathways where each pathway contributes less than 30% of the overall metabolic clearance is always preferable to flux through a unique CYP isoform since inhibition of any one of the pathways by a coadministered drug would not significantly increase the systemic exposure of the NCE, and consequently pose no significant risk. Terbinafine, an oral antifungal agent, is an example of such a drug where multiple pathways for its metabolism predict lack of significant clinical DDI for its clearance [132], which has been clinically confirmed [133]. However, terbinafine is an inhibitor of CYP2D6 and has been shown to increase the AUC of desipramine when coadministered to CYP2D6 EMs, suggesting that caution must be used with coadministration of terbinafine to patients on medications with narrow therapeutic index (TI), and its metabolism is mediated primarily by CYP2D6. Thus, terbinafine can be considered as an example of a drug that is only a perpetrator for a drug interaction that involves coadministered drugs that are metabolized primarily by CYP2D6.

If the metabolic flux of an NCE is greater than 60% through a unique CYP isoform, especially by a polymorphic enzyme (CYP2D6), the potential for DDI is extreme and careful consideration has to be given to the forward development strategy for such an NCE [134]. The case of the antihistaminic drug terfenadine stands out as an excellent example of this process. Terfenadine is primarily metabolized to its pharmacologically active form by CYP3A4. Inhibition of CYP3A4 by a variety of drugs, best exemplified by ketoconazole, results in dramatic increases in the AUC of terfenadine to toxic levels [135] that can prolong the QT interval and produce potentially fatal ventricular arrhythmias. This example shows how ketoconazole functions as the perpetrator of the DDI while terfenadine, because of its major flux through CYP3A4, becomes victim. When the TI for the NCE is narrow, knowledge of the CYP isoform(s) involved is critical and close attention is needed to the potential for DDI by coadministered drugs.

Cyclosporine, paclitaxel, tacrolimus, and warfarin are examples of drugs with narrow TIs where even dietary habits can play significant roles in toxicity. For example, in the late 1980s clinical observations noted a dramatic effect of grapefruit juice consumption on the levels of dihydropyridine calcium ion channel blockers. Flavanoids present in grapefruit juice were speculated on the basis of studies that showed Naringenin, the aglycone component of the flavanoid naringin, which is abundant in grapefruit juice, as inhibiting the *in vitro* metabolism of dihydropyridines by human liver microsomes [136]. Clinical studies on renal transplant patients showed that grapefruit juice significantly enhanced the systemic exposure of cyclosporine, while having no significant effects on prednisone or prednisolone [137]. Further investigations have established that the effect is on drugs that are primarily cleared by CYP3A4 and show low bioavailability due to presystemic metabolism by CYP3A4 in the entrocytes. The increase in systemic exposure arises from the inactivation of CYP3A4 in the entrocytes.

Epoxyfuranocoumarins formed from grapefruit juice components like bergamottin and related natural products have been identified as responsible for self-inactivation of CYP3A4 [138]. The inhibitory effect of grapefruit juice on enteric metabolism by CYP3A4 is temporal and its duration is limited to the duration of consumption [139]. Dietary effects from the consumption of other foods and natural remedies on the metabolism of some drugs have also been reported [131–143]. Examples of such processes with other CYP450 enzymes abound in the literature with over 500 published articles since 1990 that examine *in vitro*, *in vitro–in vivo* correlations, and *in vivo* drug interactions via CYP450 enzymes demonstrating the importance of such studies in drug development. Thus, early phenotyping of metabolic pathways for an NCE and the inhibitory effect of the NCE on CYP isoforms are important aspects of the development process for NCEs as drug candidates that provide an understanding of potential clinical DDI issues that may arise for which early clinical evaluations can be conducted. Reaction phenotyping is done by several methods including the use of specific inhibitors for some CYP isoforms, recombinant CYP isoforms and relative activity factors, inhibitory antibodies, and interindividual correlations [130, 144–148].

As with CYP isoforms, UGTs are also more commonly encountered in the metabolism of drugs and xenobiotics. There are several drugs and xenobiotics where glucuronidation contributes to the overall metabolic clearance mechanism and in some instances is the predominant metabolic pathway. Examples include zidovudine (AZT) [149], mycophenolate [150], retigabine [151], mitiglinide [152], fenamates [153], and gemcabene [154]. In contrast to extensive DDIs caused by binding of drugs to CYP isoforms, DDIs with the UGT family enzymes are relatively few, and are confined to compounds where UGTs are either the primary clearance mechanism or contribute substantially to it. The cancer drug irinotecan is metabolized to its active form SN-38 (7-ethyl-10-hydroxy-camptothecin), which is cleared by glucuronidation by UGT1A1. Ketoconazole increases the circulating levels of SN-38 due to potent inhibition of its glucuronidation by UGT1A1 [155]. Since irinotecan has a small TI, patients should be genotyped for UGT1A1 prior to commencement of therapy. Another example of a UGT-dependent drug interaction is that of statins and the lipid-lowering drug, gemfibrozil [156]. A significant component of simvastatin's clearance is via an unusual lactonization mechanism involving an intermediate acyl glucuronide catalyzed by UGTs [157]. Gemfibrozil, which is also metabolized to its glucuronide by UGTs, inhibits the glucuronidation of simvastatin with the resultant pharmacokinetic effect of an increase in simvastatin's AUC and decrease in biliary levels of simvastatin lactone and glucuronide [158]. UGT1A1 has a critical physiological function in the elimination of bilirubin as its glucuronide. Surprisingly, no reports on drug interactions with UGT1A1 such as that of ketoconazole have been reported to show significant increases in the blood levels of bilirubin. A possible explanation for this may be that the *in vivo* concentrations of substrates or inhibitors cleared primarily by the particular UGT under consideration may often be below their corresponding K_m and K_i values, resulting in partial site occupancy and consequently insignificant *in vivo* interactions. Hence, when considering the potential for drug interactions with the UGTs, dose and concentrations in the blood and in the liver are important parameters to consider. Reaction phenotyping of UGTs is less developed than that for CYPs. Few high specificity low K_m or K_i substrates or inhibitors have been identified, except for UGT1A1 where bilirubin is a highly specific substrate. However, it is only useful for *in vitro* inhibition assays. Phenotyping of UGT reactions is generally conducted *in vitro* using recombinant forms of the enzymes.

Drug interactions with MAOs are particularly relevant when patients are on therapy with MAO inhibitors and the MAOs are significant contributors to the clearance of a drug. As with UGTs, the K_m or K_i for MAO substrates tend to be generally high and drug interactions at therapeutic levels are less pronounced than with CYP enzymes. MAO-A and MOA-B can be selectively inactivated by Chlorgyline and Deprenyl, respectively [68]. Thus, distinguishing between these enzymes can be readily achieved using either liver microsomal suspension where MAO-A and B are present as contaminants from fractured mitochondrial membrane particles or recombinant enzymes [68].

AO has been shown to be responsible for the primary metabolism of several drugs that contain an aromatic nitrogen heterocycle [72, 88, 159, 160]. In general, when drug structures contain aromatic nitrogen heterocycles, a role for AO should be considered in preclinical and *in vitro* tests. The contribution of this enzyme to drug clearance can be defined by examining metabolism of an NCE by cytosolic preparations and inhibition by either menadione or raloxifene [161, 162]. DDIs with AO have not yet been reported in the literature.

Several *in vitro* models including cell-based, cell-free, and yeast systems as well as *in vivo* models such as genetic knock-out, gene-deficient, and chemical knock-out animals have recently been developed in order to understand the interaction between drugs and transports [163]. However, unlike the CYP-mediated drug interactions, the evidence for the transport-mediated drug interactions is often less conclusive due to a broad overlap of substrate specificities between transporters and drug-metabolizing enzymes. Some evidences for the interaction of digoxin, a good P-gp substrate, when coadministered with valspodar or verapamil (potent and selective P-gp inhibitors) have been reported [164].

1.6.2.2 Metabolism-Based Inactivation/Bioactivation.

1.6.2.2 Metabolism-Based Inactivation/Bioactivation. The metabolic activation of an NCE to a reactive intermediate is an important consideration in the development of NCEs as new drug candidates. Metabolic activation can be effected by phase 1 or phase 2 enzymes. Depending on the mechanism and the substrate, oxidative reactions can yield reactive intermediates such as epoxides, quinone methides, quinone imines, and quinones, which are alkylating agents, which react with cellular constituents such as DNA and proteins. A comprehensive listing of known bioactivation reactions and mechanisms has been presented [83, 165]. These modifications are thought to be responsible for DNA damage that can result in tumor initiation and promotion, and idiosyncratic reactions that are observed in patients post marketing of a new drug. The Ames test and micronucleus assay are routinely used to determine the carcinogenic potential of an NCE early in the development process. In addition, metabolism-based inactivation (MBI) assays have been established to detect either an enzymatic activity loss at a single concentration or an IC50 shift when an NCE is preincubated with either pooled human liver microsomes or recombinant CYP enzymes, and human hepatocytes. However, the relationships between protein covalent binding by reactive intermediates and toxicity are not well understood.

An *in vitro* test commonly used to establish reactive intermediates is the trapping of a reactive intermediate by glutathione [166–170]. While this test shows the potential for a reactive intermediate to be generated from an NCE by the oxidative enzymes in hepatic microsomes, it does not imply that toxicity will necessarily result from such an intermediate. Acetaminophen (Tylenol) and ethynylestradiol are examples of drugs that, when used in the proper manner, have proven to be extremely safe, and yet are positive for reactive intermediates when tested with the glutathione reactive intermediate screen. Accordingly, a positive reaction for an NCE in this assay should by itself not constitute a kill shot, rather it should serve as a flag for enhanced vigilance in toxicological testing, and such information should be used cautiously when selecting between NCEs for progression of compounds for drug development [171].

As described earlier, conjugation of xenobiotics to glucuronide or sulfate is generally associated with detoxification. However, phase II enzymes can also catalyze the bioactivation of drugs. For example, acyl glucuronides formed by UGTs from carboxylic acid functions present in drugs, NCEs or their metabolites can be reactive intermediates that acylate macromolecules by transferring the aglycone moiety. The modified macromolecules may be recognized by the immune system and elicit an immunological response either to the aglycone or the modified macromolecule. Consequences of such events can be hypersensitization to the drug or induction of auto immunity [80, 172–175]. Acyl glucuronides undergo intramolecular acyl group migration, which is used as a measure of reactivity. A recent NMR study provides a rapid means of assessing the migration half-life and the competing hydrolytic process. A general observation is that increased substitution at the α-carbon of the carboxylic acid increases the migration half-life and decreases toxicity [176]. Clearly, a better understanding of the relationships between reactive intermediates, macromolecular adducts, and toxicity is necessary.

NATs are also involved in bioactivation reactions via O-acetylation of N-hydroxylamines formed from CYP-mediated N-hydroxylation of arylamines. These bioactivation reactions form unstable acetoxy esters that decompose to highly reactive species, which bind to cellular DNA [83]. The O-sulfonation of compounds catalyzed by SULTs can also result in the formation reactive intermediates. Recently, it has been shown that α-hydroxytamoxifen (derived from CYP-mediated hydroxylation of tamoxifen) is bioactivated by SULTs [177].

1.6.2.3 Induction and Repression of Drug-Metabolizing Enzymes.

Among the various classes of drug-metabolizing enzymes, members of the CYP family have been the most extensively studied for susceptibility to regulation by xenobiotic and endobiotic mediators. FMOs and MAOs have not been shown to be susceptible to induction or repression by xenobiotics, and UGTs have not been sufficiently investigated despite large interindividual variability. Induction of CYPs by xenobiotics and drugs has been known since early studies on this class of enzymes. Inductions of CYP1A, 2B, and 3A families by drugs such as phenobarbital, dexamethasone, rifampin, and xenobiotics like dioxins, benzo[a]pyrene and others have been well documented. An increase in functional expression of a drug-metabolizing enzyme can result in increased clearance of a drug, resulting in reduced efficacy. Repression of expression can result in toxic accumulation of a drug above its TI. Induction mechanisms are complex and involve the proximal promoter and xenobiotic-response enhancer regions of genes to which bind the various nuclear transcription factors. The various nuclear transcription factors can interact individually or in concert, creating an interaction network that affects the magnitude of the induction response [178, 179]. Critical to the development of new drugs is the dose-dependent magnitude of the induction response to an NCE [180]. It is recognized that major differences exist in the response to inducers in humans and preclinical models as a consequence of differences in xenobiotic receptors and nuclear transcription factors including differences in the binding affinities of receptors for xenobiotics. For example, the ligand binding domains of rabbit, rodent, and human pregnane X receptor (NR1I2) share only 75–80% sequence identity. Thus, induction in preclinical species may not appropriately reflect induction in humans or vice versa. The importance of nuclear receptors in regulation of drug-metabolizing enzymes has been established. However, much effort is needed to elucidate the molecular mechanisms involved in these processes. Primary cultures of human hepatocytes have been used to examine induction of drug-metabolizing enzymes by NCEs [181]. Recently, the use of an immortalized hepatocyte cell line (Fa2N-4) has been shown to be more reproducible for examining induction [182, 183].

1.6.3 Polymorphisms

Genetic polymorphisms in the form of single nucleotide polymorphisms (SNPs) have been observed in most genes that have been examined in mammals. SNPs are the most commonly encountered mutations that are observed either in the coding or promoter regions of genes encoding drug-metabolizing enzyme. Such changes can have effects that range from innocuous to extremely deleterious. The relevant forms of polymorphism from a drug metabolism perspective are those that affect the clearance and active transport of drugs. Genetic variability in drug-metabolizing enzymes is a significant contributor to the variability in human drug pharmacokinetics. An SNP that results in a stop codon within the coding region can result in a lack of expression, or expression of a nonfunctional or unstable protein that is phenotypically observed as a lack of metabolic activity. SNPs that result in amino acid changes can alter protein structure with variable effects ranging from a lack of activity to varying degrees of impaired activity. Other SNPs can be silent, resulting in no changes in the sequences or protein expression. SNPs in the promoter regions of the gene can affect levels of protein expression. Thus, for an NCE extensively metabolized by a particular drug-metabolizing enzyme, polymorphisms in the gene could cause plasma concentrations to rise to levels far above its TI with toxic consequences. This would be particularly critical for NCEs with low TIs. Hence, knowledge of the enzymes involved in clearance of NCEs along with knowledge of polymorphisms that exist for these enzymes is critical to the development of safer drugs.

1.6.3.1 CYP450. Polymorphisms that have been discovered in the genes for CYPs, 2D6, and 2C19 are particularly good examples of the importance of knowing poly-morphisms in drug-metabolizing enzymes and the challenges in the development of NCEs as new drugs. The "Debrisoquine Polymorphism" observed for the hy-pertensive drug debrisoquin was first recognized in the mid-1970s among hyper-tensive patients as a bimodal distribution in the ratio between debrisoquine and its 4-hydroxydebrisoquine metabolite, and via familial studies was suggested to be caused by a recessive allelic gene [184–186]. Subsequently this polymorphic effect was correlated with other drugs used in the 1970s and 1980s [187], and the en-zyme responsible for the 4-hydroxylation of debrisoquine was identified [188, 189]. Extensive genotype/phenotype studies have been conducted in ethnically different populations around the globe and over 60 allelic variants have been identified for CYP2D6. The effect on enzymatic activity of these allelic forms is varied and ex-tends from increased to complete loss in metabolic capacity of this enzyme. For drugs primarily cleared by CYP2D6 the consequence of increased metabolic capac-ity would be subtherapeutic levels of drug and loss of optimal efficacy. More critically, for patients with allelic variants moderately to extensively devoid of metabolic ca-pacity, the effect would be increased drug levels with consequent toxicity. Hence, patient genotyping would appear to be critical prior to administration of drugs with low TIs and a significant contribution of CYP2D6 to their clearance. Consideration of CYP2D6 genotype can also be important when a drug such as Timolol is used top-ically. Timolol, a known CYP2D6 substrate, is known to have significant cardiopul-monary effects. The fraction absorbed systemically from a 0.5% aqueous ophthalmic application increases heart rate due to systemic absorption. With PMs the systemic exposure can reach levels sufficient to effect QT prolongation and consequent cardiac arrhythmia.

Similarly, for CYP2C19 over 20 allelic variants have been identified with sev-eral variants devoid of enzymatic activity. The activation of clopidogrel to its active metabolite is known to involve both CYP3A4 and CYP2C19. Clinical studies in healthy subjects have shown that in subjects carrying the CYP2C19*2 allelic vari-ant, platelet responsiveness was markedly decreased [190]. Similarly, in the use of the anticancer drug Indisulam, which is metabolized primarily by CYPs 2C9 and 2C19, patients carrying allelic variants CYP2C9*6 or CYP2C19*2 can be suscep-tible to neutropenia due to increased systemic exposure to the drug [191]. SNPs have been identified among all the major CYP isoforms including the CYP3A family.

The majority of allelic variants identified for CYP3A4 and 3A5 either have no phe-notypic effect or result in diminished catalytic activity, which can result in an increase in circulating drug levels to some extent. A rare allelic form of CYP3A4 recently discovered in a German subject with a midazolam clearance of $2.99 \, \text{mL} \, (\text{min kg})^{-1}$ has been shown to be due to the lack of heme incorporation into the heterologously expressed protein [192]. The frequency of this variant in the general population has not been established, and its effect on drug metabolism by such allelic carriers remains to be established. Tacrolimus, cyclosporine, and some other CYP3A cleared drugs that have low TIs are particularly sensitive to allelic variants with diminished metabolic capacity, and care in dose administration is critical. Clinical studies in renal transplant patients have shown that the CYP3A5 genotype correlated with dose, and the mean dose required to achieve efficacious concentrations was lowest for the CYP3A5*1*1 allele. Tacrolimus is nephrotoxic and hence genotyping of subjects is critical for effective dose administration [193, 194].

Allelic variants for other CYP isoforms involved in the metabolism of drugs have been documented and in some cases effects of allelic variants have been examined. A compendium of allelic variants for human CYP isoforms in families 1, 2, 3, and 4 is available on the Web at the home page of the Human Cytochrome P450 (CYP) Allele Nomenclature Committee (http://www.cypalleles.ki.se/cypalleles.html).

1.6.3.2 UGTs and SULTs. Polymorphisms have been identified in all the UGT isoforms that have been examined. The mutations are observed in the coding and promoter regions of the genes. A compendium of allelic variants in families 1 and 2 is available on the Web at http://galien.pha.ulaval.ca/alleles/alleles.html. Sixty-two allelic variants have been reported for the UGT1A1 gene. Bilirubin is cleared as its glucuronide only by UGT1A1. Several syndromes are associated with mutations in the UGT1A1 gene. The type I Crigler–Najar (CN) syndrome is lethal and is due to any of 36 allelic forms of UGT1A1 that result in complete loss of UGT1A1 activity. Systemic accumulation of bilirubin results in severe neurological toxicity. Currently, only liver transplant is the approach used to treat such patients [195]. Gene and cell therapies hold promise for future treatment of this disease [196, 197]. Sixteen allelic variants of UGT1A1 result in the less lethal, type II CN syndrome. Gilbert syndrome is a form of inherited mild hyperbilirubinemia that results from mutations in the TATA box upstream of the UGT1A1 gene resulting in altered gene expression [198]. The magnitude of severity in type II CN syndrome is determined by the intrinsic activity of allelic forms toward bilirubin and treatments vary from phenobarbital and phototherapy to dietary control [197].

The active metabolite of Irinotecan, SN-38, and the cancer drug Etoposide are drugs that are cleared primarily by UGT1A1 and have narrow TIs. To maintain drug concentrations within the therapeutic window, it would appear critical to genotype patients to be treated with such drugs. Therefore, from a drug development perspective, *in vitro* knowledge of the involvement of UGT1A1 in primary clearance of, or inhibition by, an NCE or its metabolite and their potential to limit bilirubin clearance would appear useful to conduct at an early drug discovery stage. Such knowledge is useful either to conduct early drug interaction studies to limit late stage attrition or as indicators for patient genotyping prior to commencement of therapy.

Polymorphisms have also been identified among several SULT members, and some secondary effects on metabolism have been documented. However, since this enzyme system has not yet been identified in any critical drug metabolism function, the effects of such mutations have been far less defined [199].

1.6.3.3 Other Drug-Metabolizing Enzymes. MOA-A and MOA-B are critical for maintenance of neurotransmitter homeostasis. Polymorphisms in these genes have been encountered and reports generally address polymorphisms in the promoter regions of these enzymes that result in alterations in enzyme levels [68]. No reports exist in the literature where polymorphisms encountered in the coding region of the genes have dramatically altered the catalytic activities for these enzymes.

There are five human flavin monooxygenase (FMO) gene families. Polymorphisms are known for these enzymes. However, due to a lack of significant primary roles in drug clearance or demonstrated DDIs, polymorphisms among these enzymes have not gained much attention [200]. FMO3 has received the highest attention because of its physiological role in clearance of trimethylamine. Polymorphic forms of FMO3 result in variable loss of trimethamine metabolic capacity from moderate to severe. This results in varying degrees of trimethylaminuria (fish odor syndrome) and consequent psychological effects [200].

AO/XO belongs to the family of molybdenum cofactor dependent enzymes as described earlier. XO has received the most attention because of its role in the metabolism of purines. Type I xanthinuria is caused by mutations in the XO gene that result in a lack of protein expression or inactive protein [201, 202]. Type II xanthinurias are caused by potentially inactive XO polymorphism and due to lack of the molybdenum cofactor caused by mutations in the molybdenum cofactor sulfurase gene [203, 204]. Literature on polymorphic forms of AO is limited. However, with increasing involvement of AO in drug clearance, attention needs to be focused on polymorphic forms of this enzyme. A recent report has identified Donryu rats with significant variability in clearance of the IND RS-8359. The variability was shown to be associated with mutations in the AO gene [205].

While no critical physiological function for the two forms of NAT1 and NAT2 have been identified, these enzymes have been associated with drug and xenobiotic metabolism, and with polycyclic aromatic amine-induced cancer. Humans have been classified as high, medium, and PMs with respect to these genes. Various reports have tried to link acetylator status with disease states. However, while trends have been noted, no definitive linkages have been established [206]. SNPs in the genes encoding for NAT1 and NAT2 are responsible for the high degree of variability observed in N-acetylator status. The antituberculosis drug isoniazid is cleared by NATs, and consequently the NAT genotype does influence the dose requirement for treatment of tuberculosis. Thus, from a drug metabolism perspective, it may be important to know if NATs play a role in clearance of NCEs. However, these enzymes have not become frontline screens in the drug discovery and development paradigm.

1.6.3.4 Transporters. If the NCEs are substrates for transporters, systemic or target tissue availability of the NCE may become limiting, and consequently influence the pharmacokinetics and pharmacodynamics of the NCE. Efflux transporters such as P-gp present in intestinal epithelia can have negative effects on the bioavailability of NCEs, particularly those that have poor membrane permeability. Whereas for NCEs that have good membrane permeability, efflux transporters do not play as critical role in intestinal absorption since a dose-dependent saturation of the efflux pump is possible. While systemic exposure may be achieved by overcoming the intestinal efflux barrier, target tissues that express such transporters may function as efficient barriers to tissue penetration from systemic circulation. For example, P-gp present in the blood/brain barrier and in tumor cells is an effective hindrance to brain and tumor cell access of P-gp substrates [207, 208]. Uptake and efflux transporters in the liver and kidney can have similar influences on systemic exposure of NCEs [209]. Inhibition of transporter function can, depending on the endogenous role of a transporter, result in toxicity. For example, inhibition of bile salt transporters can result in cholestasis and hepatic failure [210]. Conversely, inhibition of P-gp efflux transporters can enhance efficacy of antineoplastic agents [211]. While extensive knowledge has been gained in transporters, techniques for rapid evaluation of their roles in pharmacokinetics is still lacking and major strides are needed to effectively incorporate transporters in the drug discovery paradigm.

SNPs in many classes of transporter genes are known, particularly among those involved in bile salt homeostasis [212]. Polymorphisms in P-gp have also been identified and are thought in part to contribute to variability in P-gp function. However, insufficient studies have thus far been conducted to establish distinct roles for SNPs [213].

1.7 SUMMARY

Since the early 1980s a majority of the human enzymes involved in drug metabolism such as the CYPs, UGTs, MAOs, FMOs, and some of the lesser encountered drug-metabolizing enzymes have been isolated or heterologously expressed and characterized, and more recently some crystal structures have been solved, providing a better understanding of the structural basis for their broad specificity. Knowledge of transporters such as P-gp, MRPs, OATPs, and OCTs has expanded with construction of cell systems incorporating specific transporters for use in characterizing their role in drug efflux or uptake. Progress in molecular genetics of drug-metabolizing and related enzyme systems has enabled some understanding of the molecular basis for inherited traits such as PMs and EMs. Several CYP isoform-specific inhibitors and substrates are known that allow for the elucidation of some primary metabolic pathways and an evaluation of the potential for DDIs. The enzyme kinetic parameters (K_m, K_i, V_m) for the isoforms involved in clearance or inhibition can be accurately determined and used to assess the potential for drug interactions and issues that may be associated with metabolism by, or inhibition of, polymorphic forms that can alter projected pharmacokinetic parameters. With this information preclinical and clinical studies can be designed to address and evaluate such concerns. Expansion of our knowledge of the reaction mechanisms of CYP enzymes, their substrate specificities, and crystal structures allows for *in silico* predictive models to be constructed. This can minimize exhaustive experimental testing by selectively examining compounds in a series. Advances in structure elucidation by soft ionization mass spectroscopy allow for rapid characterization of metabolites and identification of metabolic hot spots. When done in an iterative manner, this can aid in designing compounds with better pharmacokinetic properties.

Advances in our understanding of mechanisms of induction through nuclear receptors are expanding and allow for early screening to identify compounds with the potential to induce CYP isoforms using *in vitro* ligand binding assays for nuclear receptors rather than preclinical animal models where outcomes often do not translate to humans. This knowledge allows for appropriate nonclinical or clinical testing to be conducted at an early stage to mitigate late stage attrition due to enzyme induction.

Progress in efflux and uptake transporters allows for assessment of the role of the more commonly encountered transporters in achieving targeted drug concentrations at the site of action. This is particularly important in CNS and cancer programs where P-gp plays a critical role in the blood–brain barrier and tumor cell surface by preventing drug entry. Thus if CNS or cancer drugs are P-gp substrates, depending on the intrinsic efflux parameter, TI, and safety window, such drugs may not achieve their targeted concentrations for efficacy. For drugs not targeted to organs or sites with high P-gp content, the presence of P-gp in the intestinal wall, although limiting drug entry, is not as critical since saturation of the pathway can be achieved by increasing the oral dose to overcome the efflux. Biliary efflux pumps can also play a significant role in limiting systemic exposure. As tools to assess the roles of such transporters become available, the ability to more accurately predict behavior in human subjects becomes a reality.

Knowledge of non-CYP metabolic pathways has also expanded. Tools are available to assess if enzymes, such as MAOs, UGTs, aldehyde or xanthine oxidases, contribute to the clearance. Lacking are means of assessing the magnitude of the contribution made by these enzymes to the overall clearance. As medicinal chemists become more effective in limiting CYP-mediated metabolism of NCEs, non-CYP metabolic pathways will become increasingly involved in drug clearance and consequently better tools are necessary to evaluate the role such enzymes may play in metabolism of NCEs early in the drug development process.

Most drug testing is conducted in adult humans between the ages of 18 and 60 years. Pediatric use of drugs has generally been limited to allometric scaling from adult pharmacokinetic studies. Developmental changes in the hepatic expression of some drug-metabolizing enzymes have been noted; however, far greater understanding and statistical knowledge is needed for projection of pediatric pharmacokinetics and simulation thereof by predictive tools such as SimCYP can be effectively employed.

Despite major strides in understanding of drug metabolic processes, knowledge in several areas is still lacking, and predictive tools, while significantly improved, need further enhancement. *In silico* tools are improving with better ability to predict metabolic hot spots (Meteor and Metasite). As technologies evolve, strategies used in drug metabolism studies in early drug discovery will also evolve to fit changing paradigms.

REFERENCES

1. Grabowski, H., Vernon, J., and DiMasi, J. (2002) Returns on research and development for 1990 drug introductions. *Pharmacoeconomics* **20**, 11–29.

2. Caldwell, J. (1996) The role of drug metabolism studies for efficient drug discovery and development: opportunities to enhance time- and cost-efficiency. *J. Pharm. Sci.* **2**, 117–119.

3. DiMasi, J. A., Hansen, R. W., and Grabowski, H. G. (2003) The price of innovation: new estimates of drug development costs. *J. Health Econ.* **22**, 151–185.

4. Alavijeh, M. S. and Palmer, A. M. (2004) The pivotal role of drug metabolism and pharmacokinetics in the discovery and development of new medicines. *Invest. Drugs* **7**, 755–763.

5. Segall, M. D., Beresford, A. P., Gola, J. M. R., Hawksley, D., and Tarbit, M. H. (2006) Focus on success: using a probabilistic approach to achieve an optimal balance of compound properties in drug discovery. *Expert Opin. Drug Metab. Toxicol.* **2** (2), 325–337.

6. Kola, I. and Landis, J. (2004) Can the pharmaceutical industry reduce attrition rats? *Nat. Rev. Drug Discov.* **3**, 711–715.

7. Baranczewski, P., Stanczak, A., Kautianinen, A., Sandin, P., and Edlund, P.-O. (2006) Introduction to early *in vitro* identification of metabolites of new chemical entities in drug discovery and development. *Pharmacol. Rep.* **58**, 341–352.

8. Bachmann, K. A. and Ghosh, R. (2001) The use of *in vitro* methods to predict *in vivo* pharmacokinetics and drug interactions. *Curr. Drug Metab.* **3**, 299–314.

9. Hewitt, N. J., Gómez Lechón, M. J., Houston, B. J., Hallifax, D., Brown, H. S., Maurel, P., Kenna, J. G., Gustavsson, L., Lohmann, C., Skonberg, C., et al. (2007) Primary hepatocytes: current understanding of the regulation of metabolic enzymes and transporter proteins, and pharmaceutical practice for the use of hepatocytes in metabolism, enzyme induction, transporter, clearance, and hepatotoxicity studies. *Drug Metab. Rev.* **39**, 159–234.

10. Beresford, A. P., Segall, M. D., and Tarbit, M. H. (2004) In silico prediction of ADME properties: are we making progress? *Curr. Opin. Drug Discov. Dev.* **7**, 36–42.

11. Noller, J., Schmitt, J., Hartmann, T., and Loffler, T. (2005) Just married-in vitro assays and in silico ADME prediction. *Screening* **1**, 32–33.

12. Bajorath, J. (2002) Integration of virtual and high-throughput screening. *Nat. Rev. Drug Discov.* **1**, 882–894.

13. Mahmood, I. (1998) Interspecies scaling predicting volumes, mean residence time and elimination half-life. Some considerations. *J. Pharm. Pharmacol.* **50**, 493–499.

14. Jang, J. R., Harris, R. Z., and Lau, D. T. (2001) Pharmacokinetics and its role in small molecule drug research. *Med. Res. Rev.* **21**, 382–396.

15. Fura, A., Shu, Y., Zhu, M., Hanson, R. L., Roongta, V., and Humphreys, W. G. (2004) Discovering drugs through biological transformation: role of pharmacologically active metabolites in drug discovery. *J. Med. Chem.* **47**, 4339–4351.

16. Baillie, T. A. (2003) *Drug Metabolizing Enzymes: Cytochrome P450 and Other Enzymes in Drug Discovery and Development*, Marcel Dekker, New York, pp. 147–154.

17. Evans, D. C., Watt, A. P., Nicoll-Griffith, D. A., and Baillie, T. (2004) Drug-protein adducts: an industry perspective on minimizing the potential for drug bioactivation in drug discovery and development. *Chem. Res. Toxicol.* **17**, 3–6.

18. Kalgutkar, A. S., Gardner, I., Obach, R. S., Shaffer, C. L., Callegari, E., Henne, K. R., Mutlib, A. E., Dalvie, D. K., Lee, J. S., Nakai, Y., et al. (2005) A comprehensive listing of bioactivation pathways of organic functional groups. *Curr. Drug Metab.* **6**, 161–225.

19. Clark, S. E. and Jones, B. C. (2002) Human cytochrome P450 and their role in metabolism-based drug–drug interactions. In: *Drug and Pharmaceutical Sciences: Drug–Drug Interactions*, Rodrigues, A. D., ed., Marcel Dekker, New York, pp. 55–88.

20. Obach, R. S. (2003) Drug–drug interactions: an important negative attribute in drugs. *Drug Discov. Today* **39**, 301–338.

21. Watt, A. P., Mortishire-Smith, R. J., Gerhard, U., and Thomas, T. (2003) Metabolite identification in drug discovery. *Drug Discov. Dev.* **6**, 57–65.

22. Baillie, T. A., Cayen, M. N., Fouda, H. G., Gerson, R. J., Green, J. D., Grossman, S. J., Klunk, L. J., LeBlanc, B., Perkins, D. G., and Shipley, L. A. (2002) Drug metabolites in safety testing. *Toxicol. Appl. Pharmacol.* **182**, 188–196.

23. Smith, D. A. and Obach, R. S. (2005) Seeing through the MIST: abundance versus percentage. Commentary on metabolites in safety testing. *Drug Metab. Dispos.* **33**, 1409–1417.

24. Guengerich, F. P. (2003) Cytochrome P450, drugs, and diseases. *Mol. Interv.* **3** (4), 194–204.

25. Rendic, S. (2002) Summary of information on human CYP enzymes: human P450 metabolism data. *Drug Metab. Rev.* **34**, 83–448.

26. Shou, M., Korzekwa, K. R., Crespi, C. L., Gonzalez, F. J., and Gelboin, H. V. (1994) The role of 12 cDNA-expressed human, rodent, and rabbit cytochrome P450 in the metabolism of benzo[a]pyrene and benzo[a]pyrene trans-7,8-dihydrodiol. *Mol. Carcinog.* **10** (3), 159–68.

27. Shou, M., Krausz, K. W., Gonzalez, F. J., and Gelboin, H. V. (1996) Metabolic activation of the potent carcinogen dibenzo[a,l]pyrene by human recombinant cytochrome P450, lung and liver microsomes. *Carcinogenesis* **17**, 2429–2433.

28. Baldwin, S. J., Bloomer, J. C., Smith, G. J., Ayrton, A. D., Clarke, S. E., and Chenery, R. J. (1995) Ketoconazole and sulphaphenazole as the respective selective inhibitors of P4503A and 2C9. *Xenobiotica* **25**, 261–270.

29. Sutter, T. R., Tang, Y. M., Hayes, C. L., Wo, Y., Jabs, E. W., Li, X., Yin, H., Cody, C. W., and Greenlee, W. F. (1994) Complete cDNA sequence of a human dioxin-inducible mRNA identifies a new gene subfamily of cytochrome P450 that maps to chromosome 2. *J. Biol. Chem.* **269**, 13092–13099.

30. Murray, G. I., Taylor, M. C., Mcfadyen, M. C., McKay, J. A., Greenlee, W. F., Burke, M. D., and Melvin, W. T. (1997) Tumor-specific expression of cytochrome P450 CYP1B1. *Cancer Res.* **57**, 3026–3031.

31. Belous, A. R., Hachey, D. L., Dawling, S., Roodi, N., and Parl, F. F. (2007) Cytochrome P 450 1B1-mediated estrogen metabolism results in estrogen-deoxyribonucleoside adduct formation. *Cancer Res.* **67** (2), 812–817.

32. Hukkanen, J., Pelkonen, O., Hakkola, J., and Raunio, H. (2002) Expression and regulation of xenobiotic-metabolizing cytochrome P450 (CYP) enzymes in human lung. *Crit. Rev. Toxicol.* **32** (5), 391–411.

33. Nebert, D. W., Dalton, T. P., Okey, A. B., and Gonzalez, F. J. (2004) Role of aryl hydrocarbon receptor-mediated induction of the CYP1 enzymes in environmental toxicity and cancer. *J. Biol. Chem.* **279** (23), 23847–23850.

34. Guengerich, F. P. (1997) Comparisons of catalytic selectivity of cytochrome P450 subfamily enzymes from different species. *Chem. Biol. Interact.* **106** (3), 161–82.

35. Honkakoski, P. and Negish, M. (1997) The Structure, function, and regulation of cytochrome P450 2A enzymes. *Drug Metab. Rev.* **29**, 977–996.

36. Smith, G. B. J., Bend, J. R., Bedard, J. J., Reid, K. R., Petsikas, D., and Massey, T. E. (2003) Biotransformation of 4-(methylnitrosamino)-1-(3-pyridyl)-1-butanone (NNK) in peripheral human lung microsomes. *Drug Metab. Dispos.* **31** (9), 1134–1141.

37. Ruangyuttikarn, W., Appleton, M. L., and Yost, G. S. (1991) Metabolism of 3-methylindole in human tissues. *Drug Metab. Dispos.* **19** (5), 977–84.

38. He, X. Y., Tang, L., Wang, S. L., Cai, Q. S., Wang, J. S., and Hong, J. Y. (2006) Efficient activation of aflatoxin B1 by cytochrome P450 2A13, an enzyme predominantly expressed in human respiratory tract. *Int. J. Cancer* **118**, 2665–2671.

39. Rivera, S. P., Saarikoski, S. T., and Hankinson, O. (2002) Identification of a novel dioxin-inducible cytochrome P450. *Mol. Pharmacol.* **61** (2), 255–259.

40. Court, M. H., Duan, S. X., Hesse, L. M., Venkatakrishnan, K., and Greenblatt, D. J. (2001) Cytochrome P450 2B6 is responsible for interindividual variability of propofol hydroxylation by human liver microsomes. *Anesthesiology* **94**, 110–119.

41. Lang, T. A., Klein, K. A., Fischer, J. A., Nussler, A. K. B., Neuhaus, P. B., Hofmann, U. A., Eichelbaum, M. A., Schwab, M. A., and Zanger, U. M. A. (2001) Extensive genetic polymorphism in the human CYP2B6 gene with impact on expression and function in human liver. *Pharmacogenetics* **11**, 399–415.

42. Code, E. L., Crespi, C. L., Penman, B. W., Gonzalez, F. J., Chang, T. K., and Waxman, D. J. (1997) Human cytochrome P4502B6: interindividual hepatic expression, substrate specificity, and role in procarcinogen activation. *Drug Metab. Dispos.* **25**, 985–993.

43. Lamba, V., Lamba, J., Yasuda, K., Strom, S., Davila, J., Hancock, M. L., Fackenthal, J. D., Rogan, P. K., Ring, B., Steven, A., et al. (2003) Hepatic CYP2B6 expression: gender and ethnic differences and relationship to CYP2B6 genotype and CAR (constitutive androstane receptor) expression. *J. Pharmacol. Exp. Ther.* **307**, 906–922.

44. Romkes, M., Faletto, M. B., Blaisdell, J. A., Raucy, J. L., and Goldstein, J. A. (1991) Cloning and expression of complementary DNAs for multiple members of the human cytochrome P450IIC subfamily. *Biochemistry* **30**, 3247–3255.

45. Zuber, R., Anzenbacherova, E., and Anzenbacher, P. (2002) Cytochrome P450 and experimental models of drug metabolism. *J. Cell. Mol. Med.* **6**, 189–198.

46. Hanioka, N., Kimura, S., Meyer, U. A., and Gonzalez, F. J. (1990) The human CYP2D6 locus associates with a common genetic defect in drug oxidation: a G1934 to a base change in intron 3 of a mutant CYP2D6 allele results in an aberrant 3′ splice recognition site. *Am. J. Hum. Genet.* **47**, 994–1001.

47. Mulder, H., Herder, A., Wilmink, F. W., Tamminga, W. J., Belitser, S. V., and Egberts, A. C. G. (2006) The impact of cytochrome P450-2D6 genotype on the use and interpretation of therapeutic drug monitoring in long-stay patients treated with antidepressant and antipsychotic drugs in daily psychiatric practice. *Pharmacoepidemiol. Drug Saf.* **15** (2), 107–114.

48. Lofgren, S., Hagbjork, A., Ekman, S., Fransson-Steen, R., and Terelius, Y. (2004) Metabolism of human cytochrome P450 marker substrates in mouse: a strain and gender comparison. *Xenobiotica* **34**, 811–834.

49. Ronis, M. J. J., Lindros, K. O., Ingelman-Sundberg, M., Ioannides, C., and Parke, D. V., eds. (1996) *Cytochromes P450: Metabolic and Toxicological Aspects. 9, The CYP2E Subfamily*, CRC Press, Boca Raton, FL, pp. 211–239.

50. Carr, B. A., Wan, J., Hines, R. N., and Yost, G. S. (2003) Characterization of the human lung CYP2F1 gene and identification of a novel lung-specific binding motif. *J. Biol. Chem.* **278** (18), 15473–15483.

51. Lanza, D. L., Code, E., Crespi, C. L., Gonzalez, F. J., and Yost, G. S. (1999) Specific dehydrogenation of 3-methylindole and epoxidation of naphthalene by recombinant human CYP2F1 expressed in lymphoblastoid cells. *Drug Metab. Dispos.* **27**, 798–803.

52. Thornton-Manning, J., Appleton, M. L., Gonzalez, F. J., and Yost, G. S. (1996) Metabolism of 3-methylindole by vaccinia-expressed P450 enzymes: correlation of 3-methyleneindolenine formation and protein-binding. *J. Pharmacol. Exp. Ther.* **276** (1), 21–29.

53. Zeldin, D. C., Foley, J., Ma, J., Boyle, J. E., Pascual, J. M., Moomaw, C. R., Tomer, K. B., Steenbergen, C., and Wu, S. (1996) CYP2J subfamily P450s in the lung: expression, localization, and potential functional significance. *Mol. Pharmacol.* **50** (5), 1111–1117.

54. Matsumoto, S., Hirama, T., Matsubara, T., Nagata, K., and Yamazoe, Y. (2002) Involvement of CYP2J2 on the intestinal first-pass metabolism of antihistamine drug, astemizole. *Drug Metab. Dispos.* **30** (11), 1240–1245.

55. Hoffman, S. M., Nelson, D. R., and Keeney, D. S. (2001) Organization, structure and evolution of the CYP2S gene cluster on human chromosome 19. *Pharmacogenetics* **11**, 687–698.

56. Karlgren, M., Miura, S., and Ingelman-Sundberg, M. (2005) Novel extrahepatic cytochrome P450s. *Toxicol. Appl. Pharmacol.* **207** (2, Suppl.), S57–S61.

57. Buckpitt, A., Boland, B., Isbell, M., Morin, D., Shultz, M., Baldwin, R., Chan, K., Karlson, A., Lin, C., Taff, A., et al. (2002) Naphthalene-induced respiratory tract toxicity: metabolic mechanisms of toxicity. *Drug Metab. Rev.* **34** (4), 791–820.

58. Wrighton, S. A., VandenBranden, M., and Ring, B. J. (1996) The human drug metabolizing cytochrome P450. *J. Pharmacokinet. Biopharmacol.* **24**, 461–473.

59. Li, A. P., Kaminsky, D. L., and Rasmussen, A. (1995) Substrate of human hepatic cytochrome P450 3A4. *Toxicol.* **104**, 1–8.

60. Domanski, D. L., Finta, C., Halpert, J. R., and Zaphiropoulos, P. G. (2001) cDNA cloning and initial characterization of CYP3A43, a novel human cytochrome P450. *Mol. Pharmacol.* **59**, 386–392.

61. Williams, J. A., Ring, B. J., Cantrell, V. E., Jones, D. R., Eckstein, J., Ruterbories, K., Hamman, M. A., Hall, S. D., and Wrighton, S. A. (2002) Comparative metabolic capabilities of CYP3A4, CYP3A5, and CYP3A7. *Drug Metab. Dispos.* **30**, 883–891.

62. Gellner, K., Eiselt, R., Hustert, E., Arnold, H., Koch, I., Haberl, M., Deglmann, C. J., Burk, O., Buntefuss, D., Escher, S., et al. (2001) Genomic organization of the human CYP3A locus: identification of a new inducible CYP3A gene. *Pharmacogenetics* **11**, 111–121.

63. Nebert, D. W. and Russell, D. W. (2002) Clinical importance of the cytochromes P450. *Lancet* **360**, 1155–1162.

64. Marill, J., Cresteil, T., Lanotte, M., and Chabot, G. G. (2000) Identification of human cytochrome P450s involved in the formation of all-*trans*-retinoic acid principal metabolites. *Mol. Pharmacol.* **58**, 1341–1348.

65. Cashman, J. R. (2000) Human flavin-containing monooxygenase: substrate specificity and role in drug metabolism. *Curr. Drug Metab.* **1**, 1037–1045.

66. Zhang, J. and Cashman, J. R. (2006) Quantitative analysis of FMO gene mRNA levels in human tissues. *Drug Metab. Dispos.* **34** (1), 19–26.

67. Shih, J. C., Chen, K., and Ridd, M. J. (1999) Monoamine oxidase from genes to behavior. *Annu. Rev. Neurosci.* **22**, 197–217.

68. Kalgutkar, A. S., Dalvie, D. K., Castagnoli, N., Jr., and Taylor, T. J. (2001) Interactions of nitrogen-containing xenobiotics with monoamine oxidase (MAO) isozymes A and B: SAR studies on MAO substrates and inhibitors. *Chem. Res. Toxicol.* **14** (9), 1139–1162.

69. Rashidi, M. R., Smith, J. A., Clarke, S. E., and Beedham, C. (1997) In vitro oxidation of famciclovir and 6-deoxypenciclovir by aldehyde oxidase from human, guinea pig, rabbit, and rat liver. *Drug Metab. Dispos.* **25**, 805–813.

70. Lake, B. G., Ball, S., Kao, J., Renwick, A. B., Price, R. J., and Scatina, J. A. (2002) Metabolism of zaleplon by human liver: evidence for involvement of aldehyde oxidase. *Xenobiotica* **32**, 835–847.

71. Sugihara, K., Kitamura, S., and Tatsumi, K. (1996) Involvement of mammalian liver cytosols and aldehyde oxidase in reductive metabolism of zonisamide. *Drug Metab. Dispos.* **24**, 199–202.

72. Kamel, A., Prakash, C., and Obach, S. (2002) Determination of the enzyme involved in the reductive metabolism of ziprasidone to dihydroziprasidone and the formation of S-methyl dihydroziprasidone using human in vitro systems and electrospray ionization tandem mass spectrometry. *Proceedings of the 50th ASMS Conference*, Orlando, USA.

73. Pristos, C. A. and Gustafson, D. L. (1994) Xanthine dehydrogenase and its role in cancer chemotherapy. *Oncol. Res.* **6**, 477.

74. Benhamou, S., Reinikanen, M., Bouchardy, C., Dayer, P., and Hirvonen, A. (1998) Association between lung cancer and microsomal epoxide hydrolase genotypes. *Cancer Res.* **58**, 5291–5293.

75. Razonable, R. R. and Paya, C. V. (2004) Valganciclovir for the prevention and treatment of cytomegalovirus disease in immunocompromised hosts. *Expert Rev. Anti–infect. Ther.* **2** (1), 27–42.

76. McClellan, K. and Perry, C. M. (2001) Oseltamivir: a review of its use in influenza. *Drugs* **61** (2), 261–283.

77. Tukey, R. H. and Strassburg, C. P. (2000) Human UDP-glucuronosyltransferases: metabolism, expression, and disease. *Annu. Rev. Pharmacol. Toxicol.* **40**, 581–616.

78. Evans, R. W. and Relling, M. V. (1999) Pharmacogenomics: translating functional genomics into rational therapeutics. *Science* **286**, 487–491.

79. Hawes, E. (1998) N^+-glucuronidation, a common pathway in human metabolism of drugs with tertiary amine group. *Drug Metab. Dispos.* **26** (9), 830–837.

80. Benet, L. Z. and Spahn, H. (1988) Acyl migration and covalent binding of drug glucuronides—potential toxicity mediators. *Colloq. INSERM* **173**, 261–269.

81. Kauffman, F. C. (2004) Sulfonation in pharmacology and toxicology. *Drug Metab. Rev.* **36**, 823–843.

82. Hayes, J. D. and Pulford, D. J. (1995) The glutathione S-transferase supergene family: regulation of GST and the contribution of the isoenzymes to cancer chemoprotection and drug resistance. *Crit. Rev. Biochem. Mol. Biol.* **30**, 445–600.

83. Zhou, S., Chan, E., Duan, W., Huang, M., and Chen, Y.-Z. (2005) Drug bioactivation: covalent binding to target proteins and toxicity relevance. *Drug Metab. Rev.* **37** (1), 41–213.

84. Grant, D. M. (1993) Molecular genetics of N-acetyltransferases. *Pharmacogenetics* **3**, 45–50.

85. Hanna, P. E. (1996) Metabolic activation and detoxification of arylamines. *Curr. Med. Chem.* **3**, 195–210.

86. Tenhunen, J., Salminen, M., Lundström, K., Kiviluoto, T., Savolainen, R., and Ulmanen, I. (1994) Genomic organization of the human catechol-O-methyltransferase gene and its expression from two distinct promoters. *Eur. J. Biochem.* **223**, 1049–1059.

87. Weinshilboum, R. M., Otterness, D. M., and Szumlanski, C. L. (1999) Methylation pharmacogenetics: catechol O-methyltransferase, thiopurine methyltransferase, and histamine N-methyltransferase. *Annu. Rev. Pharmacol. Toxicol.* **39**, 19–52.

88. Prakash, C., Kamel, A., Gummerus, J., and Wilner, K. (1997) Metabolism and excretion of the antipsychotic drug, ziprasidone, in humans. *Drug Metab. Dispos.* **25**, 863–872.

89. Pacifici, G. M., Donatelli, P., and Giuliani, L. (1992) Histamine N-methyl transferase: inhibition by drugs. *Br. J. Clin. Pharmacol.* **34** (4), 322–327.

90. Pahlich, S., Zakaryan, R. P., and Gehring, H. (2006) Protein arginine methylation: cellular functions and methods of analysis. *Biochim. Biophys. Acta, Protein. Proteomics* **1764** (12), 1890–1903.

91. Beringer, P. M. and Salughter, R. L. (2005) Transporters and their impact on drug disposition. *Ann. Pharmacother.* **39**, 1097–1108.

92. van Montfoort, J. E., Hagenbuch, B., Groothuis, H., Koepsell, H., Meier, P. J., and Meijer, D. K. (2003) Drug uptake systems in liver and kidney. *Curr. Drug Metab.* **4**, 185–211.

93. Chen, C. J., Chin, J. E., Ueda, K., Clark, D. P., Pastan, I., Gottesman, M. M., and Robinson, I. B. (1986) Internal duplication and homology with bacterial transport proteins in the mdr1 (P-glycoprotein) gene from multidrug-resistant human cells. *Cell* **47**, 381–389.

94. Ambudkar, S. V., Kimchi-sarfaty, C., Sauna, Z. E., and Gottesman, M. M. (2003) P-glycoprotein: from genomics to mechanism. *Oncogene* **22**, 7468–7485.

95. Marzolini, C., Paus, E., Buclin, T., and Kim, R. B. (2004) Polymorphisms in human MDR1 (P-glycoprotein): recent advances and clinical relevance. *Clin. Pharmacol. Ther.* **75**, 13–33.

96. Deeley, R. G. and Cole, S. P. C. (2006) Substrate recognition and transport by multidrug resistance protein 1 (ABCC1). *FEBS Lett.* **580**, 1103–1111.

97. Sandusky, G. E., Mintze, K. S., Pratt, S. E., and Dantzig, A. H. (2002) Expression of multidrug resistance-associated protein 2 (MRP2) in normal human tissues and carcinomas using tissue microarrays. *Histopathology* **41**, 65–74.

98. Hirohashi, T., Suzuki, H., Chu, X. Y., Tamai, I., Tsuji, A., and Sugiyama, Y. (2000) Function and expression of multidrug resistance-associated protein family in human colon adenocarcinoma cells (Caco-2). *J. Pharmacol. Exp. Ther.* **292**, 265–270.

99. Loscher, W. and Potschka, H. (2005) Drug resistance in brain diseases and the role of drug efflux transporters. *Nat. Rev. Neurosci.* **6**, 591–602.

100. Kartenbeck, J., Leuschner, U., Mayer, R., and Keepler, D. (1996) Absence of the canalicular isoform of the MRP gene-encoded conjugate export pump from the hepatocytes in Dubin–Johnson Syndrome. *Hepatology* **23**, 106–066.

101. Doyle, L. A., Yang, W., Abruzzo, L. V., Krogmann, T., Gao, Y., Rishi, A. K., and Ross, D. D. (1998) A multidrug resistance transporter from human MCF-7 breast cancer cells. *Proc. Natl Acad. Sci. USA* **95**, 15665–15670.

102. Miyake, K., Mickley, L., Litman, T., Zhan, Z., Robey, R., Cristensen, B., Brangi, M., Greenberger, L., Dean, M., Fojo, T., et al. (1999) Molecular cloning of CDNAs which are highly over expressed in mitoxantrone-resistant cells: demonstration of homology to ABC transport genes. *Cancer Res.* **59**, 8–13.

103. Maliepaard, M., Scheffer, G. L., Faneyte, I. F., van Gastelen, M. A., Pijnenborg, A. C., Schinkel, A. H., van De Vijver, M. J., Scheper, R. J., and Schellens, J. H. (2001) Subcellular localization and distribution of the breast cancer resistance protein transporter in normal human tissues. *Cancer Res.* **61**, 3458–3464.

104. Jonker, J. W., Smit, J. W. M., Brinkhuis, R. F., Maliepaard, M., Beijnen, J. H., Schellens, J. H., and Schinkel, A. H. (2000) Role of breast cancer resistance protein in the bioavailability and fetal penetration of topotecan. *J. Natl. Cancer Inst.* **92**, 1651–1656.

105. Ifergan, A., Shafran, G., Jansen, J., Hooijberg, H., Scheffer, G. L., and Assaraf, Y. G. (2004) Folate deprivation results in the loss of breast cancer resistance protein (BCRP/ABCG2) expression. A role for BCRP in cellular folate homeostasis. *J. Biol. Chem.* **279**, 25527–25534.

106. Suzuki, H. and Sugiyama, Y. (2000) Transport of drugs across the hepatic sinusoidal membrane: sinusoidal drug influx and efflux in the liver. *Semin. Liver Dis.* **20**, 251–263.

107. Hagenbuch, B. and Meier, P. J. (2003) The superfamily of organic anion transporting polypeptides. *Biochim. Biophys. Acta* **1**, 1–18.

108. Lee, W. and Kim, R. B. (2004) Transporter and renal drug elimination. *Annu. Rev. Pharmacol. Toxicol.* **44**, 137–166.

109. Sweet, D. H. and Pritchard, J. B. (1999) The molecular biology of renal organic anion and organic cation transporters. *Cell. Biochem. Biophys.* **31**, 89–118.

110. De Graff, C., Vermeulen, N. P. E., and Feenstra, K. A. (2005) Cytochrome P450 in silico: an integrative modeling approach. *J. Med. Chem.* **48** (8), 2725–2755.

111. Ekins, S., Nikolsky, Y., Nikolskaya, T., and GeneGo, S. J. (2005) Techniques: application of systems biology to absorption, distribution, metabolism, excretion and toxicity. *Trends Pharmacol. Sci.* **26** (4), 202–209.

112. Afzelius, L., Arnby, C. H., Broo, A., Carlsson, L., Isaksson, C., Ulrik Jurva, U., Kjellander, B., Kolmodin, K., Nilsson, K., Raubacher, F., et al. (2007) State of the art tools for computational site of metabolism predictions: comparative analysis, mechanistically insights, and future applications. *Drug Metab. Rev.* **39**, 61–86.

113. De Groot, M. J., Auckland, M. J., Horne, V. A., Alex, A. A., and Jones, B. C. (1999) A novel approach to predicting P450-mediated drug metabolism. CYP2D6 catalyzed N-dealkylation reactions and qualitative metabolite predictions using a combined protein and pharmacophore model for CYP2D6. *J. Med. Chem.* **42**, 4062–4070.

114. Afzelius, L., Zamora, I., Masimirembwa, C. M., Karle, A., Anderson, T. B., Mecucci, S., Baroni, M., and Cruciani, G. (2004) Conformer- and alignment independent model for predicting structurally diverse competitive CYP2C9 inhibitors. *J. Med. Chem.* **47**, 907–914.

115. Boxenbaum, H. (1984) Interspecies pharmacokinetic scaling and the evolutionary-comparative paradigm. *Drug Metab. Rev.* **15**, 1071–1121.

116. Mahmood, I. and Balian, J. D. (1996) Interspecies scaling: prediction clearance of drugs in humans: three different approaches. *Xenobiotica* **26**, 887–895.

117. Feng, M. R., Lou, X., Brown, R. R., and Hutchaleelaha, A. (2000) Allometric pharmacokinetic scaling: towards the prediction of human oral pharmacokinetics. *Pharm. Res.* **17**, 410–418.

118. Nagilla, R. and Ward, K. W. (2004) A comprehensive analysis of the role of correction factors in the allometric predictivity of clearance from rat, dog, and monkey to humans. *J. Pharm. Sci.* **93** (10), 2522–2534.

119. Obach, R. S., Baxter, J. G., Liston, T. E., Silber, B. M., Jones, B. C., Macintyre, F., Rance, D. J., and Wastall, P. (1997) The prediction of human pharmacokinetic parameters from preclinical and in vitro metabolism data. *J. Pharmacol. Exp. Ther.* **283** (1), 46–58.

120. Lave, T., Dupin, S., Schmitt, C., Chou, R. C., Jaeck, D., and Coassolo, P. F. (1997) Integration of in vitro data into allometric scaling to predict hepatic metabolic clearance in man: application to 10 extensively metabolized drugs. *J. Pharm. Sci.* **86** (5), 584–590.

121. Caldwell, G. W., Masucci, J. A., Yan, Z., and Hageman, W. (2004) Allometric scaling of pharmacokinetic parameters in drug discovery: can human CL, Vss and t1/2 be predicted from in-vivo rat data? *Eur. J. Drug Metab. Pharmacokinet.* **29** (2), 133–143.

122. McLean, M. A., Tam, C. J., Baratta, M. T., Holliman, C. L., Ings, R. M., and Galluppi, G. R. (2007) Accelerating drug development: methodology to support first-in-man pharmacokinetic studies by the use of drug candidate microdosing. *Drug Dev. Res.* **68** (1), 14–22.

123. Balani, S. K., Nagaraja, N. V., Qian, M. G., Costa, A. O., Daniels, J. S., Yang, H., Shimoga, P. R., Wu, J. T., Gan, L., Lee, F. W., et al. (2006) Evaluation of microdosing to assess pharmacokinetic linearity in rats using liquid chromatography-tandem mass spectrometry. *Drug Metab. Dispos.* **34** (3), 384–388.

124. Lappin, G., Kuhnz, W., Jochemsen, R., Kneer, J., Chaudhary, A., Oosterhuis, B., Drijfhout, W., Rowland, M., and Garner, R. C. (2006) Use of microdosing to predict pharmacokinetics at the therapeutic dose: experience with 5 drugs. *Clin. Pharmacol. Ther.* **80** (3), 203–215.

125. Sarapa, N. (2007) Exploratory IND: a new regulatory strategy for early clinical drug development in the United States. *Ernst Schering Res. Found. Workshop* **59**, 151–163, 116b.

126. Bertino, J. S., Jr., Greenberg, H. E., and Reed, M. D. (2007) American College of Clinical Pharmacology position statement on the use of microdosing in the drug development process. *J. Clin. Pharmacol.* **47** (4), 418–422.

127. Prakash, C., Shaffer, C., Tremaine, L., Liberman, R., Skipper, P., Flarakos, J., and Tannenbaum, S. (2007) Application of liquid chromatography-accelerator mass spectrometry (LC-AMS) to evaluate the metabolic profiles of a drug candidate in human urine and plasma. *Drug Metab. Lett.*, **1** (3), 226–231.

128. Souhaili, E. A., Batt, A. M., and Siest, G. (1986) Comparison of cytochrome P-450 content and activities in liver microsomes of seven animal species, including man. *Xenobiotica* **16** (4), 351–358.

129. Labedzki, A., Buters, J., Jabrane, W., and Fuhr, U. (2002) Differences in caffeine and paraxanthine metabolism between human and murine CYP1A2. *Biochem. Pharmacol.* **63** (12), 2159–2167.

130. Bjornsson, T. D., Callaghan, J. T., Einolf, H. J., Fischer, V., Gan, L., Grimm, S., Kao, J., King, S. P., Miwa, G., Ni, L., et al. (2003) The conduct of in vitro and in vivo drug–drug interaction studies: a PhRMA perspective. *J. Clin. Pharmacol.* **43**, 443–469.

131. Wienkers, L. C. and Heath, H. G. (2005) Risk assessment for drug–drug interactions caused by metabolism-based inhibition of CYP3A4 using automated in vitro assay systems and its application in early discovery process. *Drug Metab. Dispos.* **35** (7), 1232–1238.

132. Vickers, A. E. M., Sinclair, J. R., Zollinger, M., Heitz, F., Glanzel, U., Johanson, L., and Fischer, V. (1999) Multiple cytochrome P-450s involved in the metabolism of terbinafine suggest a limited potential for drug–drug interactions. *Drug Metab. Dispos.* **27** (9), 1029–1038.

133. Elewski, B. and Tavakkol, A. (2005) Safety and tolerability of oral antifungal agents in the treatment of fungal nail disease: a proven reality. *Ther. Clin. Risk Manag.* **1** (4), 299–306.

134. Gibbs, J. P., Hyland, R., and Youdim, K. (2006) Minimizing polymorphic metabolism in drug discovery: evaluation of the utility of in vitro methods for predicting pharmacokinetic consequences associated with CYP2D6 metabolism. *Drug Metab. Dispos.* **34**, 1516–1522.

135. Honig, P. K., Wortham, D. C., Zamani, K., Conner, D. P., Mullin, J. C., and Cantilena, L. R. (1993) Terfenadine–ketoconazole interaction. Pharmacokinetic and electrocardiographic consequences. *J. Am. Med. Assoc.* **269** (12), 1513–1518.

136. Guengerich, F. P. and Kim, D. H. (1990) In vitro inhibition of dihydropyridine oxidation and aflatoxin B1 activation in human liver microsomes by naringenin and other flavonoids. *Carcinogenesis* **11** (12), 2275–2279.

137. Hollander, A. A., van Rooij, J., Lentjes, G. W., Arbouw, F., van Bree, J. B., Schoemaker, R. C., van Es, L. A., Van Der Woude, F. J., and Cohen, A. F. (1995) The effect of grapefruit juice on cyclosporine and prednisone metabolism in transplant patients. *Clin. Pharmacol. Ther.* **57** (3), 318–324.

138. Bailey, D. G., Malcolm, J., Arnold, O., and Spence, J. D. (1998) Grapefruit juice–drug interactions. *Br. J. Clin. Pharmacol.* **46** (2), 101–110.

139. Greenblatt, D. J., von Moltke, L. L., Harmatz, J. S., Chen, G., Weemhoff, J. L., Jen, C., Kelley, C. J., LeDuc, B. W., and Zinny, M. A. (2003) Time course of recovery of cytochrome P450 3A function after single doses of grapefruit juice. *Clin. Pharmacol. Ther.* **74** (2), 121–129.

140. Hidaka, M., Okumura, M., Fujita, K., Ogikubo, T., Yamasaki, K., Iwakiri, T., Setoguchi, N., and Arimori, K. (2005) Effects of pomegranate juice on human cytochrome P450 3A (CYP3A) and carbamazepine pharmacokinetics in rats. *Drug Metab. Dispos.* **33** (5), 644–648.

141. Egashira, K., Ohtani, H., Itoh, S., Koyabu, N., Tsujimoto, M., Murakami, H., and Sawada, Y. (2004) Inhibitory effects of pomelo on the metabolism of tacrolimus and the activities of CYP3A4 and P-glycoprotein. *Drug Metab. Dispos.* **32** (8), 828–833.

142. Harris, R. Z., Jang, G. R., and Tsunoda, S. (2003) Dietary effects on drug metabolism and transport. *Clin. Pharmacokinet.* **42** (13), 1071–1088.

143. Guo, L., Taniguchi, M., Xiao, Y., Baba, K., Ohta, T., and Yamazoe, Y. (2000) Inhibitory effect of natural furanocoumarins on human microsomal cytochrome P450 3A activity. *Japan. J. Pharmacol.* **82** (2), 122–129.

144. Venkatakrishnan, K., von Moltke, L. L., and Greenblatt, D. J. (1998) Relative quantities of catalytically active CYP 2C9 and 2C19 in human liver microsomes: application of the relative activity factor approach. *J. Pharm. Sci.* **87** (7), 845–853.

145. Gelboin, H. V., Krausz, K. W., Gonzalez, F. J., and Yang, T. J. (1999) Inhibitory mono-clonal antibodies to human cytochrome P450 enzymes: a new avenue for drug discovery. *Trends Pharmacol. Sci.* **20** (11), 432–438.

146. Lu, A. Y. H., Wang, R. W., and Lin, J. H. (2003) Cytochrome P450 in vitro reaction phenotyping: a re-evaluation of approaches used for P450 isoform identification. *Drug Metab. Dispos.* **31** (4), 345–350.

147. Rodrigues, D. (1999) Integrated cytochrome P450 reaction phenotyping: attempting to bridge the gap between cDNA-expressed cytochromes P450 and native human liver microsomes. *Biochem. Pharmacol.* **57** (5), 465–480.

148. Venkatakrishnan, K., von Moltke, L. L., and Greenblatt, D. J. (2001) Human drug metabolism and the cytochromes P450: application and relevance of in vitro models. *J. Clin. Pharmacol.* **41** (11), 1149–1179.

149. Trapnell, C. B., Klecker, R. W., Jamis-Dow, C., and Collins, J. M. (1998) Glucuronida-tion of 3′-azido-3′-deoxythymidine (zidovudine) by human liver microsomes: relevance to clinical pharmacokinetic interactions with atovaquone, fluconazole, methadone, and valproic acid. *Antimicrob. Agents Chemother.* **42** (7), 1592–1596.

150. Bernard, O., Tojcic, J., Journault, K., Perusse, L., and Guillemette, C. (2006) Influence of nonsynonymous polymorphisms of UGT1A8 and UGT2B7 metabolizing enzymes on the formation of phenolic and acyl glucuronides of mycophenolic acid. *Drug Metab. Dispos.* **34** (9), 1539–1545.

151. Borlak, J., Gasparic, A., Locher, M., Schupke, H., and Hermann, R. (2006) N-Glucuronidation of the antiepileptic drug retigabine: results from studies with hu-man volunteers, heterologously expressed human UGTs, human liver, kidney, and liver microsomal membranes of Crigler–Najjar type II. *Metab. Clin. Exp.* **55** (6), 711–721.

152. Yu, L., Lu, S., Lin, Y., and Zeng, S. (2007) Carboxyl-glucuronidation of mitiglinide by human UDP-glucuronosyltransferases. *Biochem. Pharmacol.* **73** (11), 1842–1851.

153. Gaganis, P., Miners, J. O., and Knights, K. M. (2007) Glucuronidation of fena-mates: kinetic studies using human kidney cortical microsomes and recombinant UDP-glucuronosyltransferase (UGT) 1A9 and 2B7. *Biochem. Pharmacol.* **73** (10), 1683–1691.

154. Bauman, J. N., Goosen, T. C., Tugnait, M., Peterkin, V., Hurst, S. I., Menning, L. C., Milad, M., Court, M. H., and Williams, J. A. (2005) UDP-glucuronosyltransferase 2B7 is the major enzyme responsible for gemcabene glucuronidation in human liver microsomes. *Drug Metab. Dispos.* **33** (9), 1349–1354.

155. Kehrer, D. F. S., Mathijssen, R. H. J., Verweij, J., de Bruijn, P., and Sparreboom, A. (2002) Modulation of irinotecan metabolism by ketoconazole. *J. Clin. Oncol.* **20** (14), 3122–3129.

156. Federman, D. G., Hussain, F., and Walters, A. B. (2001) Fatal rhabdomyolysis caused by lipid-lowering therapy. *South. Med. J.* **94** (10), 1023–1026.

157. Prueksaritanont, T., Zhao, J. J., Ma, B., Roadcap, B. A., Tang, C., Qiu, Y., Liu, L., Lin, J. H., Pearson, P. G., and Baillie, T. A. (2002) Mechanistic studies on metabolic interac-tions between gemfibrozil and statins. *J. Pharmacol. Exp. Ther.* **301** (3), 1042–1051.

158. Prueksaritanont, T., Qiu, Y., Mu, L., Michel, K., Brunner, J., Richards, K. M., and Lin, J. H. (2005) Interconversion pharmacokinetics of simvastatin and its hydroxy acid in dogs: effects of gemfibrozil. *Pharm. Res.* **22** (7), 1101–1109.

159. Beedham, C., Critchley, D. J. P., and Rance, D. J. (1995) Substrate specificity of human liver aldehyde oxidase toward substituted quinazolines and phthalazines: a comparison with hepatic enzyme from guinea pig, rabbit, and baboon. *Arch. Biochem. Biophys.* **319** (2), 481–490.

160. Rashidi, M. R., Smith, J. A., Clarke, S. E., and Beedham, C. (1997) In vitro oxidation of famciclovir and 6-deoxypenciclovir by aldehyde oxidase from human, guinea pig, rabbit, and rat liver. *Drug Metab. Dispos.* **25** (7), 805–813.

161. Schofield, P. C., Robertson, I. G., and Paxton, J. W. (2000) Inter-species variation in the metabolism and inhibition of N-[(2''-dimethylamino)ethyl]acridine-4-carboxamide (DACA) by aldehyde oxidase. *Biochem. Pharmacol.* **59** (2), 161–165.

162. Obach, R. S. (2004) Potent inhibition of human liver aldehyde oxidase by raloxifene. *Drug Metab. Dispos.* **32** (1), 89–97.

163. Xia, C. Q., Milton, M. N., and Gan, L.-S. (2007) Evaluation of drug-transporter interactions using in vitro and in vivo models. *Curr. Drug Metab.* **8**, 341–363.

164. Li, J. H. (2007). Transporter-mediated drug interactions: clinical implications and in vitro assessment. *Expert Opin. Drug Metab. Toxicol.* **3** (1), 81–92.

165. Kalgutkar, A. S., Obach, R. S., and Maurer, T. S. (2007) Mechanism-based inactivation of cytochrome P450 enzymes: chemical mechanisms, structure-activity relationships and relationship to clinical drug–drug interactions and idiosyncratic adverse drug reactions. *Curr. Drug Metab.* **8** (5), 407–447.

166. Mutlib, A., Lam, W., Atherton, J., Chen, H., Galatsis, P., and Stolle, W. (2005) Application of stable isotope labeled glutathione and rapid scanning mass spectrometers in detecting and characterizing reactive metabolites. *Rapid Commun. Mass Spectrom.* **19** (23), 3482–3492.

167. Argoti, D., Liang, L., Conteh, A., Chen, L., Bershas, D., Yu, C., Vouros, P., and Yang, E. (2005) Cyanide trapping of iminium ion reactive intermediates followed by detection and structure identification using liquid chromatography-tandem mass spectrometry (LC-MS/MS). *Chem. Res. Toxicol.* **18** (10), 1537–1544.

168. Castro-Perez, J., Plumb, R., Liang, L., and Yang, E. (2005) A high-throughput liquid chromatography/tandem mass spectrometry method for screening glutathione conjugates using exact mass neutral loss acquisition. *Rapid Commun. Mass Spectrom.* **19** (6), 798–804.

169. Yan, Z., Maher, N., Torres, R., and Huebert, N. (2007) Use of a trapping agent for simultaneous capturing and high-throughput screening of both "soft" and "hard" reactive metabolites. *Anal. Chem.* **79** (11), 4206–4214.

170. Zheng, J., Ma, L., Xin, B., Olah, T., Humphreys, G. W., and Zhu, M. (2007) Screening and identification of GSH-trapped reactive metabolites using hybrid triple quadruple linear ion trap mass spectrometry. *Chem. Res. Toxicol.* **20** (5), 757–766.

171. Doss, G. A. and Baillie, T. A. (2006) Addressing metabolic activation as an integral component of drug design. *Drug Metab. Rev.* **38** (4), 641–649.

172. Kroemer, H. K. and Klotz, U. (1992) Glucuronidation of drugs. A re-evaluation of the pharmacological significance of the conjugates and modulating factors. *Clin. Pharmacokinet.* **23**, 292–310.

173. Boelsterli, U. A., Zimmerman, H. J., and Kretz-Rommel, A. (1995) Idiosyncratic liver toxicity of nonsteroidal anti-inflammatory drugs: molecular mechanisms and pathology. *Crit. Rev. Toxicol.* **25** (3), 207–235.

174. Bailey, M. J. and Dickinson, R. G. (2003) Acyl glucuronide reactivity in perspective: biological consequences. *Chem. Biol. Interact.* **145**, 117–137.

175. Boelsterli, U. A. (2003) Diclofenac-induced liver injury: a paradigm of idiosyncratic drug toxicity. *Toxicol. Appl. Pharmacol.* **192**, 307–322.

176. Walker, G. S., Atherton, J., Bauman, J., Kohl, C., Lam, W., Reily, M., Lou, Z., and Mutlib, A. (2007) Determination of degradation pathways and kinetics of acyl glucuronides by NMR spectroscopy. *Chem. Res. Toxicol.* **20**, 876–886.

177. Chen, G., Yin, S., Maiti, S., and Shao, X. (2002) 4-Hydroxytamoxifen sulfation metabolism. *J. Biochem. Mol. Toxicol.* **16**, 279–285.

178. Plant, N. (2007) The human cytochrome P450 subfamily: transcriptional regulation, inter-individual variation and interaction networks. *Biochim. Biophys. Acta* **1770** (3), 478–488.

179. Dickins, M. (2004) Induction of cytochromes P450. *Curr. Top. Med. Chem.* **4** (16), 1745–1766.

180. Smith, D. A., Dickins, M., Fahmi, O. A., Iwasaki, K., Lee, C., Obach, R. S., Padbury, G., De Morais, S. M., Ripp, S. L., Stevens, J., et al. (2007) The time to move cytochrome P450 induction into mainstream pharmacology is long overdue. *Drug Metab. Dispos.* **35** (4), 697–698.

181. Hewitt, N. J., Lechon, M. J. G., Houston, J. B., Hallifax, D., Brown, H. S., Maurel, P., Kenna, J. G., Gustavsson, L., Lohmann, C., Skonberg, C., et al. (2007) Primary hepatocytes: current understanding of the regulation of metabolic enzymes and transporter proteins, and pharmaceutical practice for the use of hepatocytes in metabolism, enzyme induction, transporter, clearance, and hepatotoxicity studies. *Drug Metab. Rev.* **39** (1), 159–234.

182. Mills, J. B., Rose, K. A., Sadagopan, N., Sahi, J., and De Morais, S. M. (2004) Induction of drug metabolism enzymes and MDR1 using a novel human hepatocyte cell line. *J. Pharmacol. Exp. Ther.* **309**, 303–309.

183. Youdim, K. A., Tyman, C. A., Jones, B. C., and Hyland, R. (2007) Induction of cytochrome P450: assessment in an immortalized human hepatocyte cell line (Fa2N4) using a novel higher throughput cocktail assay. *Drug Metab. Dispos.* **35**, 275–282.

184. Mahgoub, A., Idle, J. R., Dring, L. G., Lancaster, R., and Smith, R. L. (1977) Polymorphic hydroxylation of debrisoquine in man. *Lancet* **2** (8038), 584–586.

185. Tucker, G. T., Silas, J. H., Iyun, A. O., Lennard, M. S., and Smith, A. J. (1977) Polymorphic hydroxylation of debrisoquine. *Lancet* **2** (8040), 718.

186. Evans, D. A., Mahgoub, A., Sloan, T. P., Idle, J. R., and Smith, R. L. (1980) A family and population study of the genetic polymorphism of debrisoquine oxidation in a white British population. *J. Med. Genet.* **17** (2), 102–105.

187. Eichelbaum, M. (1984) Polymorphic drug oxidation in humans. *Fed. Proc.* **43** (8), 2298–2302.

188. Gut, J., Gasser, R., Dayer, P., Kronbach, T., Catin, T., and Meyer, U. A. (1984) Debrisoquine-type polymorphism of drug oxidation: purification from human liver of a cytochrome P450 isozyme with high activity for bufuralol hydroxylation. *FEBS Lett.* **173** (2), 287–290.

189. Inaba, T., Nakano, M., Otton, S. V., Mahon, W. A., and Kalow, W. (1984) A human cytochrome P-450 characterized by inhibition studies as the sparteine-debrisoquine monooxygenase. *Can. J. Physiol. Pharmacol.* **62** (7), 860–862.

190. Hulot, J.-S., Bura, A., Villard, E., Azizi, M., Remones, V., Goyenvalle, C., Aiach, M., Lechat, P., and Gaussem, P. (2006) Cytochrome P450 2C19 loss-of-function polymorphism is a major determinant of clopidogrel responsiveness in healthy subjects. *Blood* **108** (7), 2244–2247.

191. Zandvliet, A. S., Huitema, A. D. R., Copalu, W., Yamada, Y., Tamura, T., Beijnen, J. H., and Schellens, J. H. M. (2007) CYP2C9 and CYP2C19 polymorphic forms are related to increased indisulam exposure and higher risk of severe hematologic toxicity. *Clin. Cancer Res.* **13** (10), 2970–2976.

192. Westlind-Johnsson, A., Hermann, R., Huennemeye, A., Hauns, B., Lahu, G., Nassr, N., Zech, K., Ingelman-Sundberg, M., and von Richter, O. (2008) Identification and characterization of CYP3A4*20, a novel rare CYP3A4 allele without functional activity. *Clin. Pharmacol. Ther.* **79** (4), 339–349.

193. Thervet, E., Anglicheau, D., King, B., Schlageter, M.-H., Cassinat, B., Beaune, P., Legendre, C., and Daly, A. K. (2003) Impact of cytochrome P450 3A5 genetic polymorphism on tacrolimus doses and concentration-to-dose ratio in renal transplant recipients. *Transplantation* **76** (8), 1233–1235.

194. MacPhee, I. A. M., Fredericks, S., and Holt, D. W. (2005) Does pharmacogenetics have the potential to allow the individualization of immunosuppressive drug dosing in organ transplantation? *Expert Opin. Pharmacother.* **6** (15), 2593–2605.

195. Rela, M., Muiesan, P., Vilca-Melendez, H., Dhawan, A., Baker, A., Mieli-Vergani, G., and Heaton, N. D. (1999) Auxiliary partial orthotopic liver transplantation for Crigler–Najjar syndrome type I. *Ann. Surg.* **229** (4), 565–569.

196. Nguyen, T. H., Birraux, J., Wildhaber, B., Myara, A., Trivin, F., Le Coultre, C., Trono, D., and Chardot, C. (2006) Ex vivo lentivirus transduction and immediate transplantation of uncultured hepatocytes for treating hyperbilirubinemic Gunn rat. *Transplantation* **82** (6), 794–803.

197. Li, Q., Murphree, S. S., Willer, S. S., Bolli, R., and French, B. A. (1998) Gene therapy with bilirubin-UDP-glucuronosyltransferase in the Gunn rat model of Crigler–Najjar syndrome type 1. *Hum. Gene Ther.* **9** (4), 497–505.

198. Hirschfield, G. M. and Alexander, G. J. (2006) Gilbert's syndrome: an overview for clinical biochemists. *Ann. Clin. Biochem.* **43** (5), 340–343.

199. Nowell, S. and Falany, C. N. (2006) Pharmacogenetics of human cytosolic sulfotransferases. *Oncogene* **25** (11), 1673–1678.

200. Hisamuddin, I. M. and Yang, V. W. (2007) Genetic polymorphisms of human flavin-containing monooxygenase 3: implications for drug metabolism and clinical perspectives. *Pharmacogenomics* **8** (6), 635–643.

201. Ichida, K., Amaya, Y., Kamatani, N., Nishino, T., Hosoya, T., and Sakai, O. (1997) Identification of two mutations in human xanthine dehydrogenase gene responsible for classical type I xanthinuria. *J. Clin. Invest.* **99** (10), 2391–2397.

202. Yamamoto, T., Moriwaki, Y., Shibutani, Y., Matsui, K., Ueo, T., Takahashi, S., Tsutsumi, Z., and Hada, T. (2001) Human xanthine dehydrogenase cDNA sequence and protein in an atypical case of type I xanthinuria in comparison with normal subjects. *Clin. Chim. Acta* **304** (1–2), 153–158.

203. Yamamoto, T., Moriwaki, Y., Takahashi, S., Tsutsumi, Z., Tuneyoshi, K., Matsui, K., Cheng, J., and Hada, T. (2003) Identification of a new point mutation in the human molybdenum cofactor sulfurase gene that is responsible for xanthinuria type II. *Metab. Clin. Exp.* **52** (11), 1501–1504.

204. Peretz, H., Naamati, M. S., Levartovsky, D., Lagziel, A., Shani, E., Horn, I., Shalev, H., and Landau, D. (2007) Identification and characterization of the first mutation (Arg776Cys) in the C-terminal domain of the human molybdenum cofactor sulfurase (HMCS) associated with type II classical xanthinuria. *Mol. Genet. Metab.* **91** (1), 23–29.

205. Itoh, K., Masubuchi, A., Sasaki, T., Adachi, M., Watanabe, N., Nagata, K., Yamazoe, Y., Hiratsuka, M., Mizugaki, M., and Tanaka, Y. (2007) Genetic polymorphism of aldehyde oxidase in donryu rats. *Drug Metab. Dispos.* **35** (5), 734–739.

206. Brockton, N., Little, J., Sharp, L., and Cotton, S. C. (2000) N-acetyltransferase polymorphisms and colorectal cancer: a HuGE review. *Am. J. Epidemiol.* **151** (9), 846–861.

207. Benedetti, M. S., Whomsley, R., Espie, P., and Baltes, E. (2004) The role of the efflux transporter P-glycoprotein (P-gp) on the disposition of antiepileptic drugs: implications for drug interactions. In: *Focus on Epilepsy Research*, Benjamin Shawn, M., (ed.), Nova Science, New York, pp. 199–220.

208. Robey, R. W., Polgar, O., Deeken, J., To, K. W., and Bates, S. E. (2007) ABCG2: determining its relevance in clinical drug resistance. *Cancer Metastasis Rev.* **26** (1), 39–57.

209. Geier, A., Wagner, M., Dietrich, C. G., and Trauner, M. (2007) Principles of hepatic organic anion transporter regulation during cholestasis, inflammation and liver regeneration. *Biochim. Biophys. Acta Mol. Cell Res.* **1773** (3), 283–308.

210. Sakurai, A., Kurata, A., Onishi, Y., Hirano, H., and Ishikawa, T. (2007) Prediction of drug-induced intrahepatic cholestasis: in vitro screening and QSAR analysis of drugs inhibiting the human bile salt export pump. *Expert Opin. Drug Saf.* **6** (1), 71–86.

211. Mahadevan, D. and Shirahatti, N. (2005) Strategies for targeting the multidrug resistance-1 (MDR1)/P-gp transporter in human malignancies. *Curr. Cancer Drug Targets* **5** (6), 445–455.

212. Jansen, P. L. M. and Muller, M. (2001) Genetic transport defects as causes of cholestasis. *Falk Symposium.* (Hepatology) **Vol. 117**, pp. 27–33.

213. Leschziner, G. D., Andrew, T., Pirmohamed, M., and Johnson, M. R. (2007) ABCB1 genotype and P-gp expression, function and therapeutic drug response: a critical review and recommendations for future research. *Pharmacogenomics J.* **7** (3), 154–179.

2

ESTABLISHING ORPHAN NUCLEAR RECEPTORS PXR AND CAR AS XENOBIOTIC RECEPTORS

Tao Li, Junichiro Sonoda, and Ronald M. Evans

Gene Expression Laboratory, Howard Hughes Medical Institute, The Salk Institute for Biological Studies, La Jolla, CA, USA

2.1 INTRODUCTION

In the process of consuming food and drink and breathing polluted air, the human body is continuously exposed to numerous foreign compounds (xenobiotics), which are neither used as dietary energy sources nor as building blocks for biological matrices. These substances may be acutely or chronically harmful unless they are metabolized and eliminated. It has been shown that many of the adverse effects to living organisms by these compounds are caused by a variety of mechanisms including interaction with hormone receptors (endocrine active substances, i.e., EAS) or reaction with nucleic acids (genotoxic carcinogens) and membrane solubilization (cytoxic lipids). Accordingly, the mammalian body has evolved an efficient hepatic detoxification system, consisting of microsomal cytochrome P450 enzymes (CYPs) and other oxidative and hydroxylative enzymes (phase I), conjugation enzymes such as glucuronosyl- and sulfotransferases (phase II), and membrane-bound drug transporters such as multiple drug resistance I (MDR1) (phase III), inactivate and eliminate toxic substances. In addition to xenobiotics, some endogenous compounds such as steroids, bile acids, thyroid hormone, retinoids, cytokines, and fatty acids can also be metabolized by the same pathway.

Insight into the mechanisms by which xenobiotics activate hepatic drug metabolism and how they are eliminated by this system was revealed when two closely

Nuclear Receptors in Drug Metabolism   Edited by Wen Xie
Copyright © 2009 John Wiley & Sons, Inc.

related nuclear receptors, the pregnane X receptor (PXR) and the constitutive androstane receptor (CAR), were identified as xenobiotic sensors. PXR and CAR can be activated by a large number of structurally diverse compounds, resulting in the induction of their downstream target genes in the drug clearance pathways and conferring protection from foreign chemicals and endobiotics. Because of the enormous diversity of PXR/CAR agonists and antagonists, this nuclear receptor mediated xenobiotic regulatory pathway also contributes to drug–drug interactions and endocrine disruption.

2.2   NUCLEAR RECEPTOR AND ORPHAN NUCLEAR RECEPTOR SUPERFAMILY

Nuclear receptors are members of a superfamily of compound-inducible transcription factors mediating response to steroids, retinoids, and thyroid hormones. They regulate specific target genes involved in metabolism, development, reproduction, and other physiological processes (for reviews, see [1–3]). Using epitope selection, the first identified nuclear hormone receptor—human glucocorticoid receptor (hGR)—was cloned in 1985 [4]. Shortly after that, a number of hormone receptors were identified, such as the estrogen receptor (ER), progesterone receptor (PR), thyroid hormone receptor (TR), retinoic acid receptor (RAR), vitamin D receptor (VDR), mineralocorticoid receptor (MR), and androgen receptor (AR) [5–11]. In the late 1980s, the first nuclear receptors with unknown physiological function, estrogen-related receptors (ERR) α and β, were identified during a search for genes related to the ERs [12]. Since then, a large group of homologous proteins that lack previously identified physiological ligands or activators have been unearthed, and are classified as "orphan nuclear receptors." To date a total of 48 human nuclear receptors have been cloned (for a summary, see Table 2.1). The discovery of these orphan receptors sheds light on many unknown aspects of physiology.

Molecular cloning and structure/function analyses have revealed that the members of this superfamily have a common functional domain structure. The amino terminal region contains an activation domain (AF)-1. This abuts the DNA binding domain (DBD), followed by a hinge region and then the ligand binding domain (LBD). Close to carboxy terminus there is another transcriptional activation region (AF-2) that is ligand dependent (Figure 2.1a). A general concept for NR signaling is that in the absence of a ligand, the NR is often associated with a corepressor complex. Ligand binding to the LBD induces conformational changes that lead to the release of the corepressor complex and recruitment of the coactivator complex. Coactivator recruitment contributes to chromatin remodeling and subsequently transcriptional activation through hormone response elements (HREs) that consist of a minimal core hexameric consensus sequence, 5′ AG(G/T)TCA 3′, that can be configured into a variety of structured motifs.

The classical sex hormone receptors such as ER, PR, and AR form homodimers upon ligand binding and then interact with target DNA via HREs to control target gene expression (Figure 2.1b). In contrast, the orphan receptors or adopted orphan receptors form heterodimers with retinoid X receptors (RXRs) on direct repeat

TABLE 2.1   List of Mammalian Nuclear Receptors

	Trivial Names	Official Gene Name	Natural Ligands
Endocrine receptors (high-affinity hormonal ligands)	TRα, c-erbA-1, THRA	*NR1A1*	Thyroid hormone
	TRβ, c-erbA-2, THRB	*NR1A2*	Thyroid hormone
	RARα	*NR1B1*	Retinoic acids
	RARβ, HAP	*NR1B2*	Retinoic acids
	RARγ, RARD	*NR1B3*	Retinoic acids
	VDR	*NR1I1*	Vitamin D
	ERα	*NR3A1*	Estrogen
	ERβ	*NR3A2*	Estrogen
	GR	*NR3C1*	Glucocorticoid
	MR	*NR3C2*	Mineralocorticoid
	PR	*NR3C3*	Progesterone
	AR	*NR3C4*	Testosterone
Adopted orphan receptors (low-affinity dietary ligands)	PPARα	*NR1C1*	Fatty acids
	PPARδ, PPARβ, NUC1, FAAR	*NR1C2*	Fatty acids
	PPARγ	*NR1C3*	Fatty acids
	LXRβ, UR, OR-1, NER1, RIP15	*NR1H2*	Oxysterols
	LXRα, RLD1, LXR	*NR1H3*	Oxysterols
	FXR, RIP14, HRR1	*NR1H4*	Bile acids
	PXR, ONR1, SXR, BXR	*NR1I2*	Xenobiotics
Orphan receptors (unknown ligands)	CAR, MB67, CAR1	*NR1I3*	Xenobiotics
	RXRα	*NR2B1*	9-*cis*-RA
	RXRβ, H-2RIIBP, RCoR-1	*NR2B2*	9-*cis*-RA
	RXRγ	*NR2B3*	9-*cis*-RA
	RORα, RZRα	*NR1F1*	
	RORβ, RZRβ	*NR1F2*	
	RORγ, TOR	*NR1F3*	
	ERR1, ERRα	*NR3B1*	
	ERR2, ERRβ	*NR3B2*	
	ERR3, ERRγ	*NR3B3*	
	NGFIB, TR3, N10, NUR77, NAK1	*NR4A1*	
	NURR1, NOT, RNR1, HZF-3, TINOR	*NR4A2*	

TABLE 2.1   (*continued*)

Trivial Names	Official Gene Name	Natural Ligands
NOR1, MINOR	*NR4A3*	
SF1, ELP, FTZ-F1, AD4BP	*NR5A1*	Phospholipids
LRH1, xFF1rA, xFF1rB, FFLR, PHR	*NR5A2*	
REVERBα, EAR1, EAR1A	*NR1D1*	Heme
REVERBβ, EAR1β, BD73, RVR	*NR1D2*	Heme
TR2, TR2-11	*NR2C1*	
TR4, TAK1	*NR2C2*	
TLL, TLX, XTLL	*NR2E1*	
COUP-TFI, COUPTFA, EAR3, SVP44	*NR2F1*	
COUP-TFII, COUPTFB, ARP1, SVP40	*NR2F2*	
COUP-TFIII, EAR2	*NR2F6*	
SHP	*NR0B2*	
HNF4	*NR2A1*	
HNF4G	*NR2A2*	
PNR	*NR2E3*	
GCNF1, RTR	*NR6A1*	
DAX1, AHCH	*NR0B1*	

HREs or act independent of the ligand as monomers on half-site HREs. Over the last two decades, the biological role of several of these orphans, in particular those that act as heterodimers with RXR (Figure 2.1*b*), has been revealed through the isolation of relevant endogenous ligands as well as by generation of knock-out mouse models that lack the functional receptors. These studies have led to the understanding

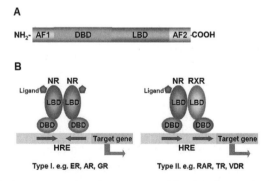

FIGURE 2.1   Nuclear receptor domain structure and its signaling mode. (*a*) Modular structure of nuclear receptors. (*b*) Signaling mode of the type I and type II nuclear receptors. See color insert.

that many orphans act as sensors for dietary lipids as opposed to high affinity endogenous hormones. For example, the liver X receptors (LXRs), peroxisome proliferator-activated receptors (PPARs), and farnesoid X receptor (FXR) have been identified as sensors for cholesterol, fatty acids, and bile acids, respectively, and shown to cooperatively regulate lipid homeostasis (reviewed in [13]). In addition, two closely related receptors, PXR and CAR, have been identified as xenobiotic sensors that mediate induction of drug clearance pathways to ensure rapid detoxification of potentially harmful substances. In this chapter we will focus on PXR and CAR, their role in drug clearance, the molecular mechanisms of their action, and the implication of xenobiotic regulation in pathophysiological conditions and endocrine disruption.

2.3 ORPHAN NUCLEAR RECEPTORS AS XENOBIOTIC RECEPTORS AND THEIR IMPLICATIONS IN PHASE I ENZYME REGULATION

In the hepatic drug clearance system, phase I enzymes, especially members of the CYP1–4 families, play important roles in xenobiotic detoxification and survival of organisms [14]. For example, the human CYP3A4 isoenzyme alone is involved in the metabolism of 50–60% of clinical drugs, whereas CYP2B metabolizes an additional 25–30% of these compounds (reviewed in [15]). One important characteristic of these CYP enzymes is their inducibility by their substrates, which allows enhanced production of these proteins as needed. For example, CYP3A is induced upon treatment with the antibiotic rifampicin, whereas CYP2B production is increased by the treatment of the antiepileptic drug phenobarbital (PB) or the planar hydrocarbon 1,4-bis[2-(3,5-dichloropyridyloxy)]benzene (TCPOBOP). The induction of CYP enzymes by these compounds can further induce a variety of other phase I enzymes and some phase II metabolic enzymes and transporters (see Chapter 1). However, the mechanism of the induction of these CYP genes by xenobiotics was unclear until PXR and CAR were initially defined as xenobiotic receptors on the basis of their activation of phase I CYP enzymes in response to a wide array of xenobiotics.

2.3.1 Cloning of PXR as a Xenobiotic Receptor and Its Regulation of P450s

2.3.1.1 Cloning and Characterization of PXR. The concept that exogenous steroids and pharmacological substances may modulate the expression of a set of enzymes to protect against subsequent exposure to toxic xenobiotic substances came up in the early 1970s [16]. Catatoxic compounds were found to induce the proliferation of the hepatic endoplasmic reticulum and the expression of cytochrome P450 genes, conferring nonspecific protection or immunity against numerous xenobiotic compounds, presumably by increasing their catabolism [17, 18]. However, it was unclear at the time which receptor(s), presuming any were involved, mediate the induction of P450 CYP genes upon the stimulation of these foreign chemicals. Given the fact that both steroid receptor agonists (e.g., dexamethasone) and antagonists (e.g., pregnenolone-16α-carbonitrile, PCN) induce CYP3A genes [17], it indicated that the induced protection was independent of the known steroid receptors.

In 1998, Kliewer et al. and Blumberg et al. identified a novel orphan nuclear receptor from mouse and human, respectively, on the basis of homology to other nuclear hormone receptors. The mouse receptor was termed as the PXR, because it was found to be activated by naturally occurring steroids such as pregnenolone and progesterone, and synthetic glucocorticoids and antiglucocorticoids [19, 28]. It was also shown that PXR binds to the CYP3A2 promoter and activates transcription in response to potent CYP3A2 inducers such as PCN or dexamethasone [19]. Thus, the response profile of PXR offered a possible explanation to the behavior of catatoxic compounds and the induction of CYP3A genes (for reviews, see [2, 20, 21]).

Shortly after the cloning of PXR, significant evidence emerged demonstrating that PXR mediates CYP3A regulation by xenobiotics. Both PXR and CYP3A are highly expressed in the liver and intestine, where drug clearance mainly occurs. PXR and its partner RXR bind to a DR3 (direct repeats of AGGTCA or closely related sequences with a spacing of three nucleotides) or ER6 (everted repeats spaced by six nucleotides) sites in the CYP3A promoter. The binding of numerous structurally unrelated drugs to PXR, including those known to induce CYP3A expression, appears to dissociate corepressor molecules such as the silencing mediator for retinoid and thyroid hormone receptor (SMRT) [22] and the nuclear receptor corepressor (NCoR) from PXR. This is followed by simultaneous recruitment of coactivator molecules, including members of the p160 family (SRC-1, GRIP, and ACTR), RIP140, and PBP (DRIP205 or TRAP220) [23] (for details, see Chapter 6). PXR null mice are both viable and fertile, indicating that in the absence of toxic insults the xenobiotic response is not required. However, PXR null mice completely lack inducibility of CYP3A by PCN or PCN-mediated induction of drug resistance [24–27]. Together these observations clearly establish PXR as the central mediator of CYP3A induction (Figure 2.2). There is also evidence showing PXR can also directly regulate phase II and phase III gene expression, which will be discussed in detail in Chapters 3 and 4, respectively.

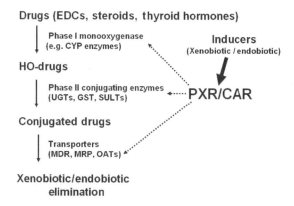

FIGURE 2.2   PXR and CAR control phases I, II, and III drug metabolism genes.

2.3.1.2 The Human Ortholog of PXR. The human homologue of PXR (hPXR) was first isolated as the steroid and xenobiotic receptor (SXR) [28] and the pregnane-activated receptor (PAR) [29]. Notably, the PXR LBD from different species is considerably divergent. Within the LBD, the amino acid identity of human PXR with mouse PXR is only 76%, whereas the DNA binding domain is highly conserved (96% identity). Thus the species specificity of the induction of CYP3A enzymes [30] might be due to the pharmacological distinction of human and rodent PXRs. For example, corticosterone and rifampicin are potent hPXR but poor mPXR activators. In contrast, PCN and dexamethasone are strong mPXR but poor hPXR activators. Despite their differences in pharmacology, the two receptors appear to act through a common metabolic pathway. From this viewpoint, perhaps the structural and pharmacological differences between human and mouse PXRs are more reflective of the differences in the diets of rodents and primates and the need to respond to a different set of xenobiotics (for review, see [2]). Xie et al. [24] created a humanized PXR mouse model and found that it activated downstream target genes but only responded to human-specific inducers such as rifampicin but not PCN. The xenobiotic response in this mouse model offers a standardized *in vivo* system for predicting potential human drug–drug interactions and is beneficial for development of new therapeutic drugs. Historically, the induction of CYP3A by drugs had been considered an unexplained adverse side effect associated with drug–drug interactions. For decades, rodent models have been standard components in the assessment of potential toxicities in the development of candidate human drugs. However, the reliability of rodents as predictors of the human xenobiotic response is compromised due to species variation. Cultured human primary hepatocytes are valuable alternative tools, but are compromised by interindividual variability, limited and unpredictable availability as well as high cost. Thus, the generation of the humanized PXR mice represents a major step toward generating a standardized humanized toxicological model (for review, see [31, 32]).

2.3.1.3 Diversity of PXR Modulators. As a xenobiotic receptor, PXR unsurprisingly binds a diverse range of structurally unrelated chemicals. In fact, x-ray crystal structures of the PXR LBD have revealed that its ligand binding pocket is relatively large compared to most other nuclear receptor LBDs, and can accommodate a hydrophobic ligand in multiple configurations [33]. A feature of its flexible binding pocket allows for molecular plasticity of ligand recognition, consistent with the low substrate specificity of xenobiotic enzymes. As a result, PXR activators include antibiotics such as rifampicin, cholesterol-lowering drugs such as SR12813 and statins, antidepressants like St. John's wort, the antineoplastic drug paclitaxel, the antimycotic clotrimazole (reviewed in [20]), bisphenol A [34], organochlorine pesticides such as chlordane, Cafestol [35], dieldrin, and endosulfan [36]. Other environmental contaminants including endocrine disrupting chemicals such as nonylphenol and phthalic acid, nonplanar polychlorinated biphenyls (PCBs), and organochloride pepticides such as *trans*-nonachlor and chlordane have all been shown to activate mouse PXR [37]. In addition to its activators, there are also a number of PXR antagonistic ligands. For example, ecteinascidin-734 blocks PXR-mediated induction of

CYP3A [23], and arsenite inhibits both untreated and rifampicin-induced CYP3A transcription in primary human hepatocytes by decreasing the activity of PXR, as well as expression of its heterodimeric nuclear receptor partner RXRα [38, 39]. Thus, in principle, it should be possible to design specific drugs that could selectively inhibit or promote the xenobiotic response. With relevance to endocrine disruption, human PXR is also activated by numerous endobiotics including bile acids, corticosterone, and estradiol as well as other estrogenic chemicals including diethylstilbestrol, and the phytoestrogen coumestrol [28, 40].

2.3.1.4 How Is PXR Regulated?

On the one hand PXR is capable of regulating drug clearance related genes, yet on the other, it can also be regulated by other proteins, suggesting that xenobiotic metabolism is involved in other physiological events. Several compounds are known to induce PXR or CAR mRNA, indicating another level of control. For example, the activation of the glucocorticoid receptor (GR) can induce expression of PXR, CAR and their heterodimeric partner RXR in cultured cells [41, 42]. In addition, in the rodent liver PXR expression is autoinduced by PCN- and PPARα-specific drugs such as perfluorodecanoic acid and clofibrate [43]. Kamiya et al. also showed that HNF4α is the key transcription factor regulating responses to xenobiotics through activation of the PXR gene during fetal liver development [44]. In theory, induction of the xenobiotic receptors could potentiate the induction of downstream target genes. Further studies are expected to reveal the relevance of xenobiotic receptor regulation and its impact on drug metabolism.

It is known that hPXR activators such as phenytoin and RU486 cause immunosuppressive side effects; on the other hand, inflammation and infection reduce hepatic CYP expression [45–47]. In addition, the levels of hepatic PXR and CAR mRNA have also been reported to be down-regulated in response to inflammatory signals [48, 49]. This broaches the question of whether xenobiotic receptors communicate with the immune system. Recently, PXR has been reported to cross talk with NF-κB signaling pathways, which regulate inflammation and the immune response [50]. The activation of hPXR inhibits NF-κB activity, whereas NF-κB target genes are up-regulated and small bowel inflammation is significantly increased in PXR null mice. On the other hand, NF-κB activation reciprocally inhibits hPXR and its target genes. This may provide a molecular explanation for the suppression of hepatic CYP mRNAs by inflammatory stimuli.

2.3.1.5 PXR Also Plays a Role in Endobiotic Metabolism.

In addition to environmental toxins, our body is continuously exposed to a variety of endogenous chemicals. For example, the secondary bile acid, lithocholic acid (LCA), is generated from nontoxic bile acids by intestinal bacteria and its elevation is implicated in pathogenesis of cholestatic liver disease and colon cancer. The observation that LCA can induce CYP3A expression led to the suggestion that PXR may modulate this induction to reduce hepatotoxicity [25, 26]. Three lines of evidence support this notion. First, LCA and its direct metabolite 3-keto LCA directly bind to and activate PXR. Second, *in vivo* activation of PXR by administration of PCN or by expression of a constitutively active form of PXR in the liver of transgenic mice results in marked resistance to LCA toxicity in rodents. Finally, a potent CYP3A inducer and agonist for human

PXR, rifampicin, has been reported to be effective in treating pruritus associated with chronic cholestasis. Detoxification of LCA by PXR appears to be mediated by the combined induction of CYP3A and the cytosolic sulfotransferase ST2A, both of which convert LCA to nontoxic metabolites [25, 26, 51]. Thus, the drug clearance pathway regulated by PXR can be utilized to detoxify endogenously produced toxins. For example, PXR may regulate inducible nitric oxide synthase (iNOS) involved in the inflammatory response [52]. Recently, Zhou et al. reported the accumulation of lipids and increased expression of the free fatty acid transporter CD36 in the livers of mice that express a constitutively activated PXR or are treated with a PXR activator. This cross-regulation of CD36 by PXR suggests that PXR plays an endobiotic role by influencing lipid homeostasis [53]. In addition, PXR takes part in steroid homeostasis and drug–hormone interactions. The activation of PXR in mice can increase plasma concentrations of corticosterone and aldosterone, which are associated with activation of cytochrome P450 CYP11 genes and 3β-hydroxysteroid dehydrogenase [54].

2.3.2 Characterization of CAR as a Xenobiotic Receptor

2.3.2.1 Cloning of CAR. The nuclear receptor CAR was identified as a xenobiotic receptor that mediated the induction of CYP2B by PB-type inducers [55] (for review, see [56]). Expressed abundantly in the liver, CAR and its partner RXR bind to the DR4 element in the CYP2B promoter, which is known to mediate inducibility by PB. CAR shows constitutive activity and induces expression of endogenous CYP2B upon transient or stable transfection in most cells. This (constitutive) CAR activity can be repressed by compounds such as androstanol and androstenol [57], whereas PB and PB-type compounds reactivate it and induce its target genes [58]. The CYP2B gene fails to respond to PB in CAR null mice, which confirms the role of CAR in mediating CYP2B induction [59]. Another example is that inducibility of CYP2B by PB-type inducers such as TCPOBOP is completely lost in CAR knock-out mice, as well as PB-mediated resistance to zoxazolamine and sensitivity to cocaine or acetaminophen [60, 61]. CAR null mice are viable and fertile, indicating that CAR function is also dispensable in the absence of toxic insults similar to PXR. Together, these results clearly show that CAR is a mediator of CYP2B induction.

2.3.2.2 CAR Activators. Similar to PXR, the LBD of CAR is also divergent among species (72% identity between human and mouse). Not surprisingly, CAR also shows strong species specificity for activators. For example, the potent mouse CAR agonist TCPOBOP and the reverse agonist androstanol do not affect human CAR, whereas the potent human CAR reverse agonist clotrimazol does not activate rodent CAR [61]. Another example is Meclizine, which acts as an agonist ligand for mouse CAR and an inverse agonist for human CAR [62]. The number of compounds that elicit CAR activity is less compared to the multitude of ligands that activate PXR. Other CAR activators include the antipsychotic chlorpromazine, plant products picrotoxin and camphor, and pesticides including PCBs, dieldrin, DDT, DDE, 6-(4-chlorophenyl)imidazo[2,1-*b*][1,3]thiazole-5-carbaldehyde

O-(3,4-dichlorobenzyl)oxime (CITCO), Methoxychlor, and its metabolites [56, 63, 64]. Ketoconazole is an antagonist for both human PXR and CAR, and can be used as a pan-antagonist of NRs involved in xenobiotic metabolism *in vivo* [65]. To date only TCPOBOP and androstanes are known to bind mouse CAR directly, whereas CITCO is the only known agonistic ligand that binds the human CAR.

2.3.2.3 A Working Model of CAR. A major distinction of CAR from other nuclear receptors is its constitutive activity in the absence of ligand in nonhepatic cells. As mentioned above, few chemicals are known to bind mouse CAR directly and modulate its interaction with coactivator molecules such as SRC-1, PGC-1, Sp1, ASC-2 and GRIP1 [57, 61, 66–69]. Rather, an alternative activation pathway appears to be engaged that is indirect and does not involve ligand binding [70]. Without the stimulation of compounds, CAR is maintained in an inactive state by being localized to the cytoplasm with its chaperone molecules such as CAR retention protein (CCRP) and heat shock protein 90 (HSP90) (reviewed in [71]). The inducing compound, while not binding, dissociates CAR from CCRP and HSP90 and triggers a cytoplasm to nuclear translocation, resulting in activation of target genes [72, 73]. Although, the mechanism of how and where CAR interacts with corepressors and coactivators is not yet clear, it was shown that this translocation depends on the activity of the protein phosphatase PP2A that modulates the LBD or an associated protein, followed by the binding of CAR/RXR to PBREs of target genes [70, 74]. It is not clear if PXR works in the same fashion as CAR, but there is also evidence showing that PXR forms a protein complex with CCRP and HSP90 and translocates from the cytoplasm into the nucleus upon PCN treatment [75]. The precise molecular mechanism of this process is important in understanding the signaling cascade of CAR/PXR xenobiotic activators.

2.3.2.4 How Is CAR Regulated? CAR is also regulated by other proteins and involved in other biochemical pathways, which makes the xenobiotic response mediated by CAR even more sophisticated. Inoue et al. reported cohesin protein SMC1 (structural maintenance of chromosomes 1) as a CAR binding protein, repressing CAR-mediated synergistic activation of CYP2B by xenobiotics such as TCPOBOP [76]. CAR activity and CYP2B gene expression are also regulated by phosphorylation through AMP-activated protein kinase (AMPK), which functions as an energy sensor. AMPK activators such as 5-amino-4-imidazolecarboxamide riboside (AICAR) induce CYP2B6 gene expression in human hepatocytes. Expression of a constitutively active form of AMPK mimics the PB induction of CYP2B gene expression. On the other hand, an AMPK inhibitor 8-bromo-AMP or the expression of a dominant negative form of AMPK abolishes the induction of CYP2B by PB, which implicates the involvement of AMPK signaling in liver drug responses mediated by CAR [77–79]. In addition, it has been shown that interleukin 1β (IL-1β) and lipopolysaccharides (LPS) decrease CAR expression and PB- or bilirubin-mediated induction of CYP2B in human hepatocytes [80], which is consistent with the observation that inflammation and sepsis inhibit drug metabolism [45–47]. These observations provide new insights into the xenobiotic metabolism pathway though CAR.

2.4 PERSPECTIVE

The number of PXR and CAR activators will continue to increase and the identification of more endogenous as well as environmental chemicals that modulate the xenobiotic receptors will improve our understanding of the mechanisms by which environmental compounds affect our endocrine balance and may offer novel strategies for preventing chemical toxicity.

Accordingly, the number of known PXR/CAR target genes has been expanding. PXR and CAR were originally characterized as independent regulators of the CYP3A and 2B genes, respectively, presumably through distinct categories of drugs. However, later observations suggest that there is significant cross-regulation of CY2Bs and 3As by these two receptors [24]. Analysis of the promoter regions of PXR/CAR target genes and identification of receptor binding sites reveal that both PXR and CAR can adaptively bind to common response elements [52, 81, 82]. In addition to CYP3A, PXR regulates a large number of genes involved in xenobiotic metabolism. These include cytochrome P450 enzymes, CYP2C, CYP1A, CYP1B, CYP2A and CYP4F, and other phase I reductases and hydrolases such as carboxylesterase, monoamine oxidase, catalase, and flavin-containing mono-oxygenases (FMO); phase II conjugating enzymes such as UDP-glucuronosyltransferase (UGT), cytosolic sulfotransferase (SULT), and GST, which solubilize hydrophobic compounds to prepare for clearance; phase III membrane-bound transporters such as MDR1 and MRP2, which act as efflux pumps to clear drugs and drug conjugates (reviewed in [20], also see [23, 51, 83, 84]). CAR has also been shown to regulate a similar array of xenobiotic genes including several CYP enzymes, aldehyde dehydrogenase, esterase, FMO, methyltransferase, GST [85], SULT [86], UGT [87], and MRP2 [81] as well as iNOS [52]. The ability of the xenobiotic receptors to respond to a numerous yet overlapping set of drugs and regulate a sophisticated network of metabolic genes suggests the existence of a well-coordinated molecular cascade of drug clearance.

In summary, we believe that the xenobiotic regulation of drug clearance by nuclear receptors will be an emerging and exciting field of research in the coming years. The results of these studies will greatly advance our understanding of the complexity of xenobiotic regulation and their implication in human physiology, pathology, pharmaceutical development as well as enable a broad assessment of environmental risks.

REFERENCES

1. McKenna, N. J. and O'Malley, B. W. (2005) An interactive course in nuclear receptor signaling: concepts and models. *Sci. STKE*, **229**, tr22.
2. Blumberg, B. and Evans, R. M. (1998) Orphan nuclear receptors—new ligands and new possibilities. *Genes Dev.* **12**, 3149–3155.
3. Mangelsdorf, D. J., Thummel, C., Beato, M., Herrlich, P., Schutz, G., Umesono, K., Blumberg, B., Kastner, P., Mark, M., Chambon, P., et al. (1995) The nuclear receptor superfamily: the second decade. *Cell* **83**, 835–839.

4. Weinberger, C., Hollenberg, S. M., Ong, E. S., Harmon, J. M., Brower, S. T., Cidlowski, J., Thompson, E. B., Rosenfeld, M. G., and Evans, R. M. (1985) Identification of human glucocorticoid receptor complementary DNA clones by epitope selection. *Science* **228**, 740–742.

5. Green, S., Walter, P., Kumar, V., Krust, A., Bornert, J. M., Argos, P., and Chambon, P. (1986) Human oestrogen receptor cDNA: sequence, expression and homology to v-erb-A. *Nature* **320**, 134–139.

6. Conneely, O. M., Sullivan, W. P., Toft, D. O., Birnbaumer, M., Cook, R. G., Maxwell, B. L., Zarucki-Schulz, T., Greene, G. L., Schrader, W. T., and O'Malley, B. W. (1986) Molecular cloning of the chicken progesterone receptor. *Science* **233**, 767–770.

7. Weinberger, C., Thompson, C. C., Ong, E. S., Lebo, R., Gruol, D. J., and Evans, R. M. (1986) The c-erb-A gene encodes a thyroid hormone receptor. *Nature* **324**, 641–646.

8. Petkovich, M., Brand, N. J., Krust, A., and Chambon, P. (1987) A human retinoic acid receptor which belongs to the family of nuclear receptors. *Nature* **330**, 444–450.

9. McDonnell, D. P., Mangelsdorf, D. J., Pike, J. W., Haussler, M. R., and O'Malley, B. W. (1987) Molecular cloning of complementary DNA encoding the avian receptor for vitamin D. *Science* **235**, 1214–1217.

10. Arriza, J. L., Weinberger, C., Cerelli, G., Glaser, T. M., Handelin, B. L., Housman, D. E., and Evans, R. M. (1987) Cloning of human mineralocorticoid receptor complementary DNA: structural and functional kinship with the glucocorticoid receptor. *Science* **237**, 268–275.

11. Lubahn, D. B., Joseph, D. R., Sullivan, P. M., Willard, H. F., French, F. S., and Wilson, E. M. (1988) Cloning of human androgen receptor complementary DNA and localization to the X chromosome. *Science* **240**, 327–330.

12. Giguere, V., Yang, N., Segui, P., and Evans, R. M. (1988) Identification of a new class of steroid hormone receptors. *Nature* **331**, 91–94.

13. Chawla, A., Repa, J. J., Evans, R. M., and Mangelsdorf, D. J. (2001) Nuclear receptors and lipid physiology: opening the X-files. *Science* **294**, 1866–1870.

14. Waxman, D. J. (1999) P450 gene induction by structurally diverse xenochemicals: central role of nuclear receptors CAR, PXR, and PPAR. *Arch. Biochem. Biophys.* **369**, 11–23.

15. Xie, W. and Evans, R. M. (2001) Orphan nuclear receptors: the exotics of xenobiotics. *J. Biol. Chem.* **276**, 37739–37742.

16. Selye, H. (1971) Hormones and resistance. *J. Pharm. Sci.* **60**, 1–28.

17. Burger, H. J., Schuetz, J. D., Schuetz, E. G., and Guzelian, P. S. (1992) Paradoxical transcriptional activation of rat liver cytochrome P-450 3A1 by dexamethasone and the antiglucocorticoid pregnenolone 16 alpha-carbonitrile: analysis by transient transfection into primary monolayer cultures of adult rat hepatocytes. *Proc. Natl Acad. Sci. USA* **89**, 2145–2149.

18. Gonzalez, F. J., Song, B. J., and Hardwick, J. P. (1986) Pregnenolone 16 alpha-carbonitrile-inducible P-450 gene family: gene conversion and differential regulation. *Mol. Cell Biol.* **6**, 2969–2976.

19. Kliewer, S. A., Moore, J. T., Wade, L., Staudinger, J. L., Watson, M. A., Jones, S. A., McKee, D. D., Oliver, B. B., Willson, T. M., Zetterstrom, R. H., et al. (1998) An orphan nuclear receptor activated by pregnanes defines a novel steroid signaling pathway. *Cell* **92**, 73–82.

20. Goodwin, B., Redinbo, M. R., and Kliewer, S. A. (2002) Regulation of cyp3a gene transcription by the pregnane X receptor. *Annu. Rev. Pharmacol. Toxicol.* **42**, 1–23.

21. Sonoda, J. and Evans, R. M. (2003) Biological function and mode of action of nuclear xenobiotic receptors. *Pure Appl. Chem.* **75**, 1733–1742.

22. Johnson, D. R., Li, C. W., Chen, L. Y., Ghosh, J. C., and Chen, J. D. (2006) Regulation and binding of pregnane X receptor by nuclear receptor corepressor silencing mediator of retinoid and thyroid hormone receptors (SMRT). *Mol. Pharmacol.* **69**, 99–108.

23. Synold, T. W., Dussault, I., and Forman, B. M. (2001) The orphan nuclear receptor SXR coordinately regulates drug metabolism and efflux. *Nat. Med.* **7**, 584–590.

24. Xie, W., Barwick, J. L., Downes, M., Blumberg, B., Simon, C. M., Nelson, M. C., Neuschwander-Tetri, B. A., Brunt, E. M., Guzelian, P. S., and Evans, R. M. (2000) Human-ized xenobiotic response in mice expressing nuclear receptor SXR. *Nature* **406**, 435–439.

25. Staudinger, J. L., Goodwin, B., Jones, S. A., Hawkins-Brown, D., MacKenzie, K. I., LaTour, A., Liu, Y., Klaassen, C. D., Brown, K. K., Reinhard, J., et al. (2001) The nuclear receptor PXR is a lithocholic acid sensor that protects against liver toxicity. *Proc. Natl Acad. Sci. USA* **98**, 3369–3374.

26. Xie, W., Radominska-Pandya, A., Shi, Y., Simon, C. M., Nelson, M. C., Ong, E. S., Waxman, D. J., and Evans, R. M. (2001) An essential role for nuclear receptors SXR/PXR in detoxification of cholestatic bile acids. *Proc. Natl Acad. Sci. USA* **98**, 3375–3380.

27. Staudinger, J., Liu, Y., Madan, A., Habeebu, S., and Klaassen, C. D. (2001) Coordinate regulation of xenobiotic and bile acid homeostasis by pregnane X receptor. *Drug Metab. Dispos.* **29**, 1467–1472.

28. Blumberg, B., Sabbagh, W., Jr., Juguilon, H., Bolado, J., Jr., van Meter, C. M., Ong, E. S., and Evans, R. M. (1998) SXR, a novel steroid and xenobiotic-sensing nuclear receptor. *Genes Dev.* **12**, 3195–3205.

29. Bertilsson, G., Heidrich, J., Svensson, K., Asman, M., Jendeberg, L., Sydow-Backman, M., Ohlsson, R., Postlind, H., Blomquist, P., and Berkenstam, A. (1998) Identification of a human nuclear receptor defines a new signaling pathway for CYP3A induction. *Proc. Natl Acad. Sci. USA* **95**, 12208–12213.

30. Gonzalez, F. J. and Gelboin, H. V. (1991) Human cytochromes P450: evolution, catalytic activities and interindividual variations in expression. *Prog. Clin. Biol. Res.* **372**, 11–20.

31. Xie, W. and Evans, R. M. (2001) Orphan nuclear receptors: the exotics of xenobiotics. *J. Biol. Chem.* **276**, 37739–37742.

32. Sonoda, J., Rosenfeld, J. M., Xu, L., Evans, R. M., and Xie, W. (2003) A nuclear receptor-mediated xenobiotic response and its implication in drug metabolism and host protection. *Curr. Drug Metab.* **4**, 59–72.

33. Watkins, R. E., Wisely, G. B., Moore, L. B., Collins, J. L., Lambert, M. H., Williams, S. P., Willson, T. M., Kliewer, S. A., and Redinbo, M. R. (2001) The human nuclear xenobiotic receptor PXR: structural determinants of directed promiscuity. *Science* **292**, 2329–2333.

34. Takeshita, A., Koibuchi, N., Oka, J., Taguchi, M., Shishiba, Y., and Ozawa, Y. (2001) Bisphenol-A, an environmental estrogen, activates the human orphan nuclear receptor, steroid and xenobiotic receptor-mediated transcription. *Eur. J. Endocrinol.* **145**, 513–517.

35. Ricketts, M. L., Boekschoten, M. V., Kreeft, A. J., Hooiveld, G. J., Moen, C. J., Muller, M., Frants, R. R., Kasanmoentalib, S., Post, S. M., Princen, H. M., et al. (2007) The cholesterol-raising factor from coffee beans, cafestol, as an agonist ligand for the farnesoid and pregnane X receptors. *Mol. Endocrinol.* **21**, 1603–1616.

36. Coumoul, X., Diry, M., and Barouki, R. (2002) PXR-dependent induction of human CYP3A4 gene expression by organochlorine pesticides. *Biochem. Pharmacol.* **64**, 1513–1519.

37. Schuetz, E. G., Brimer, C., and Schuetz, J. D. (1998) Environmental xenobiotics and the antihormones cyproterone acetate and spironolactone use the nuclear hormone pregnenolone X receptor to activate the CYP3A23 hormone response element. *Mol. Pharmacol.* **54**, 1113–1117.

38. Noreault, T. L., Jacobs, J. M., Nichols, R. C., Trask, H. W., Wrighton, S. A., Sinclair, P. R., Evans, R. M., and Sinclair, J. F. (2005) Arsenite decreases CYP3A23 induction in cultured rat hepatocytes by transcriptional and translational mechanisms. *Toxicol. Appl. Pharmacol.* **209**, 174–182.

39. Noreault, T. L., Kostrubsky, V. E., Wood, S. G., Nichols, R. C., Strom, S. C., Trask, H. W., Wrighton, S. A., Evans, R. M., Jacobs, J. M., Sinclair, P. R., et al. (2005) Arsenite decreases CYP3A4 and RXRalpha in primary human hepatocytes. *Drug Metab. Dispos.* **33**, 993–1003.

40. Sonoda, J., Chong, L. W., Downes, M., Barish, G. D., Coulter, S., Liddle, C., Lee, C. H., and Evans, R. M. (2005) Pregnane X receptor prevents hepatorenal toxicity from cholesterol metabolites. *Proc. Natl Acad. Sci. USA* **102**, 2198–2203.

41. Pascussi, J. M., Gerbal-Chaloin, S., Fabre, J. M., Maurel, P., and Vilarem, M. J. (2000) Dexamethasone enhances constitutive androstane receptor expression in human hepatocytes: consequences on cytochrome P450 gene regulation. *Mol. Pharmacol.* **58**, 1441–1450.

42. Pascussi, J. M., Drocourt, L., Fabre, J. M., Maurel, P., and Vilarem, M. J. (2000) Dexamethasone induces pregnane X receptor and retinoid X receptor-alpha expression in human hepatocytes: synergistic increase of CYP3A4 induction by pregnane X receptor activators. *Mol. Pharmacol.* **58**, 361–372.

43. Zhang, H., LeCulyse, E., Liu, L., Hu, M., Matoney, L., Zhu, W., and Yan, B. (1999) Rat pregnane X receptor: molecular cloning, tissue distribution, and xenobiotic regulation. *Arch. Biochem. Biophys.* **368**, 14–22.

44. Kamiya, A., Inoue, Y., and Gonzalez, F. J. (2003) Role of the hepatocyte nuclear factor 4alpha in control of the pregnane X receptor during fetal liver development. *Hepatology* **37**, 1375–1384.

45. Antonakis, N., Markogiannakis, E., Theodoropoulou, M., Georgoulias, V., Stournaras, C., and Gravanis, A. (1991) The antiglucocorticoid RU486 down regulates the expression of interleukin-2 receptors in normal human lymphocytes. *J. Steroid Biochem. Mol. Biol.* **39**, 929–935.

46. Riddick, D. S., Lee, C., Bhathena, A., Timsit, Y. E., Cheng, P. Y., Morgan, E. T., Prough, R. A., Ripp, S. L., Miller, K. K., Jahan, A., et al. (2004) Transcriptional suppression of cytochrome P450 genes by endogenous and exogenous chemicals. *Drug Metab. Dispos.* **32**, 367–375.

47. Morgan, E. T. (1997) Regulation of cytochromes P450 during inflammation and infection. *Drug Metab. Rev.* **29**, 1129–1188.

48. Pascussi, J. M., Gerbal-Chaloin, S., Pichard-Garcia, L., Daujat, M., Fabre, J. M., Maurel, P., and Vilarem, M. J. (2000) Interleukin-6 negatively regulates the expression of pregnane X receptor and constitutively activated receptor in primary human hepatocytes. *Biochem. Biophys. Res. Commun.* **274**, 707–713.

49. Beigneux, A. P., Moser, A. H., Shigenaga, J. K., Grunfeld, C., and Feingold, K. R. (2002) Reduction in cytochrome P-450 enzyme expression is associated with repression of CAR (constitutive androstane receptor) and PXR (pregnane X receptor) in mouse liver during the acute phase response. *Biochem. Biophys. Res. Commun.* **293**, 145–149.

50. Zhou, C., Tabb, M. M., Nelson, E. L., Grun, F., Verma, S., Sadatrafiei, A., Lin, M., Mallick, S., Forman, B. M., Thummel, K. E., et al. (2006) Mutual repression between steroid and xenobiotic receptor and NF-kappaB signaling pathways links xenobiotic metabolism and inflammation. *J. Clin. Invest.* **116**, 2280–2289.

51. Sonoda, J., Xie, W., Rosenfeld, J. M., Barwick, J. L., Guzelian, P. S., and Evans, R. M. (2002) Regulation of a xenobiotic sulfonation cascade by nuclear pregnane X receptor (PXR). *Proc. Natl Acad. Sci. USA* **99**, 13801–13806.

52. Toell, A., Kroncke, K. D., Kleinert, H., and Carlberg, C. (2002) Orphan nuclear receptor binding site in the human inducible nitric oxide synthase promoter mediates responsiveness to steroid and xenobiotic ligands. *J. Cell Biochem.* **85**, 72–82.

53. Zhou, J., Zhai, Y., Mu, Y., Gong, H., Uppal, H., Toma, D., Ren, S., Evans, R. M., and Xie, W. (2006) A novel pregnane X receptor-mediated and sterol regulatory element-binding protein-independent lipogenic pathway. *J. Biol. Chem.* **281**, 15013–15020.

54. Zhai, Y., Pai, H. V., Zhou, J., Amico, J. A., Vollmer, R. R., and Xie, W. (2007) Activation of pregnane X receptor disrupts glucocorticoid and mineralocorticoid homeostasis. *Mol. Endocrinol.* **21**, 138–147.

55. Honkakoski, P., Zelko, I., Sueyoshi, T., and Negishi, M. (1998) The nuclear orphan receptor CAR-retinoid X receptor heterodimer activates the phenobarbital-responsive enhancer module of the CYP2B gene. *Mol. Cell Biol.* **18**, 5652–5658.

56. Sueyoshi, T. and Negishi, M. (2001) Phenobarbital response elements of cytochrome P450 genes and nuclear receptors. *Annu. Rev. Pharmacol. Toxicol.* **41**, 123–143.

57. Forman, B. M., Tzameli, I., Choi, H. S., Chen, J., Simha, D., Seol, W., Evans, R. M., and Moore, D. D. (1998) Androstane metabolites bind to and deactivate the nuclear receptor CAR-beta. *Nature* **395**, 612–615.

58. Sueyoshi, T., Kawamoto, T., Zelko, I., Honkakoski, P., and Negishi, M. (1999) The repressed nuclear receptor CAR responds to phenobarbital in activating the human CYP2B6 gene. *J. Biol. Chem.* **274**, 6043–6046.

59. Wei, P., Zhang, J., Egan-Hafley, M., Liang, S., and Moore, D. D. (2000) The nuclear receptor CAR mediates specific xenobiotic induction of drug metabolism. *Nature* **407**, 920–923.

60. Zhang, J., Huang, W., Chua, S. S., Wei, P., and Moore, D. D. (2002) Modulation of acetaminophen-induced hepatotoxicity by the xenobiotic receptor CAR. *Science* **298**, 422–424.

61. Moore, L. B., Parks, D. J., Jones, S. A., Bledsoe, R. K., Consler, T. G., Stimmel, J. B., Goodwin, B., Liddle, C., Blanchard, S. G., Willson, T. M., et al. (2000) Orphan nuclear receptors constitutive androstane receptor and pregnane X receptor share xenobiotic and steroid ligands. *J. Biol. Chem.* **275**, 15122–15127.

62. Huang, W., Zhang, J., Wei, P., Schrader, W. T., and Moore, D. D. (2004) Meclizine is an agonist ligand for mouse constitutive androstane receptor (CAR) and an inverse agonist for human CAR. *Mol. Endocrinol.* **18**, 2402–2408.

63. Wyde, M. E., Kirwan, S. E., Zhang, F., Laughter, A., Hoffman, H. B., Bartolucci-Page, E., Gaido, K. W., Yan, B., and You, L. (2005) Di-n-butyl phthalate activates constitutive

androstane receptor and pregnane X receptor and enhances the expression of steroid-metabolizing enzymes in the liver of rat fetuses. *Toxicol. Sci.* **86**, 281–290.

64. Maglich, J. M., Parks, D. J., Moore, L. B., Collins, J. L., Goodwin, B., Billin, A. N., Stoltz, C. A., Kliewer, S. A., Lambert, M. H., Willson, T. M., et al. (2003) Identification of a novel human constitutive androstane receptor (CAR) agonist and its use in the identification of CAR target genes. *J. Biol. Chem.* **278**, 17277–17283.

65. Huang, H., Wang, H., Sinz, M., Zoeckler, M., Staudinger, J., Redinbo, M. R., Teotico, D. G., Locker, J., Kalpana, G. V., and Mani, S. (2007) Inhibition of drug metabolism by blocking the activation of nuclear receptors by ketoconazole. *Oncogene* **26**, 258–268.

66. Tzameli, I., Pissios, P., Schuetz, E. G., and Moore, D. D. (2000) The xenobiotic compound 1,4-bis[2-(3,5-dichloropyridyloxy)]benzene is an agonist ligand for the nuclear receptor CAR. *Mol. Cell Biol.* **20**, 2951–2958.

67. Min, G., Kemper, J. K., and Kemper, B. (2002) Glucocorticoid receptor-interacting protein 1 mediates ligand-independent nuclear translocation and activation of constitutive androstane receptor in vivo. *J. Biol. Chem.* **277**, 26356–26363.

68. Choi, E., Lee, S., Yeom, S. Y., Kim, G. H., Lee, J. W., and Kim, S. W. (2005) Characterization of activating signal cointegrator-2 as a novel transcriptional coactivator of the xenobiotic nuclear receptor constitutive androstane receptor. *Mol. Endocrinol.* **19**, 1711–1719.

69. Shiraki, T., Sakai, N., Kanaya, E., and Jingami, H. (2003) Activation of orphan nuclear constitutive androstane receptor requires subnuclear targeting by peroxisome proliferator-activated receptor gamma coactivator-1 alpha. A possible link between xenobiotic response and nutritional state. *J. Biol. Chem.* **278**, 11344–11350.

70. Zelko, I., Sueyoshi, T., Kawamoto, T., Moore, R., and Negishi, M. (2001) The peptide near the C terminus regulates receptor CAR nuclear translocation induced by xenochemicals in mouse liver. *Mol. Cell Biol.* **21**, 2838–2846.

71. Timsit, Y. E. and Negishi, M. (2007) CAR and PXR: the xenobiotic-sensing receptors. *Steroids* **72**, 231–246.

72. Kawamoto, T., Sueyoshi, T., Zelko, I., Moore, R., Washburn, K., and Negishi, M. (1999) Phenobarbital-responsive nuclear translocation of the receptor CAR in induction of the CYP2B gene. *Mol. Cell Biol.* **19**, 6318–6322.

73. Kobayashi, K., Sueyoshi, T., Inoue, K., Moore, R., and Negishi, M. (2003) Cytoplasmic accumulation of the nuclear receptor CAR by a tetratricopeptide repeat protein in HepG2 cells. *Mol. Pharmacol.* **64**, 1069–1075.

74. Koike, C., Moore, R., and Negishi, M. (2005) Localization of the nuclear receptor CAR at the cell membrane of mouse liver. *FEBS Lett.* **579**, 6733–6736.

75. Squires, E. J., Sueyoshi, T., and Negishi, M. (2004) Cytoplasmic localization of pregnane X receptor and ligand-dependent nuclear translocation in mouse liver. *J. Biol. Chem.* **279**, 49307–49314.

76. Inoue, K., Borchers, C. H., and Negishi, M. (2006) Cohesin protein SMC1 represses the nuclear receptor CAR-mediated synergistic activation of a human P450 gene by xenobiotics. *Biochem. J.* **398**, 125–133.

77. Blattler, S. M., Rencurel, F., Kaufmann, M. R., and Meyer, U. A. (2007) In the regulation of cytochrome P450 genes, phenobarbital targets LKB1 for necessary activation of AMP-activated protein kinase. *Proc. Natl Acad. Sci. USA* **104**, 1045–1050.

78. Rencurel, F., Stenhouse, A., Hawley, S. A., Friedberg, T., Hardie, D. G., Sutherland, C., and Wolf, C. R. (2005) AMP-activated protein kinase mediates phenobarbital induction of CYP2B gene expression in hepatocytes and a newly derived human hepatoma cell line. *J. Biol. Chem.* **280**, 4367–4373.

79. Shindo, S., Numazawa, S., and Yoshida, T. (2007) A physiological role of AMP-activated protein kinase in phenobarbital-mediated constitutive androstane receptor activation and CYP2B induction. *Biochem. J.* **401**, 735–741.

80. Assenat, E., Gerbal-Chaloin, S., Larrey, D., Saric, J., Fabre, J. M., Maurel, P., Vilarem, M. J., and Pascussi, J. M. (2004) Interleukin 1beta inhibits CAR-induced expression of hepatic genes involved in drug and bilirubin clearance. *Hepatology* **40**, 951–960.

81. Kast, H. R., Goodwin, B., Tarr, P. T., Jones, S. A., Anisfeld, A. M., Stoltz, C. M., Tontonoz, P., Kliewer, S., Willson, T. M., and Edwards, P. A. (2002) Regulation of multidrug resistance-associated protein 2 (ABCC2) by the nuclear receptors pregnane X receptor, farnesoid X-activated receptor, and constitutive androstane receptor. *J. Biol. Chem.* **277**, 2908–2915.

82. Xie, W., Yeuh, M. F., Radominska-Pandya, A., Saini, S. P., Negishi, Y., Bottroff, B. S., Cabrera, G. Y., Tukey, R. H., and Evans, R. M. (2003) Control of steroid, heme, and carcinogen metabolism by nuclear pregnane X receptor and constitutive androstane receptor. *Proc. Natl Acad. Sci. USA* **100**, 4150–4155.

83. Rae, J. M., Johnson, M. D., Lippman, M. E., and Flockhart, D. A. (2001) Rifampin is a selective, pleiotropic inducer of drug metabolism genes in human hepatocytes: studies with cDNA and oligonucleotide expression arrays. *J. Pharmacol. Exp. Ther.* **299**, 849–857.

84. Maglich, J. M., Stoltz, C. M., Goodwin, B., Hawkins-Brown, D., Moore, J. T., and Kliewer, S. A. (2002) Nuclear pregnane x receptor and constitutive androstane receptor regulate overlapping but distinct sets of genes involved in xenobiotic detoxification. *Mol. Pharmacol.* **62**, 638–646.

85. Ueda, A., Hamadeh, H. K., Webb, H. K., Yamamoto, Y., Sueyoshi, T., Afshari, C. A., Lehmann, J. M., and Negishi, M. (2002) Diverse roles of the nuclear orphan receptor CAR in regulating hepatic genes in response to phenobarbital. *Mol. Pharmacol.* **61**, 1–6.

86. Garcia-Allan, C., Lord, P. G., Loughlin, J. M., Orton, T. C., and Sidaway, J. E. (2000) Identification of phenobarbitone-modulated genes in mouse liver by differential display. *J. Biochem. Mol. Toxicol.* **14**, 65–72.

87. Sugatani, J., Kojima, H., Ueda, A., Kakizaki, S., Yoshinari, K., Gong, Q. H., Owens, I. S., Negishi, M., and Sueyoshi, T. (2001) The phenobarbital response enhancer module in the human bilirubin UDP-glucuronosyltransferase UGT1A1 gene and regulation by the nuclear receptor CAR. *Hepatology* **33**, 1232–1238.

3

NUCLEAR RECEPTOR-MEDIATED REGULATION OF PHASE II CONJUGATING ENZYMES

OLIVIER BARBIER

Faculty of Pharmacy, Laval University and Research Center of the "Centre Hospitalier de l'Université Laval—CRCHUL", Québec, Canada

3.1 INTRODUCTION

Almost all drugs that undergo metabolic transformation are converted to metabolites, which are more polar than the parent molecules. Biotransformation not only promotes drug elimination but also often results in inactivation of pharmacological activity, thereby changing the overall biological properties of the drug [1]. In general, drugs are metabolized by sequential reactions involving phase I and II enzymes, and the functional groups added by the phase I P450 (NADPH-cytochrome P450) enzymes are then used as acceptors for a polar group incorporated by phase II conjugating enzymes [2]. Thus sequential biotransformations by the phase I and II drug-/xenobiotic-metabolizing enzymes (D/XMEs) constitute the major routes for drug metabolism. Phase I metabolism usually does not result in a large change in molecular weight or water solubility, but is of great importance because oxidative reactions add or expose sites where the phase II metabolism can subsequently occur [3]. The phase II reactions generally produce nontoxic polar metabolites and facilitate biliary and/or urinary excretion. Most drugs undergo phase I reaction prior to phase II metabolism, but molecules with sites amenable to conjugation may undergo conjugation reactions directly. Furthermore, molecules that are directly conjugated may also sustain competing or additional phase I oxidation [3]. The metabolites from

Nuclear Receptors in Drug Metabolism   Edited by Wen Xie
Copyright © 2009 John Wiley & Sons, Inc.

phases I and II reactions, being hydrosolubles, are excreted from the cells with the aid of membrane efflux pumps (phase III).

It is increasingly accepted that all phases of xenobiotic metabolism may be coordinately regulated by a series of xenobiotic-activated transcription factors (also termed xenosensors), namely the nuclear receptors (NRs), constitutive androstane receptor (CAR), pregnane X receptor (PXR), and aryl hydrocarbon receptor (AhR). Such coordinate regulation of DMEs results in increased protection of cells against toxic levels of hydrophobic xeno- or endobiotics. Therefore, large parts of the present chapter will summarize the current knowledge on the tight interactions between these transcription factors and phase II DMEs. In some cases, metabolism of chemicals generates more toxic species making the liver and gastrointestinal tract particularly susceptible to oxidative-type processes, such as chemical toxicity and carcinogenesis [4]. Both organs are equipped with defense mechanisms to detoxify reactive oxygen intermediates and minimize oxidative stress, and recent observations suggest that the nuclear factor erythroid 2-related factor 2 (Nrf2) transcription factor is critical in such protection by regulating phase II DMEs as well as cellular antioxidant defenses [4]. These regulatory pathways and their consequences will also be addressed in this chapter. Finally, recent observations also indicate that a number of other nuclear transcription factors, such as the farnesoid X receptor (FXR), the liver X receptor (LXR), the peroxisome proliferator-activated receptors (PPARs), and the hepatic nuclear factors (HNFs), are important modulators of DME encoding genes. The role of these receptors will be briefly addressed at the end of this chapter.

3.2   PHASE II DRUG METABOLIZING ENZYMES

The major phase II metabolizing enzymes include the UDP-glucuronosyltransferases (UGTs), sulfotransferases (SULTs), glutathione S-transferases (GSTs), arylamine N-acetyltransferases (NATs), and epoxide hydrolases (EPHXs) enzymes. All together, members belonging to these five enzyme families contribute to the phase II metabolism of more than 90% of drugs sustaining conjugation [5]. Among the drugs conjugated, ~35% are substrates for UGTs, ~20% for SULTs, ~15% for GSTs, and ~10% for NAT [5]. In addition to drugs and other xenobiotics, these enzymes also conjugate a large variety of endogenous molecules (at least in humans), and thus play also a significant role in the inactivation and excretion of active endobiotics, such as bilirubin, bile acids, or steroid and thyroid hormones [6]. Thus, modulating the expression and activity of these metabolizing enzymes will affect both xeno- and endobiotics levels.

3.2.1   Glucuronidation and UGTs

Glucuronidation is a major drug-metabolizing reaction in humans and accounts for approximately 40–70% of xenobiotic elimination [7, 8]. The UGTs catalyze the glucuronidation reaction, which corresponds to the transfer of glucuronic acid from the ubiquitous UDP-glucuronic acid (UDPGA) cofactor to hydrophobic molecules

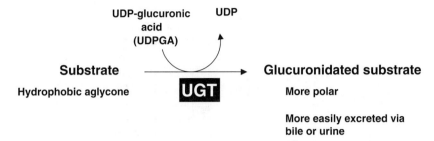

FIGURE 3.1   The glucuronidation reaction.

(aglycone) via nucleophilic attack of functional groups such as carboxyls, hydroxyls, thiols, or primary and secondary amines [2]. The UDPGA cofactor being highly hydrophilic, the resulting glucuronide conjugates have a high solubility in water and are easily eliminated from the human body into bile or urine (Figure 3.1). Some drugs are metabolized by phase I reactions to generate acceptor groups for the glucuronic acid. For these drugs both phase I and II pathways are relevant [8]. However, there are also many drugs for which glucuronidation is the major metabolic pathway [8]. In this case, the glucuronides formed can potentially interfere with the pharmacokinetics of the parent compound. For example, after oral administration, first-pass conjugation in the gut and the liver, two organs with high UGT expression, may lead to a low bioavailability and impaired efficiency of the drug. If the glucuronides are formed in the liver, they can be excreted by either renal or biliary elimination. Whereas molecules with a molecular weight of $450 \, \mathrm{g \, mol^{-1}}$ and greater are better substrates for canalicular secretion, those with a lower molecular weight are predominantly transported across the sinusoidal membrane into the blood and are then eliminated by the kidney [9].

On the basis of amino acid sequence homology, human UGTs have been classified into four families: UGT1, UGT2, UGT3, and UGT8 [10]. The most important drug-conjugating UGTs belong to UGT1 and UGT2 families (Figure 3.2). While the UGT1 family is composed by only one subfamily, i.e., UGT1A, the members of the UGT2 are further divided into UGT2A and UGT2B subfamilies. In contrast to the UGT2 genes that are encoded by different genes clustered on chromosome 4q13, the human UGT1As are produced from a single gene locus located on chromosome 2q37 [10]. This gene contains 13 individual promoters and exons and a shared set of exons 2–5. The presence of missense mutations within three exons 1 of the UGT1A gene results in the formation of only nine proteins. However, recent studies have identified an additional common exon 5 (named 5b) that can also be alternatively shared with other exons to form an additional set of nine UGT1A, designated with the symbol i2 (Figure 3.3) [11, 12]. With the exception of UGT1A7, 1A8, and 1A10, which are expressed in the stomach (UGT1A7, 1A8, and 1A10) and/or gastrointestinal tract (UGT1A8 and 1A10), all other human UGT1A and UGT2B enzymes are expressed in the liver and kidney, which are considered as the most important organs for xenobiotic glucuronidation [6, 13]. However, elevated glucuronidation

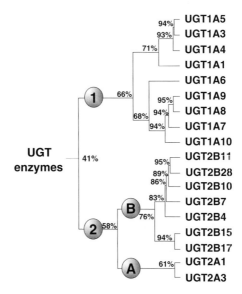

FIGURE 3.2   Human UDP-glucuronosyltransferase enzymes.

activities are also detected in peripheral tissues exposed to drugs and xenobiotics, namely the gastrointestinal tract. Of the 18 functional human UGTs, UGT1A5, 2A1, 2A3, 2B4, 2B10, 2B11, 2B17, and 2B28 appear to exhibit low or negligible activity toward drugs and other xenobiotics [14]. Thus UGT1A1, 1A3, 1A4, 1A6, 1A9, 2B7, and 2B15 are considered to be the isoforms of greater importance in hepatic drug elimination [15]. UGT1A1 is of particular importance because it is responsible for glucuronidation, and thus detoxification, of the hepatic heme breakdown product, bilirubin (Figure 3.3) [16]. Deficiency in UGT1A1 causes unconjugated hyperbilirubinemia termed Crigler–Najjar and Gilbert syndromes [17]. UGT1A6 has been recognized as the isoform responsible for the glucuronidation of serotonin and phenols (e.g., acetaminophen, APAP), but there is considerable substrate overlap between the UGT1A enzymes [3, 18, 19]. The UGT2B7 is recognized for its role in glucuronidation of morphine, one of only few pharmacologically active glucuronide conjugates, and Zidovudine (AZT) [20, 21]. In the following pages, human enzymes are designated by capital letters (i.e., UGT), whereas "Ugt" refers to rodent isoforms [10].

3.2.2   Sulfonation and SULTs

Hepatic sulfation (or more correctly sulfonation) of drugs is a common phase II metabolic mechanism for increasing hydrophilicity in preparation for biliary or urinary excretion. Sulfonation may occur directly (i.e., acetaminophen) or may follow phase I oxidation (i.e., hydroxyphenobarbital). However, hepatic sulfonation can also

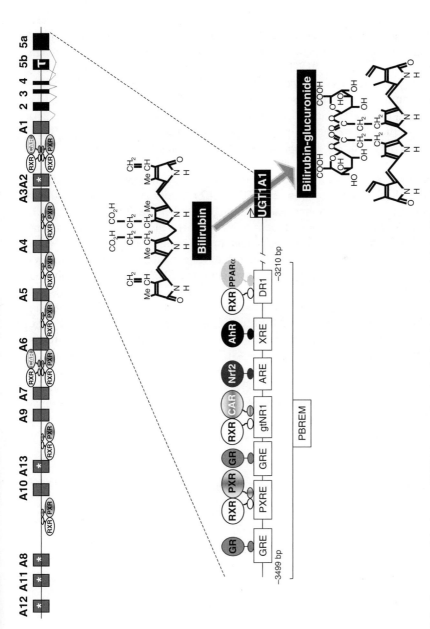

FIGURE 3.3 Structure of the human UGT1A gene showing the human genes in which PXR and CAR response elements have been identified (upper panel) and response elements of the PBREM, which control the expression of the human bilirubin-conjugating UGT1A1 enzyme (lower panel); *, pseudogenes; T, alternative exon 5.

lead to activation of hepatotoxins (i.e., troglitazone), DNA binding carcinogens (i.e., tamoxifen), or prodrugs (i.e., minoxidil) [22–25].

Cytosolic sulfotransferases are the main sulfonation enzymes and are involved in the biotransformation of a wide variety of structurally diverse endo- and xenobiotics, including many therapeutic agents and endogenous steroids [26]. SULTs involved in the phase II biotransformation of endo- and xenobiotics are cytosolic, although there are membrane-bound SULTs that are responsible for catalyzing the sulfation of protein tyrosyl residues and polysaccharides [26, 27]. Cytosolic SULTs belong to a superfamily of genes that are divided into two subfamilies, the phenol SULTs (SULT1) and the hydroxysteroid SULTs (SULT2) [26]. A third subfamily (SULT4) of brain-specific sulfotransferase enzymes has also been described while having been less characterized [26, 28]. Eleven SULTs have been identified in humans [26, 29]. SULTs have broad overlapping substrate specificities, but can be distinguished by their particular affinities for substrates, their thermal stability and their sensitivity to pharmacological inhibitors [26, 29]. The SULT1 family is subdivided into four subfamilies: (i) SULT1A comprises four human members (SULT1A1, 1A2, 1A3, and 1A4) that sulfonate phenolic-type xenobiotics; (ii) SULT1B (one human member, SULT1B1, conjugates dopa/tyrosine and thyroid hormones); (iii) SULT1C (two human enzymes, SULT1C2 and 1C4, which conjugate hydroxyarylamines); and (iv) SULT1E (human SULT1E1 conjugates estrogens). The SULT2 family is subdivided into two subfamilies: (a) SULT2A (human SULT2A1 sulfonates neutral steroids and bile acids) and (b) SULT2B (the two human SULT2B1-v1 and v2 conjugate sterols) [3, 30–34] (Figure 3.4). Sulfonation of xenobiotics is primarily mediated by the SULT1A subfamily, which catalyzes sulfonation of hydroxyl and monoamine groups on phenolic-type molecules [3]. However, SULT1C also plays a role in xenobiotic sulfonation by sulfating hydroxyarylamines, which can be activated into carcinogens by sulfonation [35]. SULT enzymes have a widespread tissue distribution and are expressed in the liver, lung, brain, skin, platelets, breast, kidney, and gastrointestinal tissue [26].

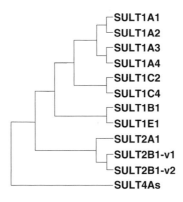

FIGURE 3.4   Cytosolic human sulfotransferase enzymes. (Adapted from [33].)

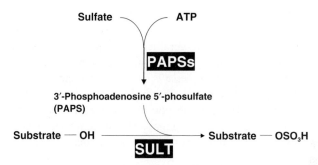

FIGURE 3.5   The sulfonation reaction.

The physiological function of SULTs is to catalyze the transfer of the sulfonyl group from the cosubstrate 3'-phosphoadenosine 5'-phosphosulfate (PAPS) to the hydroxyl, sulfhydryl, amino, or N-oxide groups of the acceptor substrates [26] (Figure 3.5). In the cytosol, inorganic sulfate is conjugated with adenosine triphosphate to form PAPS by a bifunctional enzyme, 3'-phosphoadenosine 5'-phosphosulfate synthetase (PAPSS). There are two isoforms: PAPSS1 and PAPSS2, which are highly expressed in the brain and liver, respectively [26]. Synthesis of PAPS is fast but is limited by hepatic sulfate concentrations, which are largely dependent on equilibrium with circulating inorganic sulfate [36–38].

3.2.3   Sulfonation and Glucuronidation: Two Complementary Metabolic Pathways

Sulfonation is a high affinity and low capacity phase II reaction that works in concert with glucuronidation on overlapping substrates; sulfonation predominates at low substrate concentrations and glucuronidation at high concentrations when sulfonation has been saturated [39]. In fact, hepatic PAPS concentrations ($\sim$23 nmol g^{-1} liver) are low enough to be depleted rapidly, and at substrate concentrations lower than those necessary to achieve sulfotransferase maximal velocity [37, 38]. Hence, sulfonation in many cases is a cofactor-limited instead of enzyme-limited reaction. In contrast, UDPGA is abundant in the liver ($\sim$300 nmol g^{-1} liver) and glucuronidation is generally considered as a low affinity and high capacity conjugation reaction [3]. Thus, saturation of sulfonation results in an increase in glucuronidation rate, so that the combined rate of these two reactions is linear and hepatic extraction ratio is constant, until glucuronidation is saturated at very high substrate concentrations [3, 39, 40–43].

On the other hand, sulfate conjugation is reversible, since sulfated molecules may be desulfated in the liver by sulfatase enzymes, subsequently reconjugated, deconjugated, etc., giving rise to the phenomenon of "futile cycling" between the parent compound and the sulfate metabolite [39]. Glucuronide conjugates can also be deconjugated in an acidic environment or enzymatically via β-glucuronidase. However, the liver only has very low glucuronidase activity, and the hepatic futile cycle observed for sulfate conjugate is not appreciable for the majority of glucuronide conjugates (with the exception of acyl glucuronides) [44, 45]. Nevertheless, with high β-glucuronidase activity in the gut microflora, many glucuronidated drugs excreted via bile into the duodenum are deconjugated to the parent drug, which may be reabsorbed resulting in "enterohepatic cycling" of conjugation–deconjugation [3, 46]. This enterohepatic cycle can also be an object of drug–drug interactions. Since carrier proteins are involved in the transport of glucuronides across cell membranes, such interactions at the transporter level may result in conjugate accumulation in tissues, cells, or in the circulation with corresponding pharmacological or toxicological consequences [9, 47].

Despite (and because) of these differences, glucuronidation and sulfonation often play complementary roles in drug metabolism and elimination. As detailed in the following sections, the expression and activity of drug-conjugating sulfotransferase and glucuronosyltransferase enzymes are coordinately regulated by a similar subset of NRs. Modulation of drug sulfonation and glucuronidation causing increasing or decreasing conjugation rates in the liver and intestine often results in significant pharmacodynamic changes *in vivo* [3, 48].

3.2.4   Glutathione Conjugation and GSTs

Glutathione conjugation is of particular importance because substrates of this reaction are often potent electrophiles. Conjugation with intracellular glutathione results in the detoxification of these reactive species, which could otherwise bind intracellular macromolecules. Thus, GSTs are involved in cellular protection against xenobiotics and oxidative stress as well as resistance to chemotherapeutic compounds such as doxorubicin [49, 50]. A broad spectrum of diverse electrophiles can undergo glutathione conjugation. Molecular moieties amenable for glutathione conjugation include electrophilic carbon atoms as well as electrophilic nitrogen, oxygen, and sulfur atoms [3]. Substrates of GSTs can be parent compound electrophilic, phase I metabolites, and even phase II conjugates. GST-mediated metabolism predominantly occurs in the cytosol, but some activity also resides in the endoplasmic reticulum [51]. Most GSTs catalyze the conjugation of glutathione (GSH) to a variety of electrophilic substrates [52]; however, some GSTs can function as GSH peroxidases and ligandins, making it difficult to assign specific roles for individual GST enzymes [53]. Unlike PAPS and UDPGA, which are high energy unstable cofactors, glutathione is a stable tripeptide that reacts with high energy unstable substrate [3]. Furthermore, intracellular levels of glutathione are extremely elevated ($\sim$10 mM) and difficult to deplete. Such elevated concentrations of glutathione are thought to provide the driving force for non-ATP-mediated transport processes, as well as to stimulate the function of ATP-driven transport processes [54].

The GSTs comprise a complex and widespread enzyme superfamily that has been subdivided further into an ever-increasing number of classes on the basis of a variety of criteria, including homology, as well as immunological, kinetic, and tertiary/quaternary structural properties [55]. GSTs are present in both membrane and soluble fractions, and cytosolic (or soluble) GSTs are responsible for conjugation of drugs. Approximately 20 cytosolic GSTs have been identified in vertebrates and

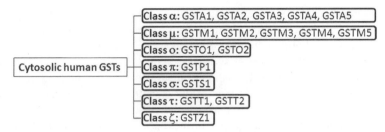

FIGURE 3.6   Human glutathione *S*-transferase enzymes. (Adapted from [58].)

categorized into seven distinct classes: α (GSTA), π (GSTP), μ (GSTM), θ (GSTT), ω (GSTO), ζ (GSTZ), and σ (GSTS) [56, 57] (Figure 3.6). Other classes of GSTs (β, δ, ε, etc.) have been identified in nonmammalian species [56]. In humans and rodents, cytosolic GST isoenzymes within a class typically share more than 40% identity, and those between classes share 25% identity [56]. Cytosolic GSTs exist as heterodimers (α and μ) and homodimers, and dimerization is necessary for GSTs to carry out their function in catalyzing glutathione conjugation [56]. The α- and μ-class GSTs are the major GST subunits expressed in the adult liver [50, 59]. The α-class GSTs exhibit selenium-independent glutathione peroxidase activity, which plays an important role in protecting cells against lipid and nucleotide hydroperoxides [50, 60]. π-class GSTs are known to be expressed in hepatocellular carcinomas and considered to be a tumor marker [50, 61]. The loss of GST protection can increase the susceptibility of preneoplastic hepatocytes to chemical-induced genotoxicity during chemical carcinogenesis. Therefore, GST induction is not only for cell detoxification and survival but also for cancer prevention [62].

An important feature of glutathione conjugates is that they typically are not deconjugated in the same manner as sulfate or glucuronide metabolites. The peptide bonds in the glutathione moiety of the molecule may be hydrolyzed sequentially to form a cysteine metabolite followed by N-acetylation to form a mercapturate metabolite. Thus, GSTs are considered as catalyzing the first of four steps required for the synthesis of mercapturic acid [56], and metabolites formed in the kidney are excreted in the urine [63]. Besides catalyzing conjugation reactions, cytosolic GSTs also bind, covalently and noncovalently, hydrophobic nonsubstrate ligands [56]. This type of activity contributes to intracellular transport, sequestration, and disposition of xenobiotics and hormones. Affinity labeling of rat class α GST revealed a high affinity nonsubstrate binding site within the cleft between the two subunits [64], indicating that there are two distinct binding sites in certain isoenzymes. The second nonsubstrate binding site form in heterodimers will be distinct from those in homodimers, and it may provide an evolutionary reason why it is beneficial for members within the α- and μ-classes to heterodimerize [56].

3.2.5   Acetylation, Methylation, and EPHXs

Arylamine NATs are a family of DMEs that catalyze the conjugation of an acetyl group from acetyl-coenzyme A to the terminal amino group of arylamines, arylhydrazides, and certain heterocyclic amines, as well as the *N*-hydroxyl group of aromatic and heterocyclic amines [65]. Three NAT reactions have been documented: *N*-acetyltransfer and *O*-acetyltransfer from an acetyl donor to an acceptor substrate and *N*-, *O*-acetyltransfer from an *N*-arylhydroxamic acid. In humans, acetylation is an important route in the biotransformation of many aromatic and heterocyclic amines drugs, and acetylation by NATs aids in the excretion of exogenous compounds, such as drugs and carcinogens [65]. The acetyltransferases relevant to human drug metabolism include two families of enzymes: *N*-acetyltransferase 1 (NAT1) and *N*-acetyltransferase 2 (NAT2) [65].

A common feature of acetylation and methylation is that these metabolic reactions decrease water solubility. A huge number of cytosolic methyltransferases are responsible for DNA, RNA, proteins, or lipids methylation [66]. The most characterized methyltransferase is the catechol-*O*-methyltransferase (COMT) enzyme, which catalyzes the biotransformation of estrogens and endogenous catecholamine neurotransmitters such as norepinephrine [66]. It also catalyzes the O-methylation of catechol drugs, including the anti-Parkinson's disease agent L-dopa and the antihypertensive methyldopa [66]. The thiopurine *S*-methyltransferase (TPMT) like COMT is an AdoMet-dependent methyltransferase, which catalyzes the AdoMet-dependent S-methylation of thiopurine drugs such as 6-mercaptopurine. Thiopurines are cytotoxic, immunosuppressant drugs that are used to treat childhood leukemia, inflammatory bowel disease, and organ transplant recipients [66]. Patients homozygous for common variant TPMT alleles that result in low levels of enzyme activity are at greatly increased risk for life-threatening thiopurine-induced toxicity [66].

EPHXs are important multifunctional enzymes from both the deactivation and activation of reactive species. Furthermore, they convert any potentially reactive epoxide formed by the P450s system into a diol metabolite, which is usually less reactive, more water soluble, and more easily cleared by GSTs. There are two major types of EPHX enzymes: the microsomal (mEPHX), which uses epoxides of polycyclic aromatics or drugs as substrates (type 1) and which controls hepatic uptake of bile acids (type 2) [67], and the soluble EPHX (sEPHX), which forms diols from many endogenous and exogenous epoxides, including fatty acids and leukotrienes [68].

3.3 THE XENOSENSORS CAR AND PXR: TWO MASTER REGULATORS OF PHASE II METABOLISM

3.3.1 General Considerations

The concept of "catatoxic steroids," such as pregnenolone-16-carbonitrile (PCN), which protect the liver against the effects of toxic substances by inducing detoxifying enzymes, has existed for long times [69]. PXR and CAR function as sensors for toxic compounds derived from endogenous metabolism or xenobiotics in order to enhance their elimination [70]. PXR and CAR regulate gene transcription through conventional mechanisms, involving xenobiotic binding, nuclear translocation, heterodimerization with their partner, the retinoid X receptor (RXR), and interactions with the

5' regulatory sequences of target genes [71]. Many xenobiotics interact with PXR and CAR either as agonists, activators, or inverse agonists [72, 73]. For example, phenobarbital (PB) activates both CAR and PXR, whereas clomitrazole and androstanol are activators of PXR but inverse agonists of CAR [74]. Similarly, bile acids, such as cholic acid and lithocholic acid (LCA), which are primary ligands of FXR, are activators of PXR and suppressors of CAR transcriptional activity [75]. 1,4-Bis[2-(3,5-dichloropyridyloxy)]benzene (TCPOBOP) is believed to be a potent ligand for mCAR but not for hCAR, and PB activates both receptors but only at high concentrations and possibly via an indirect mechanism. 6-(4-Chlorophenyl)imidazo[2,1-b][1,3]thiazole-5-carbaldehyde *O*-(3,4-dichlorobenzyl)oxime (CITCO) activates the human CAR, which is also subject to activation by pregnane, 17α-hydroxyprogesterone, pregnenolone, 17α-hydroxypregnenolone, 17ß-estradiol, RU486, and cortisol [76]. Human PXR is activated by xenobiotics such as rifampicin, clomitrazole, and hyperforin, one of the active constituents of the St. John's wort remedy [77–80]. Interestingly, ligand binding is not a prerequisite for activation of CAR [80]. This receptor seems to be constitutively active (i.e., even in the absence of ligand). This activity could be inhibited by androstanol or androstenol. It appeared that CAR adopts an active conformation in the absence of ligand and is shifted toward an inactive conformation by androstanol [81].

The tissue distribution analysis of CAR and PXR mRNA expression indicated that the major site of expression is the liver, with significant levels being detected in the intestine (small intestine-hPXR and hCAR and colon-hPXR), the kidney, stomach (hPXR), adrenal (hCAR), and testis (hCAR) [71, 78, 82]. Such tissue distribution of expression is consistent with the major role that these xenobiotic sensors exert in controlling detoxification. Overall, since CAR and PXR are activated by some of the same ligands and induce specific but overlapping sets of genes, they provide the cell with two overlapping and semiredundant mechanisms for recognizing and eliminating toxicants.

3.3.2   Phase II XMEs Identified as PXR and/or CAR Target Genes

As summarized in Tables 3.1 and 3.2, a number of genes encoding phase II DMEs have been identified as targets of the PXR and/or CAR receptors in humans and laboratory animals.

In human hepatoma HepG2 cells, UGT1A1 and 1A3 expression was markedly increased by the cotransfection of PXR and subsequent activation with rifampicin [83]. The expression of UGT1A6 and 1A4 was also increased, although to a much lesser extent compared to that of UGT1A1 and 1A3. The expression of UGT1A9 and all UGT2B members remained unchanged. In colon cancer Caco-2 cells, a similar response was also observed with induction of UGT1A1, 1A3, and 1A4 mRNA levels, whereas expression of other human UGTs remained unchanged (Table 3.1). However, in these cells, expression of UGT1A6 was not induced by PXR [83]. The lack of UGT2B response in both HepG2 and Caco-2 cells suggests that this UGT subfamily may not be transcriptional target of PXR. In human hepatocytes in primary culture, the human CAR activator, CITCO, induced P450s and a series of phase II encoding

TABLE 3.1   Human Phase II Enzymes as PXR and/or CAR Target Genes

PXR[a]	CAR
UGT1A1 ↑ (HepG2, Caco-2, L-I, Tg mice)	UGT1A1 ↑ (HH)
UGT1A3 ↑ (HepG2, Caco-2, L-I, Tg mice)	UGT1A6 ↑ (HH)
UGT1A4 ↑ (HepG2, Caco-2, L-I, Tg mice)	GSTA2 ↑ (HH)
UGT1A6 ↑ (HepG2, L-I, Tg mice)	SULT1A1 ↑ (HH)
UGT1A9 ↑ (L-I, Tg mice)	SULT2A1 ↑ (HepG2)
UGT1A5 ↑ (I, Tg mice)	SULT1A2 ↑ (HepG2)
UGT1A10 ↑ (I, Tg mice)	NAT1 ↑ (HepG2)
SULT2A1 ↑ (HepG2, HH)	

[a] ↑, induction of expression (cells or tissue); HH, human hepatocytes; L, liver; I, intestine; Tg, transgenic mice expressing the human enzymes.

genes, such as GSTA2, UGT1A1, and SULT1A1 [71]. In HepG2 cells, treatment with CITCO failed to modulate transcript levels of UGT1A6, SULT1A1, SULT1E1, GSTM1, and mEPHX1, but resulted in significant induction of UGT1A1, SULT2A1, SULT1A2, and NAT1 (Table 3.1) [84]. A similar pattern of regulation was observed with HepG2 cells treated with PB [84]. NAT1 was also significantly induced after exposure to PB and CITCO in rats and primary rat hepatocytes [85].

Biochemical induction studies in rats and primary rat hepatocytes demonstrated that digitoxigenin monodigitoxoside UGT activity could be induced by glucocorticoids, which is consistent with induction of UGT activity via PXR [71]. Similarly, hepatic UGT activity versus bilirubin, 1-naphthol, chloramphenicol, thyroxine, and trioodothyronine were induced, and mRNA levels of Ugt1a1 and 1a9 were more than 100% increased in mice following PCN treatment [86]. By contrast, the PXR

TABLE 3.2   Phase II Enzymes as PXR and/or CAR Target Genes in Rodents

PXR[a]	CAR
Ugt1a1 ↑ (L: WT, VP-hPXR, hPXR mice)	Ugt1a1 ↑ (WT, VP-hCAR mice)
Ugt1a9 ↑ (L: WT mice)	Sult1c2 ↑ (WT, ♀, L)
Ugt1a6 ↑ (L: VP-hPXR mice)	Sult1d1 ↑ (WT, ♀-♂, L)
Sult2a1 ↑ (♀, L: WT, VP-hPXR mice)	Sult1e1 ↑ (WT, ♀, L)
Sult1e1 ↑ (WT, ♀-♂, L)	Sult2a1/2 ↑ (WT, ♀, L)
PAPSs2 ↑ (WT, ♀-♂, L)	Sult3a1 ↑ (WT, ♀-♂, L)
GSTM ↑ (WT, VP-hPXR ♀-♂, L-I)	Sult4a1 ↑ (WT, ♀, L)
GSTA ↑ (WT, VP-hPXR ♀-♂, I)	PAPSs2 ↑ (WT, ♀, L)
GSTP ↑ (WT, VP-hPXR ♀, L-I)	
GSTP ↓ (WT, VP-hPXR ♂, L-I)	

[a] ↑, induction of expression; ↓, repression of expression (cells or tissue); L, liver; I, intestine; VP, transgenic mice expressing constitutively active human PXR or CAR; WT, wild-type mice; ♀, female; ♂, male.

null mice were resistant to the induction of all these UGT activities by PCN, while having slightly higher basal activities than the wild-type controls [86]. On the other hand, the expression of Ugt1a2, 1a6, and 2b5 was not affected by PCN regardless of PXR phenotype [86]. These results indicate that PCN induces glucuronidation in the mouse liver and that PXR regulates PCN-inducible expression of selected Ugt isoforms (Table 3.2). While unaffected in wild-type animals, the Ugt1a6 expression was also up-regulated in the transgenic animals expressing a constitutively activated hPXR (VP-hPXR) [87]. Similarly, Ugt1a1 mRNA and protein levels were also up-regulated in transgenic VP-hPXR mice and upon rifampicin-treated humanized hPXR transgenic mice [88, 89], as was the microsomal glucuronidation activity toward β-estradiol, thyroid hormones, corticosterone, and xenobiotics such as 4-nitrophenol and the carcinogen 2-amino-1-methyl-6-phenylimidazo[4,5-b]pyridine (4-OHPhIP) [90, 91]. In mice expressing the entire human UGT1A locus, treatment with PCN resulted in induced hepatic expression of human UGT1A1, 1A3, 1A4, 1A6, and 1A9 [92]. The same genes were also induced in the small intestine, where transcript levels of UGT1A5 and 1A10 were also induced [92]. Overall, these analyses indicate that all human UGT1A enzymes are susceptible to be up-regulated by PXR in a tissue-dependent manner (Figure 3.3).

The positive role that PXR exert on the activity of the Sult2a1 and expression of the PAPS synthesizing enzyme in mice was reported few years ago [93]. More recently, the control of the expression of 11 hepatic sulfotransferase enzymes and the two PAPS isozymes by PXR and CAR activators has been examined in mice [94]. CAR activators caused induction of Sult1c2, 1d1, 1e1, 2a1/2, 3a1, 4a1, and PAPSs2 expression in female animals, and of Sult1d1 and 3a1 in male mice (Table 3.2). PXR activators also affected Sults in a sex-dependent manner: in male, PXR ligands up-regulated Sult1e1 mRNA levels, while Sult1e1 and 2a1/2 expression was only activated in female animals and PAPSs2 mRNA levels were induced in both male and female mice [94].

Expression of some GSTs is also controlled by CAR and PXR in mice (Table 3.2) [3, 95–97]. As for Sult enzymes, PXR regulates GSTs in an isoform-, sex-, and tissue-dependent manner [98]. Thus, GSTM expression is up-regulated by PXR in both liver and intestine from male and female tissues, whereas GSTA is induced in the small intestine and not in the liver and GSTP is induced in females but repressed in males [98]. Interestingly, a similar pattern of GST regulation was seen in wild-type mice in response to pharmacological activation of PXR and the effect of PCN was abolished in PXR null animals.

3.3.3   Physiological and Pharmacological Interests of Phase II Enzymes Regulation by PXR and/or CAR

Considering the importance of UGTs, SULTs, and GSTs in the detoxification process, their PXR- and CAR-dependent regulation affects homeostasis of a number of endo- and xenobiotics. Some of these effects are of therapeutic potential, whereas others are thought to be deleterious.

The therapeutic interest of the PXR- and CAR-mediated induction of phase II enzymes is very well exemplified by the regulation of bilirubin glucuronidation. Bilirubin is an oxidative end product of heme catabolism. It is predominantly metabolized in the liver and then excreted into the bile. Accumulation of bilirubin in the blood is potentially hepato- and neurotoxic [91]. UGT1A1 is the primary enzyme responsible for bilirubin glucuronidation, a critical step for hepatic bilirubin clearance (Figure 3.3). In accordance with the positive effects that CAR exerts on the UGT1A1 gene expression, treatment with phenobarbital has been shown to dramatically alleviate the hyperbilirubinemia in Criggler–Najjar patients (who present reduced UGT1A1 activity). Accordingly, expression of the constitutively activated VP-CAR confers mice with a resistance to hyperbilirubinemia induced by bilirubin infusion [99]. In another independent report, activation of CAR also resulted in increased hepatic expression of genes known to be involved in bilirubin metabolism, including Ugt1a1, and this induction was absent in CAR-null mice [100]. However, the basal level of serum bilirubin was also not affected in mice lacking PXR alone or both PXR and CAR proteins (double knock-out). Upon bilirubin infusion, the CAR null and the double knock-out animals showed similar sensitivity as the wild type. By contrast, the PXR null mice exhibited a surprisingly complete resistance to hyperbilirubinemia [99]. These observations suggested that PXR negatively interferes with the bilirubin metabolic pathway, which was supported by induction of bilirubin detoxifying genes, such as Ugt1A1, OATP (organic anion transporting polypeptides)4/ SLC (solute carrier)21A6, GSTA2, and MRP2 in PXR null animals. But, expressing the VP-hPXR in mice also resulted in reducing plasma bilirubin after infusion [87], thus indicating that PXR plays both positive and negative roles in regulating bilirubin homeostasis. However, the exact mechanism for such differential roles remains to be clarified.

Phase II induction upon PXR and/or CAR activation can also have deleterious consequences for human health. Indeed, CAR is essential for thyroid tumorigenesis following exposure to PB and TCPOBOP [100]. Thyroid hormones (triiodothyronine, T3, and thyroxine, T4) are inactivated in the liver by UGT1As and SULTs to glucuronide and sulfate conjugates, which are eliminated via urine and bile [101]. The control of thyroid hormones metabolism by CAR is supported by the observations that patients treated with PB have reduced serum T4 levels, and that hypothyroid patients need thyroid replacement therapy after treatment with antiepileptic drugs such as PB [102, 103]. Furthermore, two recent studies [104, 105] used wild-type and CAR null mice to demonstrate that CAR mediates the induction of UGT and SULT involved in these metabolic reactions in wild type (Figure 3.7). Enzyme induction coincided with reduced levels of T4 and T3 and increased levels of the thyroid-stimulating hormone (TSH). TSH levels are thought to increase in order to compensate for the decrease in thyroid hormone levels and this leads to an increase in thyroid follicular cell proliferation and the development of thyroid tumors [105]. Thyroid follicular cell proliferation (believed to be a precursor of thyroid carcinogenesis in rodents) was observed in response to PB and TCPOBOP in wild-type mice but not CAR null mice [105]. These results were consistent with previous observations pointing to

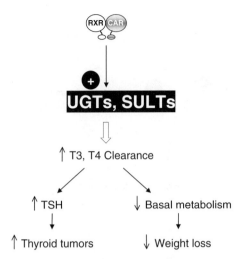

FIGURE 3.7   CAR controls thyroid hormone metabolism, an effect involved in thyroid tumorigenesis and control of obesity.

numerous thyroid hormone linked disorders during chronic treatments with drugs such as PB and phenytoin [102, 103]. These disorders are thought to be the consequence of xenobiotic-mediated enhancement of metabolism and excretion of thyroid hormones via induction of UGT and SULT enzymes through CAR activation [75, 80].

From another point of view, the CAR-mediated induction of thyroid hormones metabolism can also be considered as having elevated therapeutic potential for obesity treatment [75]. Thyroid hormones regulate basal metabolism, and T3/T4 plasma levels correlate with energy expenditure and caloric loss (Figure 3.7). Actually, T3 has been shown to be reduced by 33% in subjects following weight loss program. The concomitant decrease in basal metabolism is believed to represent a mechanism to resist weight loss. Maglich et al. [104] have reported that Cyp2b10, Sult1a1, and Ugt1a1 gene expression is significantly induced during fasting in wild type but not CAR null mice. In parallel, the same authors observed that T3 and T4 levels were greater in CAR null animals than in wild-type littermates during fasting. When mice were placed on a 40% caloric restriction diet for 12 weeks, CAR null mice lost over twice as much weight as their littermates, a phenotype attributed to the sustained thyroid hormone levels in the caloric-restricted CAR null mice [104]. These data suggest that the CAR-dependent modulation of thyroid hormones glucuronidation and sulfonation plays an important role in metabolic rate from food intake and energy homeostasis and thus has implications in obesity and is associated with metabolic disorders. In conclusion, depending on the physiopathological situation, a given regulatory process involving NRs and phase II DMEs can translate into both beneficial and deleterious effects.

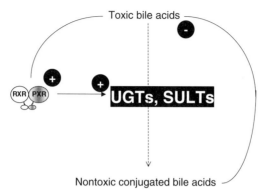

FIGURE 3.8   PXR activation is an essential wheel in the autoregulatory mechanism by which bile acid stimulate their own sulfonation and glucuronidation.

Beyond the metabolism of bilirubin and thyroid hormones, PXR plays also a major role in protection from bile acid toxicity [70]. The basis for PXR being involved in bile acid homeostasis comes from *in vivo* observations showing that PXR activation is protective against hepatotoxic bile acids [79, 106] and that accumulation of bile acid precursors leads to PXR activation [107]. Considering that PXR is a receptor for the cholestatic LCA, it is considered as an essential wheel in an autoregulatory mechanism, by which toxic bile acids stimulate their own inactivation. Among the subset of PXR target genes that encode bile acid detoxification proteins, the phase II enzymes UGT1A3 and SULT2A1 are important for cholestatic bile acid elimination (Figure 3.8) [108]. The up-regulation of bile acid glucuronidation through PXR activation is supported by the observation that urinary levels of bile acid glucuronides in healthy volunteers treated with rifampicin are increased twofold [109]. These observations are of clinical interest since rifampicin is considered as the safest treatment for cholestasis-induced pruritus [110]. In addition to bile acid, PXR- and CAR-mediated regulation of phase II conjugating enzymes also affects other endobiotics, such as the glucocorticoid hormones (i.e., corticosterone and cortisol), the glucuronidation of which is increased in constitutive hPXR transgenic mice [87]. Finally, the regulation of UGT and SULT by PXR and CAR has led to the idea that PXR- and CAR-mediated gene regulation may play a role in estrogen homeostasis. Indeed, metabolic pathways to deactivate estrogens include glucuronidation and sulfonation [111]. Estrogen is one of the preferred substrates for UGT1A1 and estrogen glucuronidation activity is increased in the VP-hPXR mice [87]. The PXR target gene SULT1E1 mediates estrogen inactivation, since the sulfonated estrogens cannot bind to and activate estrogen receptor (ER) [112]. Estrogens promote mammary epithelial proliferation and are required for breast cancer formation. These observations suggest that PXR and CAR activators may have potential therapeutic interest for clinical treatment of breast cancer [91].

CAR and PXR activators, such as PB and rifampicin, are widely used in clinic, and multiple drugs therapy is often used with a single patient [113]. For example,

in individuals infected with HIV and tuberculosis, both antituberculosis and antiretroviral agents have to be used [114]. Rifampicin is an effective antibiotic against
gram-positive bacteria including mycobacteria, being frequently used currently in
the chemotherapy of tuberculosis along with isoniazid, pyrazinamide, and ethamb-
utol/streptomycin [115]. While a number of drug–drug interactions involving the
PXR-mediated regulation of P450s by rifampicin have been reported, only few
interactions with phase II DMEs are currently listed in the literature [113]. However,
Gallicano and colleagues [116] reported that a significant pharmacokinetic interaction
between rifampicin and the antiretroviral agent zidovudin was mediated by induction
of both glucuronidation and amination pathways. Similarly, rifampicin increased the
urinary excretion of lamotrigine-glucuronide in a group of healthy volunteers [117].
By contrast, a randomized, double-blinded, placebo-controlled study has assessed
the effect of steady-state dosing of rifampicin on the pharmacokinetics of morphine.
The oral clearance of morphine was increased significantly, but the renal clearance
of morphine, morphine-3, and 6-glucuronide was not altered. Following rifampicin
ingestion, no analgesic effect of morphine was observed, as assessed by the cold
pressor test. These results suggested that a clinically significant interaction occurred
between the two drugs, but the mechanism could not be attributed to induction of
UGT enzymes [118]. Overall, when occurring, the magnitude of UGT-based drug
interactions with the PXR activators is generally smaller compared to P450-mediated
drug interactions [19, 119].

3.3.4 Functional Mechanisms of Phase II Genes Regulation by PXR and CAR: Role of the Glucocorticoid Receptor

Accordingly, with the regulation of their expression upon receptor activation, various
studies have reported the presence of functional PXR and/or CAR response elements
within phase II XMEs genes. Thus, a PXR- and CAR-inducible composite element,
containing both inverted repeat (IR2) and direct repeat (DR4) of the hexanucleotide
AGGTCA structure, has been characterized within the human SULT2A1 promoter
[120]. This composite element confers the xenobiotic response of the promoter, and
is bound by PXR and CAR in the presence of their respective ligands [120]. An
IR0 motif found just under 200 base pairs upstream from the transcriptional start
site of the mouse and rat Sult2a1 genes can bind several NRs, including the vitamin
D receptor (VDR), FXR, and PXR, but not CAR [121, 122]. Similarly, the human
UGT1A1 gene possesses a critical region between nucleotides −3483 and −3194,
which contains three CAR-responsive motifs, NR3 (a DR-3 element), NR4, and
gtNR1 (both DR-4 elements). This region, which was first identified as responding
to PB, is called phenobarbital-responsive enhancer module (PBREM) (Figure 3.3).
The gtNR1 motif was found to be the site of CAR/RXR binding, but both NR3
and NR4 were also required for full enhancer activity, both in HepG2 cells and in
primary mouse hepatocytes [87]. The importance of this multiple NRs responsive
region is such that a genetic variation (T-3279G) in the PBREM is associated with
hyperbilirubinemia [123] and drug metabolism and toxicity alterations [124].

The appropriate transcriptional activity of PXR and CAR also requires the formation of regulatory complexes that comprise a series of coactivators and corepressors, such as PPARγ coactivator (PGC)-1, steroid receptor coactivator (SRC)-1, transcriptional intermediary factor (TIF)-2, GR-interacting protein (GRIP)-1, silencing mediator of retinoic acid and thyroid hormone receptor (SMRT), nuclear receptor corepressor (NCoR), and so on. It is therefore likely that these cofactors also play important roles in the control of phase II genes transcription upon PXR and CAR activation. However, in addition to these classical NR coregulators, another ligand-activated transcription factor, the glucocorticoid receptor (GR), also plays a significant role for the integrity of the PXR- and CAR-mediated regulation of phase II enzymes expression. In human hepatocytes, hPXR mRNA levels are induced in response to the GR activators dexamethasone (DEX), prednisolone, and hydrocortisone [125], and there are a lot of evidences that this occurs as a result of direct regulation of PXR expression by the GR at the level of expression [126]. Similarly, up-regulation of hCAR mRNA also occurs in response to these compounds in human hepatocytes [127]. These results further support the concept that the GR may act by regulating indirectly PXR and CAR target genes, including those encoding phase II DMEs. In addition to regulating PXR and CAR expression, GR also influences CAR- and PXR-mediated regulation of UGT1A1, as demonstrated in HepG2 transfected with the 290-bp PBREM reporter construct and expression plasmids for CAR and PXR in the presence or absence of exogenously expressed GR. In the presence of GR, DEX enhances more prominently PXR- and CAR-dependent induction of the promoter [128]. Furthermore, a GR response element was identified in this promoter region (Figure 3.3) [128]. Such GR-dependent enhancement was further reinforced in the presence of the GRIP-1 [129].

3.4   AhR AND Nrf2, TWO IMPORTANT REGULATORS OF PHASE II ENZYMES

In addition to PXR and CAR, another xenosensor is well characterized for its role in regulating phase I and II metabolizing enzyme: the AhR. More recent observations illustrated that expression of most of the AhR target genes is also transcriptionally controlled by the Nrf2, an important controller of the cellular oxidative status.

3.4.1   AhR and Phase II XMEs Regulation

AhR was initially characterized for its implication in the vertebrate response to many planar aromatic hydrocarbons (PAHs). Indeed, the AhR binds exogenous ligands and transcriptionally activates a series of target genes (termed the "AhR battery"), including numerous phase II XME encoding genes (Table 3.3) [130, 131]. This AhR-mediated pathway is commonly viewed as an "adaptive" response toward xenobiotic agents. In addition to adaptive metabolism, the AhR also mediates a spectrum of toxic endpoints in response to high affinity agonists [132]. AhR is a member of

TABLE 3.3   AhR-Regulated Phase II Enzymes

AhR[a]

Human
    UGT1A1 ↑ (HepG2, L; Tg mice)
    UGT1A6 ↑ (HepG2, L; Tg mice)
    UGT1A3 ↑ (L-I; Tg mice)
    UGT1A4 ↑ (L-I; Tg mice)
    UGT1A5 ↑ (I; Tg mice)
    UGT1A7 ↑ (I; Tg mice)
    UGT1A8 ↑ (I; Tg mice)
    UGT1A9 ↑ (L-I; Tg mice)
    UGT1A10 ↑ (I; Tg mice)
    SULT1A1 ↑ (HepG2)
    SULT1A2 ↑ (HepG2)
    SULT2A1 ↑ (HepG2)
    GSTM1 ↑ (HepG2)
    NAT1 ↑ (HepG2)
    mEPHX ↑ (HepG2)
    COMT ↑ (mammary gland cells)
    GSTA1 ↑ (HH)
    GSTA2 ↑ (HH)

Rodents
    Ugt1a1 ↑ (L; WT mice)
    Ugt1a6 ↑ (L; WT mice)
    Ugt1a7 ↑ (L; WT mice, rats)
    Sult1a1 ↓ (L; WT mice)
    Sult1b1 ↓ (L; WT mice)
    Sult1c1 ↓ (L; WT mice)
    Sult1c2 ↓ (L; WT mice)
    Sult1d1 ↓ (L; WT mice)
    Sult1e1 ↓ (L; WT mice)
    Sult3a1 ↓ (L; WT mice)
    PAPSs2 ↓ (L; WT mice)
    Sult5a1 ↑ (L; WT mice)

[a] ↑, induction of expression; ↓, repression of expression; (cells or tissue); L, liver; I, intestine; WT, wild-type mice; Tg, transgenic mice expressing the human genes.

the basic helix–loop–helix Per–Arnt–Sim transcription factor family [91]. Without ligand stimulation, the AhR is present in the cytosol in an inactive complex [133], while ligand binding increases nuclear shuttling of the receptor and, once in the nucleus, AhR heterodimerizes with a constitutively active nuclear protein known as Arnt (aryl hydrocarbon receptor nuclear translocator) [134]. Heterodimers then binds to xenobiotic response elements (XREs, also termed dioxin response elements,

DREs), identified in the 5′ upstream region of AhR target genes [134–136]. In addition to PHAs, many structurally diverse compounds including halogenated aromatic hydrocarbons (HAHs) and flavonoids activate AhR [137]. The most potent known ligand for the AhR, 2,3,7,8-tetrachlorodibenzo-*p*-dioxin (TCDD), is highly resistant to metabolic degradation and elicits numerous AhR-dependent toxic events (thymic atrophy, chloracne, tumor promotion, hepatomegaly, cachexia, and death) [138]. Kaempferol is one of the most abundant flavonoids acting as a dietary ligand for AhR [139], and identified as a phase II inducer present in a large amount in cakes [140, 141]. The modulation of drug-metabolizing enzymes by flavonoids is important in terms of human health since these enzymes can inactivate carcinogens, which contributes to the cancer prevention properties of these compounds [142]. The regulation of UGT and GST genes is also critical for the detoxification of xenobiotics from diet. Indeed, cooking, pan frying, and grilling of meat cause the formation of mutagens or carcinogens such as PAHs and heterocyclic amines [143]. Data from both *in vitro* and *in vivo* studies suggest that exposure to polycyclic aromatic hydrocarbons, such as benzo(*a*)pyrene, may be important determinant of colorectal cancer risk [144, 145]. Interestingly, the flavonoid-enriched green tea has many anticarcinogenic effects, and these effects are partly caused by activating the AhR-UGT/GST signaling pathway (Figure 3.9).

As listed in Table 3.3, AhR activation results in the induction of a large number of phase II enzymes expression, including numerous GSTs and UGTs. In human hepatoma HepG2 cells, various AhR agonists exert differential inducing effects on the expression of UGT1A1, 1A6, SULT1A1, 1A2, 2A1, GSTM1, NAT1, and mEPHX [84]. Studies with transgenic mice expressing the human UGT1 locus have revealed that all nine expressed hepatic and intestinal isoforms are induced by TCDD [92]. With human UGT1A1, one XRE has been characterized at −3.3 kb [146], within

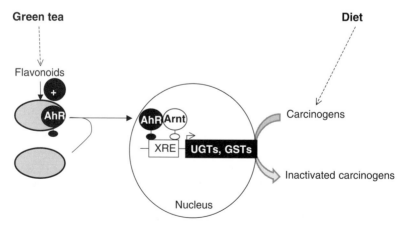

FIGURE 3.9 The anti-carcinogenic effects of green tea are partly caused by the flavonoids-dependent activation of the AhR-GST/UGT pathway.

the PBREM, which also contains binding motifs for CAR, PXR, and GR (Figure 3.3) [147]. This XRE is activated in transgenic mice expressing the human UGT1A1 enzyme [148]. A functional XRE has also been identified in the human UGT1A6 gene and AhR-deficient mice exhibit reduced basal and TCDD-inducible Ugt1a6 expression [131]. In fact, rat and murine Ugt1a6 and 1a7 appear to be coregulated by AhR [149]. When determined using 2-aminophenol as the substrate, long-term ingestion of green tea extracts increases UGT activity in rats [150, 151], and this induction is considered to contribute to the anticarcinogenic effect of green tea (Figure 3.9). On the other hand, rat Ugt1a7 was identified as a major (T4) thyroxine hormone conjugating enzyme in this species [152], and administration of TCDD increases the biliary clearance of T4 indicating that the TCDD-dependent activation of AhR affects thyroid hormones homeostasis. Thus, the elevated T4 glucuronide clearance results in a feedback control of TSH release, which correlates with growth of the thyroid gland [153, 154]. Such observations are relevant to human health, since populations exposed to TCDD is known to have increased risk for thyroid gland cancer (such as people involved in the Seveso incident in 1976, Italy) [155]. Accordingly, workers exposed to TCDD present inversely correlated plasma levels of TCDD and thyroid hormones [156]. At the opposite, AhR activators act as strong inhibitors of estrogen-dependent mammary and uterine tumor formation and growth [157], and epidemiological studies have evidenced that women exposed to TCDD in Seveso have a decreased incidence of both breast and endometrial cancer [158]. Interestingly, TCDD treatment leads to an increased COMT activity in mammary gland cancer cells [159]. COMT catalyzes the conversion of estrogens into 4-hydroxy catechol metabolites [111], which are inactive metabolites to the carcinogenic hormones. Thus, these observations suggest that induction of COMT in the presence of AhR ligands may play a role in the anticarcinogenic effects of these agonists versus estrogen-dependent cancers. However, AhR and ER regulatory pathways also share numerous cross talks and the contribution of the COMT induction within the anticancer effects of AhR activators is unclear, for instance [160].

In addition to UGTs, AhR is also an important regulator for SULT and GST enzymes. Thus, long-term ingestion of green tea extracts increases cytosolic GST activity in female rats [161]. Human GSTA1 and A2 are induced by AhR agonists in human hepatocytes [162, 163]. By contrast, AhR activators, such as TCDD and β-naphtoflavone, down-regulate Sult1a1, 1b1, 1c1, 1c2, 1d1, 1e1, 3a1, and PAPSS2 expression in mice [94]. Interestingly, in the same study, Sult5a1 mRNA expression was induced upon AhR activation [94].

3.4.2 Nrf2 and Phase II XMEs Regulation

It has recently been recognized that phase II genes of the AhR battery are linked to a second gene battery, termed "Nrf2 battery," which is involved in protection against oxidative stress [164–169]. Linkage between AhR and Nrf2 batteries is probably achieved by multiple mechanisms in a species and tissue-specific manner: Nrf2 is a target gene for AhR [170]; Nrf2 can be activated by ROS generated by the AhR target CYP1A1 gene [171, 172]; and finally a direct cross-interaction between AhR and Nrf2 has been proposed in the regulation of some target genes such as the NAD(P)H:quinone oxidoreductase 1 (NQO1) [168]. Nrf2 belongs to the basic region-leucine zipper family and is activated in response to electrophiles and reactive oxygen species [4]. Nrf2 binds to the promoters of target genes through specific response elements regulating both their constitutive and inducible expression [173–175]. Under basal conditions, Nrf2 is largely bound to the cytoskeletal anchoring protein Kelch like ECH-associated protein 1 (Keap1) in the cytoplasm [176, 177]. During period of oxidative stress or following exposure to electrophiles, Keap1 releases Nrf2 from sequestration and Nrf2 translocates to nucleus enabling gene transcription [176]. Once in the nucleus Nrf2 can heterodimerize with a variety of transcriptional regulatory proteins, such as members of the activator protein (AP)-1 family (Fos or the small Maf family of transcription factors), the nuclear factor-kappaB (NF-κB) and glucocorticoid receptor [178–181]. These protein complexes then bind to motifs known as antioxidant or electrophilic response elements (ARE/EpRE) located in the promoter regions of detoxification genes [173, 174]. ARE-containing genes identified as downstream targets of Nrf2 are involved in a variety of cellular functions including drug metabolism, ROS scavenging, GSH homeostasis, stress proteins, and efflux transport pathways [167]. Nrf2-activating chemicals that induce ARE genes have been categorized as cytoprotective agents. These include phenolic antioxidants (butylated hydroxyanisole, BHA, and tert-butyl-hydroquinone, tBHQ), synthetic antioxidant (oltipraz, 3H-1,2-dithiol-3-thione, D3T), phorbol esters (phorbol 12-myristate, PMA), triterpenoid analogues (oleanolic acid derivatives), and isothiocyanates (sulforaphane, phenethyl isothiocyanate) [167].

The nonprotein thiol (GSH) maintains intracellular redox balance and protects against oxidative insult. Additionally, GSH can detoxify chemicals through direct binding or enzymatic conjugation by GST. Decreased levels of GSH are observed in hepatocytes and fibroblasts from Nrf2 knock-out mice [182]. ARE were identified in the promoter of genes encoding glutathione synthesizing enzymes, such as the glutathione synthetase and glutamate cysteine ligase. Furthermore, livers from Nrf2 null animals exhibit lower expression of these genes corresponding with diminished GSH stores [182, 183]. Numerous GST isoforms are regulated at least in part by Nrf2 [179] (Table 3.4). For example, constitutive gene expression of GST isoforms (A1, A2, A3, A4, M1, M3, and M4) is markedly less in livers from Nrf2 null mice [183]. Basal levels of GSTA1/2, A3, and M1 are similarly decreased in the small intestine of these mutant animals [184]. Oltipraz induces GSTA2 mRNA and protein expression in a rat hepatocyte-derived cell line (H4IIE), via PI3-kinase-mediated nuclear translocation and binding to the CCAAT enhancer binding protein (C/EBP) response element in the GSTA2 gene promoter [185]. Accordingly, the Nrf2-dependent control of other GST expression involves various other transcription factor binding sites (NF-κB, AP-1, GRE, etc.); however, the understanding of GST genes induction by Nrf2 activators is still incomplete. Nevertheless, Keap1 null animals demonstrated pronounced nuclear accumulation of Nrf2 and overexpression

TABLE 3.4 Nrf2-Regulated Phase II
Enzymes

Nrf2[a]

Nrf2 null mice
  GSTA1 ↑ (liver, intestine)
  GSTA2 ↑ (liver, intestine)
  GSTA3 ↑ (liver, intestine)
  GSTA4 ↑ (liver)
  GSTM1 ↑ (liver, intestine)
  GSTM3 ↑ (liver)
  GSTM4 ↑ (liver)

Wild-type rats
  Ugt1a3 ↑ (liver, intestine)
  Ugt1a6 ↑ (liver, lung)
  Ugt1a7 ↑ (liver)
  Ugt1a2 ↑ (intestine)
  Ugt2b1 ↑ (intestine)
  Ugt2b3 ↑ (intestine)
  Ugt2b8 ↑ (intestine)
  Ugt2b12 ↑ (intestine)

Wild-type mice
  Sult1a1 ↑ (♂, liver)
  Sult1b1 ↑ (♂, liver)
  Sult1c1 ↑ (♂, liver), ↓ (♀, liver)
  Sult1c2 ↑ (♂, liver), ↓ (♀, liver)
  Sult1d1 ↑ (♂, liver)
  Sult1e1 ↑ (♂, liver)
  Sult3a1 ↑ (♂, liver)
  Sult5a1 ↑ (♂, liver), ↓ (♀, liver)
  PAPSs2 ↑ (♂, liver)

UGT transgenic mice
  UGT1A1↑ (liver, intestine)

[a] ↑, induction of expression; ↓, repression
of expression; ♀, female; ♂, male.

of Nrf2 target genes, such as GSTs [186]. Interestingly, simultaneous deletion of Nrf2 and Keap1 reversed this phenotype. Similar observations were also done with hepatocytes specific conditional Keap1 deleted animals [187]. The Nrf2-dependent regulation of GSTA4 expression is thought to be part of an autoregulatory homeostatic defense mechanism against lipid peroxidation products. Indeed, GSTA4 inactivates 4-hydroxynonenal (HNE), an endogenous lipid peroxidation product formed through membrane damage by reactive oxygen species [188]. 4-HNE favors desequestration

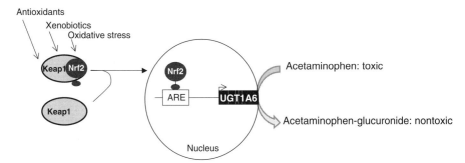

FIGURE 3.10 Nrf2-dependent activation of the UGT1A6 mediates acetaminophen detoxification.

of Nrf2, from the cytoplasmic Nrf2–Keap1 complex, thus leading to its nuclear accumulation and increasing GSTA4 expression [56].

In Nrf2-deficient mice, the basal expression of Ugt1a6 in liver and lung is depressed about 56% of wild type [189, 190]. Similarly, induction of hepatic Ugt1a6 by oltipraz is abolished in Nrf2 null animals [165]. More recently, ARE-like sequences were identified in the human UGT1A6 promoter, suggesting that both the human and mouse genes expression is dependent upon Nrf2 [191]. UGT1A6 is an important detoxification enzyme for acetaminophen, and treatment of Nrf2 null animals with APAP results in enhanced liver injury and mortality compared to wild-type and heterozygote counterparts [189, 192]. These observations correlate with the reduced expression of Ugt1a6 in Nrf2 null animals [189, 190] and point out a strong role for Nrf2 in attenuating drug-induced toxicity (Figure 3.10). Interestingly, APAP toxicity involves generation of a highly reactive eletrophile, depletion of GSH stores, protein covalent adduct formation, and oxidative stress, and reduced APAP-glucuronide formation was hypothesized to increase availability of APAP for bioactivation to these reactive metabolites.

A recent study demonstrates induction of multiple Ugt isoforms in the liver and intestinal tract in male rats given oltipraz [193]. Higher levels of hepatic Ugt1a3, 1a6, and 1a7 mRNA and intestinal Ugt1a2, 1a3, 2b1, 2b3, 2b8, and 2b12 mRNA were seen after oltipraz treatment. Up-regulation of human UGT1A1 expression by the Nrf2 pathway was recently confirmed in HepG2 cells and in transgenic mice expressing the human UGT1A locus [194]. In addition, a functional ARE was located in the PBREM region of the human UGT1A1 promoter, close to the AhR-responsive element (Figure 3.3) [194]. Microsomal epoxide hydrolase (mEH) inactivates epoxides within the cell. Basal mRNA levels of mEH are reduced in multiple tissues of Nrf2 null animals [165, 195]. Similarly, treatment of these mutant animals with prototypal Nrf2 activators fails to induce mEH. In mice, Nrf2 activation modulates sulfotransferase enzyme expression in a sex-dependent manner [94]. Indeed, activators, such as oltipraz or BHA, up-regulate Sult1a1, 1b1, 1c1, 1c2, 1d1, 1e1, 3a1, 5a1, and PAPSS2 in male animals, whereas Sult1c1, 1c2, and 5a1 are down-regulated in females [94]. However, Sult1d1 expression is induced in both sexes.

3.5 PPARs AND PHASE II XMEs REGULATION

Peroxisome proliferators (PP) are a large group of structurally diverse chemicals that include plasticizers, hypolipidemic drugs, and industrial solvents. These chemicals induce a variety of pathologies in rodent liver including hepatomegaly, hepatocyte proliferation, hepatocyte peroxisome proliferation, and alterations in the levels of lipid-metabolizing genes [196]. Responses to PP exposure in the liver are mainly controlled by a member of the NR superfamily called peroxisome proliferator-activated receptor α (PPARα). The PPAR subfamily consists of three distinct subtypes, PPARα, PPARβ/δ, and PPARγ that show high homology, common structural and functional properties but distinct expression patterns [197, 198]. Whereas PPARα and PPARβ/δ are expressed preferentially in tissues where fatty acids are catabolized namely liver, kidney, heart, skeletal muscle, and vascular endothelial cells, PPARγ is highly expressed in adipose tissue [199]. PPARs heterodimerize with RXR and then bind to PPAR response elements (PPRE) corresponding to DR-1 and 2 sequences [200]. PPARα is activated by a large number of endogenous and exogenous compounds including hydroxyeicosatetraenoic acid (HETE), prostaglandins, and polyunsaturated fatty acids, as well as structurally heterogeneous PP and the cholesterol-lowering fibrates (i.e., gemfibrozil, clofibrate, and fenofibrate) [196]. In addition to PXR, CAR, AhR, and Nrf2, PPARα has been identified as another NR that regulates various phase II enzymes [201]. Human UGT1A1, 1A3, 1A4, 1A6, and 1A9 were identified as positively regulated target genes of the ligand-activated PPARα in human hepatocytes and transgenic mice (Table 3.5) [202–204]. Functional PPREs have been identified in the promoters of the human UGT1A1, 1A3, and 1A9 genes [202, 204]. Glucuronidation constitutes the main metabolic route for fibrates and the PPARα-regulated UGTs are involved in their conjugation [201, 205], which indicates that pharmacological activators of this receptor stimulate their own glucuronidation. Similarly, UGTs also conjugate the endogenous PPARα activators HETE metabolites [201, 205, 206], which indicates that PPARα can autoregulate its own transactivational function by reducing intracellular levels of unconjugated natural and xenobiotic activators. PPARα also activates the human UGT2B4 gene transcription through direct binding to its promoter [207]. UGT2B4 exerts an important role in the detoxification of cholestatic bile acid, such as LCA [108, 201]. Since UGT1A3 also participates in bile acid glucuronidation, these observations support a potential therapeutic role for fibrates in hepatic detoxification under cholestatic conditions [108]. On the basis of the elevated number of drugs that are glucuronidated, it could have been anticipated that the fibrate–PPARα–UGT pathway may result in numerous drug–drug interactions. However, the clinical use of fibrates appears relatively safe, and very few interactions involving fibrate-dependent induction of UGT genes have been reported. Nevertheless, glucuronide derivatives of fibrates, such as gemfibrozil-glucuronide,

TABLE 3.5 XMEs as Target Genes for PPARs, LXR, FXR, and HNFs

PPARα[a]	PPARγ	LXR	FXR	HNF4α	HNF1α
UGT1A1 ↑ (HH, L: Tg mice)	SULT2B1 ↑ (HK)	SULT2B1 ↑ (HK)	UGT2B4 ↑ (HH, HepG2)	UGT1A1 ↑ (HH)	UGT1A1 ↑ (HH)
UGT1A3 ↑ (HH, L: Tg mice)	UGT1A9 ↑ (HM)[a]	UGT1A3 ↑ (HH, L: Tg mice)	UGT2B7 ↓ (Caco-2)	UGT1A9 ↑ (HH)	UGT1A6 ↑ (HH)
UGT1A4 ↑ (HH, L: Tg mice)	GSTA2 ↑ (L: rat)	Sult1e1 ↑ (L: WT mice)	SULT2A1 ↓ (HepG2)	SULT2A1 ↑ (HH)	UGT2B7 ↑ (Caco-2)
UGT1A6 ↑ (HH, L: Tg mice)		Sult2a1 ↑ (L: WT mice)	UGT2B15 ↓ (LNCaP)		UGT2B17 ↑ (Caco-2)
UGT1A9 ↑ (HH, L: Tg mice)			UGT2B17 ↓ (LNCaP)		Ugt1a7 ↑ (L: rat)
UGT2B4 ↑ (HH, HepG2)					Ugt1a6 ↑ (L: rat)
GSTP ↓ (L: WT rats)					
Sult1c1 ↓ (L: WT mice)					
Sult1c2 ↓ (L: WT mice)					
Sult3a1 ↓ (L: WT mice)					
Sult5a1 ↓ (L: WT mice)					
Sult1e1 ↓ (L: WT mice)					
SULT2B1 ↑ (HK)					

[a] ↑, induction of expression; ↓, repression of expression; L, liver; HH, human hepatocytes; HK, human keratinocytes; HM, human macrophages; WT, wild type.

have been shown to serve as inhibitor for statins oxidation and transport, thus increasing the myotoxicity risk associated with the use of such cholesterol-lowering drugs [208, 209].

Beyond UGTs, PPARα activation also modulates other phase II XMEs, such as GSTP, the expression of which is repressed [210]. This repression functions via the interactions of ligand-activated PPARα with Jun, thus inactivating the Jun-dependent induction of the GSTP gene promoter. In preneoplastic rat liver lesions, PPARα expression is decreased and this reduction correlates the increased expression of GSTP mRNA. These observations indicate that PPARα regulates, at least in part, the derepression of the GSTP gene that occurs in the early stages of hepatocarcinogenesis. PPARα activation also negatively regulates Sult1c1, 1c2, 3a1, 5a1, and 1e1 in mice liver [94]. Sult1e1 is also dramatically down-regulated in rats [211], while the SULT2B1b gene expression and enzyme activity are increased in the presence of PPARα, PPARβ/δ, and PPARγ activators in cultured human keratinocytes (Table 3.5) [212]. Interestingly, the most dramatic effect occurred with the PPARγ activator ciglitazone, and the up-regulation of SULT2B1b mRNA by ciglitazone appears to occur at a transcriptional level. The SULT2B1b gene encodes the major cholesterol sulfonating enzyme and cholesterol plays an important role in epidermal differentiation and corneocyte desquamation; thus, these effects identify PPARγ as an important regulator of skin physiology. On the other hand, PPARγ also plays a central role in lipid and glucose homeostasis and has been implicated in the pathogenesis of chronic inflammatory disorders such as arthritis, atherosclerosis, and inflammatory bowel disease [213]. PPARγ, which is activated by the insulin-sensitizers glitazones, is also an important target for the development of new drugs aimed at preventing and treating cancer [214]. As PPARα, PPARγ is also a positive regulator of human UGT1A9 expression [215]; however, in contrast to fibrates, the glitazones activators of the gamma isoform are only feebly glucuronidated [215]. Nevertheless, the co-occurrence of hypertriglyceridemia (i.e., target for PPARα activators) and type 2 diabetes (target for PPARγ agonists) in metabolic syndrome patients has led the pharmaceutical industry to generate novel molecules capable of activating both PPARα and PPARγ receptors, i.e., the glitazar series [201]. Some of these coagonists, namely muraglitazar and tesaglitazar, are excellent substrates for glucuronidation, and it is expected that these molecules will stimulate their own inactivation under the form of glucuronide conjugates by stimulating both PPARα-UGT and PPARγ-UGT pathways [216].

The PPARγ/RXR heterodimer promotes rat GSTA2 expression by inducing Nrf2 and C/EBP nuclear levels and activation [62], and evidences demonstrate that ARE and CEBP response elements are both necessary for GSTA2 induction by PPARγ and RXR ligands. However, three functional PPREs were identified to be essential for full ligand responsiveness [62]. These functional PPREs form a PPAR response module, and it is likely that this module is essential for the formation of transactivation complexes comprising Nrf2 and C/EBP [62].

Prostaglandins, such as the prostaglanding J2 (PGJ2), are endogenous activators of PPARγ [205]. Overexpression of GSTM1, A1, and P1 in breast cancer MCF7 cells significantly inhibited the PGJ2-dependent transactivation of a PPARγ responsive reporter gene [217]. The degree of inhibition correlated with the level of GST expressed.

However, only few glutathione conjugates of PGJ2 were formed in the presence of the GSTs, and mechanism by which GSTs bind to and sequester PGJ2 in the cytosol away from their nuclear target PPARγ was suggested [217]. Taken together, these observations suggest that GSTs inhibit their PPARγ-dependent up-regulation.

3.6   FXR/LXRα AND PHASE II XMEs REGULATION

The LXRs and FXR are intracellular sensors for sterols and bile acids, respectively. LXRs are activated by physiological levels of oxysterols, such as 24S-hydroxycholesterol and 24-25-epoxycholesterol, whereas FXR ligands are the primary bile acids chenodeoxycholic and cholic acids, and their tauroconjugates [218]. Consistent with their role as lipid sensors, FXR and LXRs are highly expressed in enterohepatic tissues, where they exert opposite actions on the same gene battery [218]. By functioning in a coordinate manner, these two NRs allow the maintenance of cholesterol, carbohydrate, and bile acid homeostasis [108].

Numerous studies established the role of FXR as a bile acid sensor that regulates a network of genes encoding enzymes involved in the synthesis, metabolism, or transport of bile acids [218]. Indeed, FXR regulates bile acid conjugating UGT enzymes, in a tissue- and isoform-specific manner. FXR activation in human hepatic cells leads to increased expression and activity of the 6α-hydroxylated bile acid conjugating UGT2B4 enzyme [219], without affecting the expression of UGT1A3, an enzyme that glucuronidates LCA and CDCA at their 24 carboxyl-position [203]. By contrast, in colon carcinoma Caco-2 cells, bile acid-activated FXR was identified as a negative regulator of the UGT2B7 gene, which encodes an isoform involved in the formation of 3-hydroxy-glucuronide bile acids [220]. Similarly, FXR functions as a repressor rather than an inducer of sulfonation in mice and humans. Indeed, disruption of the FXR gene in mice resulted in increased hepatic levels of SULT mRNAs and proteins, as well as in elevated bile concentrations of 3α-sulfated bile acids in animals fed with an LCA-enriched diet (Table 3.5) [221]. Human SULT2A1 protein and mRNA levels were also decreased in HepG2 cells treated with FXR agonists. These results suggest that SULT2A1 is negatively regulated through CDCA-mediated FXR activation. By contrast, LXR activation confers a female-specific resistance to bile acid toxicity and cholestasis by increasing expression of Sult2a and selected bile acid transporters in mice [222]. In LXR-deficient animals, a basal expression of these gene products was reduced, while these animals exhibited heightened cholestatic sensitivity. LXR activation also stimulates bile acid glucuronidation, since the human UGT1A3 gene was identified as a positively regulated target gene and contains a functional response element for this receptor [223]. By contrast, UGT2B4 expression remains unaffected in hepatic cells treated with the LXR agonist T0901317.

Beyond the control of bile acid levels, recent observations support a role for LXR in the control of hepatic estrogen metabolism and cholesterol sulfonation [212, 224]. LXR controls estrogen homeostasis by regulating the basal and inducible hepatic expression of estrogen sulfotransferase SULT1E1, an enzyme critical for metabolic estrogen deactivation. Genetic or pharmacological activation of LXR resulted in

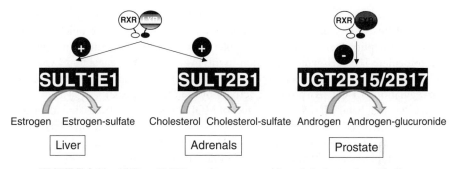

FIGURE 3.11   FXR and LXR regulate sex steroids and cholesterol metabolism.

SULT1E1 induction, which in turn inhibited estrogen-dependent uterine epithelial cell proliferation and gene expression, as well as breast cancer growth in a nude mouse model of tumorigenicity. Interestingly, SULT1E1 regulation appears to be liver specific, and this tissue is so important for estrogen metabolism that a compromised liver function has been linked to hyperestrogenism in patients. In cultured human keratinocytes, LXR activators also increase expression of SULT2B1b, a key enzyme for cholesterol sulfation, which may affect keratinocyte differentiation and desquamation, two processes regulated by cholesterol sulfate [212]. In prostate cancer cells, FXR activation results in a reduction of the expression and activity of the two androgen-conjugating UGT2B15 and 2B17 enzymes [225]. In summary, these results have demonstrated that LXR and FXR are regulators of sex steroid and cholesterol homeostasis, which may have implications for drug development in the treatment of estrogen-, androgen-, and cholesterol-related disorders (Figure 3.11).

3.7   HNF AND PHASE II XMEs REGULATION

HNFs are structurally related NRs that regulate numerous liver-specific genes either alone or as members of a complex regulatory network involving other NRs critical for the maintenance of liver phenotype [213]. In humans, HNF1 exists in two isoforms associated with alternative splicing, HNF1α and HNF1β [226]. Although both isoforms bind the same DNA sequence as homodimers, HNF1α appears to be a more dominant transactivator. Both isoforms are expressed in polarized epithelia of multiple organs, including the liver, digestive tract, pancreas, and kidney, but only HNF1α is expressed in lungs [227]. In humans, mutations in HNF1α are linked to MODY3, an autosomal dominant form of non-insulin-dependent diabetes mellitus [228, 229]. HNF1α regulates human UGT1A1 [230], 1A6 [231], 2B17 [232], and 2B7 [232] and the hepatic expression of rats Ugt1a7 (Table 3.5) [233]. The proximal 320-bp promoter of the rat Ugt2b1 gene contains several DNA binding sites for liver-specific transcription factors, including the hepatocyte HNF1 and C/EBP [234, 235]. These liver-specific transcription factors may have conferred the liver-specific expression pattern of this UGT isoform. Other transcription factors, such as the ubiquitous octamer transcription factor 1 (Oct-1), can interact with HNF1 to modulate its capacity to up-regulate UGT2B gene expression in the liver [236]. The capacity of Oct-1 to enhance the transcriptional activity of HNF1α provides a plausible explanation for the regulation of UGT2B7 promoter activity by genotoxic stress.

On the other hand, HNF4α is a product of MODY1 gene, which was first identified as a liver-enriched transcription factor that regulates a number of genes involved in the developmental program of organs such as the liver, intestine, and kidney [237]. Actually, since it physically interacts with a significantly high number of genes (∼13,000 genes) in hepatocytes, HNF4α has been proposed to act as a master regulator of gene expression in liver [238]. Among HNF4α target genes, those encoding phase II enzymes are UGT1A1, 1A9, and SULT2A1 (Table 3.5) [239]. The role of HNF4α in regulating drug-metabolizing enzymes is reinforced by the observation that HNF4α regulates the human UGT1A9 gene through a specific response element that is absent from the closely linked UGT1A7, 1A8, and 1A10 genes [215]. In addition, the proximal promoters of UGT1A1, 2B11, and 2B15 also bind HNF4α and can be regulated in a feed-forward and multi-input mode in hepatocytes [238]. Further, HNF4α controls xenobiotic responses in the fetal liver by regulating the expression of PXR during liver development via binding to HNF4 response elements in the PXR promoter [240]. When gene expression is analyzed in mice with HNF4α null livers, one of the genes whose mRNA is found to be absent is mPXR [241]. In parallel, ligand-activated PXR interferes with HNF4α by squelching the common coactivator PGC1a [242, 243]. CAR activation also inhibits HNF4α transactivation activity by mechanisms to be identified [244]. Together, these observations indicate that PXR/CAR and HNF4α are closely linked regulatory pathways and that PXR and CAR activation may also affect the HNF4α-dependent regulation of phase II enzymes. As an example, HNF4α has an essential role for hepatic SULT2A1 expression, while rifampicin-activation of PXR negatively interferes with the HNF4α-dependent induction of SULT2A1 expression [245].

3.8   REGULATION OF PHASE II CONJUGATING ENZYMES BY STEROID AND THYROID RECEPTORS

In addition to UGT1A1, steroids, and especially glucocorticoids also induce SULTs expression [48, 246, 247]. Induction of Sult1a in rat occurs by GR activation, while Sult2a transcription appears to be controlled by both GR and other transcriptional factors [248, 249]. On the other hand, rat Sult2a transcription is effectively repressed by androgens through a mechanism involving the androgen receptor (AR) [248]; however, effects of this repression on protein and functional activity remain to be demonstrated. By contrast, various studies established that the AR, which mediates the transcriptional activity of the active male hormone, dihydrotestosterone, is a negative regulator of UGT2B15 and 2B17 expression in prostate cancer cells [6, 250]. Interestingly, these two enzymes are major determinants of androgen action in prostate cells [251]. Taken together, these informations demonstrate that androgens down-regulate their own inactivation in androgen-dependent prostate cancer cells, thus further potentiating their cellular accumulation and carcinogenic effects. AR activation also produces an induction of the UGT2B11 expression within prostate cells [250]. Since UGT2B11 is specific for monohydroxy fatty acid (i.e., HETEs) glucuronidation [206], this effect is thought to reduce the angiogenic and metastatic properties of such fatty acid derivatives [250]. Recent observations indicate that the androgen-conjugating UGT2B15 enzyme is also a positively regulated ER target gene in estrogen-sensitive cancer cells [252]. Interestingly, a biphasic regulation of UGT1A10 was also reported in such cell models [253]. At low estradiol concentration (0.1 nM), UGT1A10 expression was up-regulated, whereas pharmacological concentrations (10 nM) reduced both UGT1A10 expression and estradiol glucuronidation [253]. Thus, in mammary gland cancer cells, the expression of UGTs is tightly controlled by ER. COMT is also an important inactivating and conjugating enzyme for active estrogens. In mammary gland cells, estradiol-activated ER provokes a 50% reduction of COMT expression through binding to negative ER response element within the human COMT gene. However, and on the basis of the major role that the futile cycle of sulfonation–desulfonation exerts on the maintenance of estrogen homeostasis, the regulation of the estrogen sulfotransferase enzyme (SULT1E1) has been thoroughly analyzed in breast cancer cells [254]. These analyses revealed that estrogens stimulate their own sulfonation, thus providing the cells with an efficient mean to tightly regulate their concentrations [254]. Interestingly, the estrogen-related receptor alpha (ERRα), which is functionally related to ER, also regulates steroid sulfonation by controlling the adrenal expression of SULT2A1, the human enzyme sulfating the androgen and estrogen precursor, dehydroepiandrosterone [255]. This observation illustrates the important role that ERRα exerts in controlling adrenal steroid production, which in humans corresponds to approximately 50% of the formation of active steroid hormones [256].

Altogether, these observations demonstrate that nuclear steroid sensors control the expression of enzymes involved in the metabolism of their high affinity ligands within their target tissues. Similarly, rat Ugt1a7 has recently been characterized as an important conjugating enzyme for thyroid hormones (T4), and it should be noted that Ugt1a7 is a thyroid hormone receptor (THR)-induced target gene in this species [152]. This suggests that the THR-dependent regulation of this UGT isoform is part of an autoregulatory process, in which T4 stimulates its own metabolism.

3.9   CONCLUDING REMARKS AND PERSPECTIVES

Phase II drug metabolism was initially considered as a constitutive detoxification process occurring specifically in the liver. However, an increasing body of evidences generated during the last decade clearly established that conjugating enzymes are also expressed in numerous extrahepatic tissues, and that their expression is tightly controlled by transcription factors that function as sensors for endo- and xenobiotics. Furthermore, the important role that these enzymes exert in the control of hormones levels and actions, as well as the identification of specific NRs involved in their regulation, has moved scientists' attention on these enzymes from the pure drug metabolism to concept of phase II enzymes as drug target. Signaling pathways controlling xenobiotic metabolism are embedded within a tangle of regulatory networks controlling the homeostasis of glucose, bile acids, lipids, hormones, inflammation, and others [75]. These cross talks explain how physiopathological stimuli affect drug metabolism and disposition. They also explain how drug and other xenobiotics affect physiological functions, thus opening the way to interesting pharmacological opportunities for a wide variety of pathological situations. Nevertheless, there are marked species differences in response to NRs activating agents and phase II enzymes expression [71]. For example, the rat cells are responsive to DEX but not rifampicin, whereas in the human cells rifampicin was a more potent inducer than DEX [257]. Similarly, some phase II DMEs exists in rodents while being absent in humans (i.e., Ugt1a2); the opposite is also true (i.e., UGT1A3) [10]. Overall, these differences can make it difficult to extrapolate conclusions across species boundaries.

The extent of cross talk between all the phase II genes regulators increases the complexity of the molecular interactions that need to be understood and means that the individual effects of each of the receptors are difficult to dissect. Furthermore, although NRs activate battery of target genes, coregulators control the activity of batteries of NRs and transcription factors. In addition, coactivators are subjected to transcriptional regulation, post-translational modification, controlled degradation and polymorphic activity, and these modifications may influence the activity of their NR partners. With few exceptions, the role of coregulators in the NR-dependent regulation of phase II DMEs has received only little attention. Thus, future experiments will have to identify all the components of these regulatory complexes to further increase our understanding of the manner in which drug-metabolizing enzymes respond to endo- and xenobiotics sensors, with a particular attention on tissue- and cell-specific manner in which these enzymes are regulated.

REFERENCES

1. Iyanagi, T. (2007) Molecular mechanism of phase I and phase II drug-metabolizing enzymes: implications for detoxification. *Int. Rev. Cytol.* **260**, 35–112.

2. Dutton, G. J. (1980) *Glucuronidation of Drugs and Other Compounds.* CRC Press, Boca Raton, FL.

3. Zamek-Gliszczynski, M. J., Hoffmaster, K. A., Nezasa, K., Tallman, M. N., and Brouwer, K. L. (2006) Integration of hepatic drug transporters and phase II metabolizing enzymes: mechanisms of hepatic excretion of sulfate, glucuronide, and glutathione metabolites. *Eur. J. Pharm. Sci.* **27**, 447–486.

4. Aleksunes, L. M. and Manautou, J. E. (2007) Emerging role of Nrf2 in protecting against hepatic and gastrointestinal disease. *Toxicol. Pathol.* **35**, 459–473.

5. Evans, W. E. and Relling, M. V. (1999) Pharmacogenomics: translating functional genomics into rational therapeutics. *Science* **286**, 487–491.

6. Bélanger, A., Pelletier, G., Labrie, F., Barbier, O., and Chouinard, S. (2003) Inactivation of androgens by UDP-glucuronosyltransferase enzymes in humans. *Trends Endocrinol. Metab.* **14**, 473–479.

7. Wells, P. G., Mackenzie, P. I., Chowdhury, J. R., Guillemette, C., Gregory, P. A., Ishii, Y., Hansen, A. J., Kessler, F. K., Kim, P. M., Chowdhury, N. R., et al. (2004) Glucuronidation and the UDP-glucuronosyltransferases in health and disease. *Drug Metab. Dispos.* **32**, 281–290.

8. Shipkova, M. and Wieland, E. (2005) Glucuronidation in therapeutic drug monitoring. *Clin. Chim. Acta.* **358**, 2–23.

9. Sallustio, B. C., Sabordo, L., Evans, A. M., and Nation, R. L. (2000) Hepatic disposition of electrophilic acyl glucuronide conjugates. *Curr. Drug Metab.* **1**, 163–180.

10. Mackenzie, P. I., Bock, K. W., Burchell, B., Guillemette, C., Ikushiro, S., Iyanagi, T., Miners, J. O., Owens, I. S., and Nebert, D. W. (2005) Nomenclature update for the mammalian UDP glycosyltransferase (UGT) gene superfamily. *Pharmacogenetics Genomics* **15**, 677–685.

11. Girard, H., Lévesque, É., Bellemare, J., Journault, K., Caillier, B., and Guillemette, C. (2007) Genetic diversity at the UGT1 locus is amplified by a novel 3′ alternative splicing mechanism leading to nine additional UGT1A proteins that act as regulators of glucuronidation activity. *Pharmacogenetics Genomics* **17**, 1077–1089.

12. Lévesque, É., Girard, H., Journault, K., Lépine, J., and Guillemette, C. (2007) Regulation of the UGT1A1 bilirubin-conjugating pathway: role of a new splicing event at the UGT1A locus. *Hepatology* **45**, 128–138.

13. Tukey, R. H. and Strassburg, C. P. (2000) Human UDP-glucuronosyltransferases: metabolism, expression, and disease. *Annu. Rev. Pharmacol. Toxicol.* **40**, 581–616.

14. Miners, J. O., Knights, K. M., Houston, J. B., and Mackenzie, P. I. (2006) In vitro-in vivo correlation for drugs and other compounds eliminated by glucuronidation in humans: pitfalls and promises. *Biochem. Pharmacol.* **71**, 1531–1539.

15. Miners, J. O., Smith, P. A., Sorich, M. J., McKinnon, R. A., and Mackenzie, P. I. (2004) Predicting human drug glucuronidation parameters: application of in vitro and in silico modeling approaches. *Annu. Rev. Pharmacol. Toxicol.* **44**, 1–25.

16. Bosma, P. J., Seppen, J., Goldhoorn, B., Bakker, C., Oude Elferink, R. P., Chowdhury, J. R., Chowdhury, N. R., and Jansen, P. L. (1994) Bilirubin UDP-glucuronosyltransferase 1 is the only relevant bilirubin glucuronidating isoform in man. *J. Biol. Chem.* **269**, 17960–17964.

17. Bosma, P. J., Chowdhury, J. R., Bakker, C., Gantla, S., de Boer, A., Oostra, B. A., Lindhout, D., Tytgat, G. N., Jansen, P. L., Oude Elferink, R. P., et al. (1995) The genetic basis of the reduced expression of bilirubin UDP-glucuronosyltransferase 1 in Gilbert's syndrome. *N. Engl. J. Med.* **333**, 1171–1175.

18. Court, M. H., Duan, S. X., von Moltke, L. L., Greenblatt, D. J., Patten, C. J., Miners, J. O., and Mackenzie, P. I. (2001) Interindividual variability in acetaminophen glucuronidation by human liver microsomes: identification of relevant acetaminophen UDP-glucuronosyltransferase isoforms. *J. Pharmacol. Exp. Ther.* **299**, 998–1006.

19. Williams, J. A., Hyland, R., Jones, B. C., Smith, D. A., Hurst, S., Goosen, T. C., Peterkin, V., Koup, J. R., and Ball, S. E. (2004) Drug–drug interactions for UDP-glucuronosyltransferase substrates: a pharmacokinetic explanation for typically observed low exposure (AUCi/AUC) ratios. *Drug Metab. Dispos.* **32**, 1201–1208.

20. Coffman, B. L., Rios, G. R., King, C. D., and Tephly, T. R. (1997) Human UGT2B7 catalyzes morphine glucuronidation. *Drug Metab. Dispos.* **25**, 1–4.

21. Barbier, O., Turgeon, D., Girard, C., Green, M. D., Tephly, T. R., Hum, D. W., and Bélanger, A. (2000) 3′-Azido-3′-deoxythimidine (AZT) is glucuronidated by human UDP-glucuronosyltransferase 2B7 (UGT2B7). *Drug Metab. Dispos.* **28**, 497–502.

22. Crayford, J. V. and Hutson, D. H. (1980) Comparative metabolism of phenobarbitone in the rat (CFE) and mouse (CF1). *Food Cosmet. Toxicol.* **18**, 503–509.

23. Buhl, A. E., Waldon, D. J., Baker, C. A., and Johnson, G. A. (1990) Minoxidil sulfate is the active metabolite that stimulates hair follicles. *J. Invest. Dermatol.* **95**, 553–557.

24. Glatt, H., Davis, W., Meinl, W., Hermersdorfer, H., Venitt, S., and Phillips, D. H. (1998) Rat, but not human, sulfotransferase activates a tamoxifen metabolite to produce DNA adducts and gene mutations in bacteria and mammalian cells in culture. *Carcinogenesis* **19**, 1709–1713.

25. Funk, C., Ponelle, C., Scheuermann, G., and Pantze, M. (2001) Cholestatic potential of troglitazone as a possible factor contributing to troglitazone-induced hepatotoxicity: in vivo and in vitro interaction at the canalicular bile salt export pump (Bsep) in the rat. *Mol. Pharmacol.* **59**, 627–635.

26. Nowell, S. and Falany, C. N. (2006) Pharmacogenetics of human cytosolic sulfotransferases. *Oncogene* **25**, 1673–1678.

27. Huttner, W. B. (1988) Tyrosine sulfation and the secretory pathway. *Annu. Rev. Physiol.* **50**, 363–376.

28. Liyou, N. E., Buller, K. M., Tresillian, M. J., Elvin, C. M., Scott, H. L., Dodd, P. R., Tannenberg, A. E., and McManus, M. E. (2003) Localization of a brain sulfotransferase, SULT4A1, in the human and rat brain: an immunohistochemical study. *J. Histochem. Cytochem.* **51**, 1655–1664.

29. Glatt, H. (2000) Sulfotransferases in the bioactivation of xenobiotics. *Chem. Biol. Interact.* **129**, 141–170.

30. Otterness, D. M., Wieben, E. D., Wood, T. C., Watson, W. G., Madden, B. J., McCormick, D. J., and Weinshilboum, R. M. (1992) Human liver dehydroepiandrosterone sulfotransferase: molecular cloning and expression of cDNA. *Mol. Pharmacol.* **41**, 865–872.

31. Her, C., Wood, T. C., Eichler, E. E., Mohrenweiser, H. W., Ramagli, L. S., Siciliano, M. J., and Weinshilboum, R. M. (1998) Human hydroxysteroid sulfotransferase SULT2B1, two enzymes encoded by a single chromosome 19 gene. *Genomics* **53**, 284–295.

32. Strott, C. A. (2002) Sulfonation and molecular action. *Endocr. Rev.* **23**, 703–732.

33. Gamage, N., Barnett, A., Hempel, N., Duggleby, R. G., Windmill, K. F., Martin, J. L., and McManus, M. E. (2006) Human sulfotransferases and their role in chemical metabolism. *Toxicol. Sci.* **90**, 5–22.

34. Radominska, A., Comer, K. A., Zimniak, P., Falany, J., Iscan, M., and Falany, C. N. (1990) Human liver steroid sulphotransferase sulphates bile acids. *Biochem. J.* **272**, 597–604.

35. Nagata, K., Ozawa, S., Miyata, M., Shimada, M., Gong, D. W., Yamazoe, Y., and Kato, R. (1993) Isolation and expression of a cDNA encoding a male-specific rat sulfotransferase that catalyzes activation of N-hydroxy-2-acetylaminofluorene. *J. Biol. Chem.* **268**, 24720–24725.

36. Mulder, G. J. and Scholtens, E. (1978) The availability of inorganic sulphate in blood for sulphate conjugation of drugs in rat liver in vivo. (35S)Sulphate incorporation into harmol sulphate. *Biochem. J.* **172**, 247–251.

37. Hjelle, J. J., Hazelton, G. A., and Klaassen, C. D. (1985) Acetaminophen decreases adenosine 3′-phosphate 5′-phosphosulfate and uridine diphosphoglucuronic acid in rat liver. *Drug Metab. Dispos.* **13**, 35–41.

38. Kim, H. J., Rozman, P., Madhu, C., and Klaassen, C. D. (1992) Homeostasis of sulfate and 3′-phosphoadenosine 5′-phosphosulfate in rats after acetaminophen administration. *J. Pharmacol. Exp. Ther.* **261**, 1015–1021.

39. Pang, K. S., Schwab, A. J., Goresky, C. A., and Chiba, M. (1994) Transport, binding, and metabolism of sulfate conjugates in the liver. *Chem. Biol. Interact.* **92**, 179–207.

40. Pang, K. S., Koster, H., Halsema, I. C., Scholtens, E., and Mulder, G. J. (1981) Aberrant pharmacokinetics of harmol in the perfused rat liver preparation: sulfate and glucuronide conjugations. *J. Pharmacol. Exp. Ther.* **219**, 134–140.

41. Pang, K. S., Koster, H., Halsema, I. C., Scholtens, E., Mulder, G. J., and Stillwell, R. N. (1983) Normal and retrograde perfusion to probe the zonal distribution of sulfation and glucuronidation activities of harmol in the perfused rat liver preparation. *J. Pharmacol. Exp. Ther.* **224**, 647–653.

42. Morris, M. E., Yuen, V., and Pang, K. S. (1988) Competing pathways in drug metabolism. II. An identical, anterior enzymic distribution for 2- and 5-sulfoconjugation and a posterior localization for 5-glucuronidation of gentisamide in the rat liver. *J. Pharmacokinet. Biopharm.* **16**, 633–656.

43. Ratna, S., Chiba, M., Bandyopadhyay, L., and Pang, K. S. (1993) Futile cycling between 4-methylumbelliferone and its conjugates in perfused rat liver. *Hepatology* **17**, 838–853.

44. Smith, P. C., McDonagh, A. F., and Benet, L. Z. (1990) Effect of esterase inhibition on the disposition of zomepirac glucuronide and its covalent binding to plasma proteins in the guinea pig. *J. Pharmacol. Exp. Ther.* **252**, 218–224.

45. Liu, J. H. and Smith, P. C. (2006) Predicting the pharmacokinetics of acyl glucuronides and their parent compounds in disease states. *Curr. Drug Metab.* **7**, 147–163.

46. Roberts, M. S., Magnusson, B. M., Burczynski, F. J., and Weiss, M. (2002) Enterohepatic circulation: physiological, pharmacokinetic and clinical implications. *Clin. Pharmacokinet.* **41**, 751–790.

47. Shipkova, M., Armstrong, V. W., Oellerich, M., and Wieland, E. (2003) Acyl glucuronide drug metabolites: toxicological and analytical implications. *Ther. Drug Monit.* **25**, 1–16.

48. Duanmu, Z., Dunbar, J., Falany, C. N., and Runge-Morris, M. (2000) Induction of rat hepatic aryl sulfotransferase (SULT1A1) gene expression by triamcinolone acetonide: impact on minoxidil-mediated hypotension. *Toxicol. Appl. Pharmacol.* **164**, 312–320.

49. Duvoix, A., Schnekenburger, M., Delhalle, S., Blasius, R., Borde-Chiche, P., Morceau, F., Dicato, M., and Diederich, M. (2004) Expression of glutathione S-transferase P1-1 in leukemic cells is regulated by inducible AP-1 binding. *Cancer Lett.* **216**, 207–219.

50. Kim, S. K., Abdelmegeed, M. A., and Novak, R. F. (2006) Identification of the insulin signaling cascade in the regulation of alpha-class glutathione S-transferase expression in primary cultured rat hepatocytes. *J. Pharmacol. Exp. Ther.* **316**, 1255–1261.

51. Morgenstern, R., Lundqvist, G., Andersson, G., Balk, L., and DePierre, J. W. (1984) The distribution of microsomal glutathione transferase among different organelles, different organs, and different organisms. *Biochem. Pharmacol.* **33**, 3609–3614.

52. DeRidder, B. P. and Goldsbrough, P. B. (2006) Organ-specific expression of glutathione S-transferases and the efficacy of herbicide safeners in Arabidopsis. *Plant Physiol.* **140**, 167–175.

53. Edwards, R., Dixon, D. P., and Walbot, V. (2000) Plant glutathione S-transferases: enzymes with multiple functions in sickness and in health. *Trends Plant. Sci.* **5**, 193–198.

54. Haimeur, A., Conseil, G., Deeley, R. G., and Cole, S. P. (2004) The MRP-related and BCRP/ABCG2 multidrug resistance proteins: biology, substrate specificity and regulation. *Curr. Drug Metab.* **5**, 21–53.

55. Sheehan, D., Meade, G., Foley, V. M., and Dowd, C. A. (2001) Structure, function and evolution of glutathione transferases: implications for classification of non-mammalian members of an ancient enzyme superfamily. *Biochem. J.* **360**, 1–16.

56. Hayes, J. D., Flanagan, J. U., and Jowsey, I. R. (2005) Glutathione transferases. *Annu. Rev. Pharmacol. Toxicol.* **45**, 51–88.

57. Nishinaka, T., Ichijo, Y., Ito, M., Kimura, M., Katsuyama, M., Iwata, K., Miura, T., Terada, T., and Yabe-Nishimura, C. (2007) Curcumin activates human glutathione S-transferase P1 expression through antioxidant response element. *Toxicol. Lett.* **170**, 238–247.

58. Pool-Zobel, B., Veeriah, S., and Bohmer, F. D. (2005) Modulation of xenobiotic metabolising enzymes by anticarcinogens—focus on glutathione S-transferases and their role as targets of dietary chemoprevention in colorectal carcinogenesis. *Mutat. Res.* **591**, 74–92.

59. Mannervik, B., Alin, P., Guthenberg, C., Jensson, H., Tahir, M. K., Warholm, M., and Jornvall, H. (1985) Identification of three classes of cytosolic glutathione transferase common to several mammalian species: correlation between structural data and enzymatic properties. *Proc. Natl Acad. Sci. USA* **82**, 7202–7206.

60. Sun, Q., Komura, S., Ohishi, N., and Yagi, K. (1996) Alpha-class isozymes of glutathione S-transferase in rat liver cytosol possess glutathione peroxidase activity toward phospholipid hydroperoxide. *Biochem. Mol. Biol. Int.* **39**, 343–352.

61. Yusof, Y. A., Yan, K. L., and Hussain, S. N. (2003) Immunohistochemical expression of pi class glutathione S-transferase and alpha-fetoprotein in hepatocellular carcinoma and chronic liver disease. *Anal. Quant. Cytol. Histol.* **25**, 332–338.

62. Park, E. Y., Cho, I. J., and Kim, S. G. (2004) Transactivation of the PPAR-responsive enhancer module in chemopreventive glutathione S-transferase gene by the peroxisome proliferator-activated receptor-gamma and retinoid X receptor heterodimer. *Cancer Res.* **64**, 3701–3713.

63. Salinas, A. E. and Wong, M. G. (1999) Glutathione S-transferases—a review. *Curr. Med. Chem.* **6**, 279–309.

64. Vargo, M. A. and Colman, R. F. (2001) Affinity labeling of rat glutathione S-transferase isozyme 1-1 by 17beta-iodoacetoxy-estradiol-3-sulfate. *J. Biol. Chem.* **276**, 2031–2036.

65. Sim, E., Westwood, I., and Fullam, E. (2007) Arylamine N-acetyltransferases. *Expert Opin. Drug Metab. Toxicol.* **3**, 169–184.

66. Weinshilboum, R. M. (2006) Pharmacogenomics: catechol O-methyltransferase to thiopurine S-methyltransferase. *Cell Mol. Neurobiol.* **26**, 539–561.

67. Shimada, T. (2006) Xenobiotic-metabolizing enzymes involved in activation and detoxification of carcinogenic polycyclic aromatic hydrocarbons. *Drug Metab. Pharmacokinet.* **21**, 257–276.

68. Chiamvimonvat, N., Ho, C. M., Tsai, H. J., and Hammock, B. D. (2007) The soluble epoxide hydrolase as a pharmaceutical target for hypertension. *J. Cardiovasc. Pharmacol.* **50**, 225–237.

69. Selye, H. (1972) Prevention by catatoxic steroids of lithocholic acid-induced biliary concrements in the rat. *Proc. Soc. Exp. Biol. Med.* **141**, 555–558.

70. Timsit, Y. E. and Negishi, M. (2007) CAR and PXR: the xenobiotic-sensing receptors. *Steroids* **72**, 231–246.

71. Stanley, L. A., Horsburgh, B. C., Ross, J., Scheer, N., and Wolf, C. R. (2006) PXR and CAR: nuclear receptors which play a pivotal role in drug disposition and chemical toxicity. *Drug Metab. Rev.* **38**, 515–597.

72. Ekins, S., Mirny, L., and Schuetz, E. G. (2002) A ligand-based approach to understanding selectivity of nuclear hormone receptors PXR, CAR, FXR, LXRalpha, and LXRbeta. *Pharm. Res.* **19**, 1788–1800.

73. Moore, L. B., Maglich, J. M., McKee, D. D., Wisely, B., Willson, T. M., Kliewer, S. A., Lambert, M. H., and Moore, J. T. (2002) Pregnane X receptor (PXR), constitutive androstane receptor (CAR), and benzoate X receptor (BXR) define three pharmacologically distinct classes of nuclear receptors. *Mol. Endocrinol.* **16**, 977–986.

74. Tzameli, I. and Moore, D. D. (2001) Role reversal: new insights from new ligands for the xenobiotic receptor CAR. *Trends Endocrinol. Metab.* **12**, 7–10.

75. Pascussi, J. M., Gerbal-Chaloin, S., Duret, C., Daujat-Chavanieu, M., Vilarem, M. J., and Maurel, P. (2008) The tangle of nuclear receptors that controls xenobiotic metabolism and transport: crosstalk and consequences. *Annu. Rev. Pharmacol. Toxicol.* **48**, 1–32.

76. Moore, L. B., Parks, D. J., Jones, S. A., Bledsoe, R. K., Consler, T. G., Stimmel, J. B., Goodwin, B., Liddle, C., Blanchard, S. G., Willson, T. M., et al. (2000) Orphan nuclear receptors constitutive androstane receptor and pregnane X receptor share xenobiotic and steroid ligands. *J. Biol. Chem.* **275**, 15122–15127.

77. Lehmann, J. M., McKee, D. D., Watson, M. A., Willson, T. M., Moore, J. T., and Kliewer, S. A. (1998) The human orphan nuclear receptor PXR is activated by compounds that regulate CYP3A4 gene expression and cause drug interactions. *J. Clin. Invest.* **102**, 1016–1023.

78. Moore, J. T. and Kliewer, S. A. (2000) Use of the nuclear receptor PXR to predict drug interactions. *Toxicology* **153**, 1–10.

79. Staudinger, J. L., Goodwin, B., Jones, S. A., Hawkins-Brown, D., MacKenzie, K. I., LaTour, A., Liu, Y., Klaassen, C. D., Brown, K. K., Reinhard, J., et al. (2001) The nuclear receptor PXR is a lithocholic acid sensor that protects against liver toxicity. *Proc. Natl Acad. Sci. USA* **98**, 3369–3374.

80. Kretschmer, X. C. and Baldwin, W. S. (2005) CAR and PXR: xenosensors of endocrine disrupters? *Chem. Biol. Interact.* **155**, 111–128.

81. Forman, B. M., Tzameli, I., Choi, H. S., Chen, J., Simha, D., Seol, W., Evans, R. M., and Moore, D. D. (1998) Androstane metabolites bind to and deactivate the nuclear receptor CAR-beta. *Nature* **395**, 612–615.

82. Lamba, J. K., Lamba, V., Yasuda, K., Lin, Y. S., Assem, M., Thompson, E., Strom, S., and Schuetz, E. (2004) Expression of constitutive androstane receptor splice variants in human tissues and their functional consequences. *J. Pharmacol. Exp. Ther.* **311**, 811–821.

83. Gardner-Stephen, D., Heydel, J. M., Goyal, A., Lu, Y., Xie, W., Lindblom, T., Mackenzie, P., and Radominska-Pandya, A. (2004) Human PXR variants and their differential effects

on the regulation of human UDP-glucuronosyltransferase gene expression. *Drug Metab. Dispos.* **32**, 340–347.

84. Westerink, W. M. and Schoonen, W. G. (2007) Phase II enzyme levels in HepG2 cells and cryopreserved primary human hepatocytes and their induction in HepG2 cells. *Toxicol. In Vitro* **21**, 1592–1602.

85. Schuetz, E. G., Hazelton, G. A., Hall, J., Watkins, P. B., Klaassen, C. D., and Guzelian, P. S. (1986) Induction of digitoxigenin monodigitoxoside UDP-glucuronosyltransferase activity by glucocorticoids and other inducers of cytochrome P-450p in primary monolayer cultures of adult rat hepatocytes and in human liver. *J. Biol. Chem.* **261**, 8270–8275.

86. Chen, C., Staudinger, J. L., and Klaassen, C. D. (2003) Nuclear receptor, pregnane X receptor, is required for induction of UDP-glucuronosyltransferases in mouse liver by pregnenolone-16 alpha-carbonitrile. *Drug Metab. Dispos.* **31**, 908–915.

87. Xie, W., Yeuh, M. F., Radominska-Pandya, A., Saini, S. P., Negishi, Y., Bottroff, B. S., Cabrera, G. Y., Tukey, R. H., and Evans, R. M. (2003) Control of steroid, heme, and carcinogen metabolism by nuclear pregnane X receptor and constitutive androstane receptor. *Proc. Natl Acad. Sci. USA* **100**, 4150–4155.

88. Xie, W., Barwick, J. L., Downes, M., Blumberg, B., Simon, C. M., Nelson, M. C., Neuschwander-Tetri, B. A., Brunt, E. M., Guzelian, P. S., and Evans, R. M. (2000) Humanized xenobiotic response in mice expressing nuclear receptor SXR. *Nature* **406**, 435–439.

89. Xie, W. and Evans, R. M. (2002) Pharmaceutical use of mouse models humanized for the xenobiotic receptor. *Drug Discov. Today* **7**, 509–515.

90. Xie, W., Barwick, J. L., Simon, C. M., Pierce, A. M., Safe, S., Blumberg, B., Guzelian, P. S., and Evans, R. M. (2000) Reciprocal activation of xenobiotic response genes by nuclear receptors SXR/PXR and CAR. *Genes Dev.* **14**, 3014–3023.

91. Zhou, J., Zhang, J., and Xie, W. (2005) Xenobiotic nuclear receptor-mediated regulation of UDP-glucuronosyl-transferases. *Curr. Drug Metab.* **6**, 289–298.

92. Chen, S., Beaton, D., Nguyen, N., Senekeo-Effenberger, K., Brace-Sinnokrak, E., Argikar, U., Remmel, R. P., Trottier, J., Barbier, O., Ritter, J. K., et al. (2005) Tissue-specific, inducible, and hormonal control of the human UDP-glucuronosyltransferase-1 (UGT1) locus. *J. Biol. Chem.* **280**, 37547–37557.

93. Sonoda, J., Xie, W., Rosenfeld, J. M., Barwick, J. L., Guzelian, P. S., and Evans, R. M. (2002) Regulation of a xenobiotic sulfonation cascade by nuclear pregnane X receptor (PXR). *Proc. Natl Acad. Sci. USA* **99**, 13801–13806.

94. Alnouti, Y. and Klaassen, C. D. (2008) Regulation of sulfotransferase enzymes by prototypical microsomal enzyme inducers in mice. *J. Pharmacol. Exp. Ther.* **324** (2), 612–621.

95. DePierre, J. W., Seidegard, J., Morgenstern, R., Balk, L., Meijer, J., Astrom, A., Norelius, I., and Ernster, L. (1984) Induction of cytosolic glutathione transferase and microsomal epoxide hydrolase activities in extrahepatic organs of the rat by phenobarbital, 3-methylcholanthrene and trans-stilbene oxide. *Xenobiotica* **14**, 295–301.

96. Rae, J. M., Johnson, M. D., Lippman, M. E., and Flockhart, D. A. (2001) Rifampin is a selective, pleiotropic inducer of drug metabolism genes in human hepatocytes: studies with cDNA and oligonucleotide expression arrays. *J. Pharmacol. Exp. Ther.* **299**, 849–857.

97. Shibayama, Y., Ikeda, R., Motoya, T., and Yamada, K. (2004) St. John's wort (hypericum perforatum) induces overexpression of multidrug resistance protein 2 (MRP2) in rats: a 30-day ingestion study. *Food Chem. Toxicol.* **42**, 995–1002.

98. Gong, H., Singh, S. V., Singh, S. P., Mu, Y., Lee, J. H., Saini, S. P., Toma, D., Ren, S., Kagan, V. E., Day, B. W., et al. (2006) Orphan nuclear receptor pregnane X receptor sensitizes oxidative stress responses in transgenic mice and cancerous cells. *Mol. Endocrinol.* **20**, 279–290.

99. Saini, S. P., Mu, Y., Gong, H., Toma, D., Uppal, H., Ren, S., Li, S., Poloyac, S. M., and Xie, W. (2005) Dual role of orphan nuclear receptor pregnane X receptor in bilirubin detoxification in mice. *Hepatology* **41**, 497–505.

100. Huang, W., Zhang, J., Chua, S. S., Qatanani, M., Han, Y., Granata, R., and Moore, D. D. (2003) Induction of bilirubin clearance by the constitutive androstane receptor (CAR). *Proc. Natl Acad. Sci. USA* **100**, 4156–4161.

101. Tong, Z., Li, H., Goljer, I., McConnell, O., and Chandrasekaran, A. (2007) In vitro glucuronidation of thyroxine and triiodothyronine by liver microsomes and recombinant human UDP-glucuronosyltransferases. *Drug Metab. Dispos.* **35**, 2203–2210.

102. Eiris-Punal, J., Del Rio-Garma, M., Del Rio-Garma, M. C., Lojo-Rocamonde, S., Novo-Rodriguez, I., and Castro-Gago, M. (1999) Long-term treatment of children with epilepsy with valproate or carbamazepine may cause subclinical hypothyroidism. *Epilepsia* **40**, 1761–1766.

103. Klaassen, C. D. and Hood, A. M. (2001) Effects of microsomal enzyme inducers on thyroid follicular cell proliferation and thyroid hormone metabolism. *Toxicol. Pathol.* **29**, 34–40.

104. Maglich, J. M., Watson, J., McMillen, P. J., Goodwin, B., Willson, T. M., and Moore, J. T. (2004) The nuclear receptor CAR is a regulator of thyroid hormone metabolism during caloric restriction. *J. Biol. Chem.* **279**, 19832–19838.

105. Qatanani, M., Zhang, J., and Moore, D. D. (2005) Role of the constitutive androstane receptor in xenobiotic-induced thyroid hormone metabolism. *Endocrinology* **146**, 995–1002.

106. Xie, W., Radominska-Pandya, A., Shi, Y., Simon, C. M., Nelson, M. C., Ong, E. S., Waxman, D. J., and Evans, R. M. (2001) An essential role for nuclear receptors SXR/PXR in detoxification of cholestatic bile acids. *Proc. Natl Acad. Sci. USA* **98**, 3375–3380.

107. Goodwin, B., Gauthier, K. C., Umetani, M., Watson, M. A., Lochansky, M. I., Collins, J. L., Leitersdorf, E., Mangelsdorf, D. J., Kliewer, S. A., and Repa, J. J. (2003) Identification of bile acid precursors as endogenous ligands for the nuclear xenobiotic pregnane X receptor. *Proc. Natl Acad. Sci. USA* **100**, 223–228.

108. Trottier, J., Milkiewicz, P., Kaeding, J., Verreault, M., and Barbier, O. (2006) Coordinate regulation of hepatic bile acid oxidation and conjugation by nuclear receptors. *Mol. Pharm.* **3**, 212–222.

109. Wietholtz, H., Marschall, H. U., Sjovall, J., and Matern, S. (1996) Stimulation of bile acid 6 alpha-hydroxylation by rifampin. *J. Hepatol.* **24**, 713–718.

110. Khurana, S. and Singh, P. (2006) Rifampin is safe for treatment of pruritus due to chronic cholestasis: a meta-analysis of prospective randomized-controlled trials. *Liver Int.* **26**, 943–948.

111. Lépine, J., Bernard, O., Plante, M., Tetu, B., Pelletier, G., Labrie, F., Bélanger, A., and Guillemette, C. (2004) Specificity and regioselectivity of the conjugation of estradiol,

estrone, and their catecholestrogen and methoxyestrogen metabolites by human uridine diphospho-glucuronosyltransferases expressed in endometrium. *J. Clin. Endocrinol. Metab.* **89**, 5222–5232.

112. Falany, J. L., Macrina, N., and Falany, C. N. (2002) Regulation of MCF-7 breast cancer cell growth by beta-estradiol sulfation. *Breast Cancer Res. Treat.* **74**, 167–176.

113. Chen, J. and Raymond, K. (2006) Roles of rifampicin in drug–drug interactions: underlying molecular mechanisms involving the nuclear pregnane X receptor. *Ann. Clin. Microbiol. Antimicrob.* **5**, 3.

114. Dlodlo, R. A., Fujiwara, P. I., and Enarson, D. A. (2005) Should tuberculosis treatment and control be addressed differently in HIV-infected and -uninfected individuals? *Eur. Respir. J.* **25**, 751–757.

115. Davies, P. D. and Yew, W. W. (2003) Recent developments in the treatment of tuberculosis. *Expert Opin. Investig. Drugs* **12**, 1297–1312.

116. Gallicano, K. D., Sahai, J., Shukla, V. K., Seguin, I., Pakuts, A., Kwok, D., Foster, B. C., and Cameron, D. W. (1999) Induction of zidovudine glucuronidation and amination pathways by rifampicin in HIV-infected patients. *Br. J. Clin. Pharmacol.* **48**, 168–179.

117. Ebert, U., Thong, N. Q., Oertel, R., and Kirch, W. (2000) Effects of rifampicin and cimetidine on pharmacokinetics and pharmacodynamics of lamotrigine in healthy subjects. *Eur. J. Clin. Pharmacol.* **56**, 299–304.

118. Fromm, M. F., Eckhardt, K., Li, S., Schanzle, G., Hofmann, U., Mikus, G., and Eichelbaum, M. (1997) Loss of analgesic effect of morphine due to coadministration of rifampin. *Pain* **72**, 261–267.

119. Lin, J. H. and Lu, A. Y. (1998) Inhibition and induction of cytochrome P450 and the clinical implications. *Clin. Pharmacokinet.* **35**, 361–390.

120. Echchgadda, I., Song, C. S., Oh, T., Ahmed, M., De La Cruz, I. J., and Chatterjee, B. (2007) The xenobiotic-sensing nuclear receptors pregnane X receptor, constitutive androstane receptor, and orphan nuclear receptor hepatocyte nuclear factor 4alpha in the regulation of human steroid-/bile acid-sulfotransferase. *Mol. Endocrinol.* **21**, 2099–2111.

121. Echchgadda, I., Song, C. S., Oh, T. S., Cho, S. H., Rivera, O. J., and Chatterjee, B. (2004) Gene regulation for the senescence marker protein DHEA-sulfotransferase by the xenobiotic-activated nuclear pregnane X receptor (PXR). *Mech. Ageing Dev.* **125**, 733–745.

122. Echchgadda, I., Song, C. S., Roy, A. K., and Chatterjee, B. (2004) Dehydroepiandrosterone sulfotransferase is a target for transcriptional induction by the vitamin D receptor. *Mol. Pharmacol.* **65**, 720–729.

123. Ferraris, A., D'Amato, G., Nobili, V., Torres, B., Marcellini, M., and Dallapiccola, B. (2006) Combined test for UGT1A1 -3279T–>G and A(TA)nTAA polymorphisms best predicts Gilbert's syndrome in Italian pediatric patients. *Genet. Test* **10**, 121–125.

124. Kitagawa, C., Ando, M., Ando, Y., Sekido, Y., Wakai, K., Imaizumi, K., Shimokata, K., and Hasegawa, Y. (2005) Genetic polymorphism in the phenobarbital-responsive enhancer module of the UDP-glucuronosyltransferase 1A1 gene and irinotecan toxicity. *Pharmacogenetics Genomics* **15**, 35–41.

125. Pascussi, J. M., Drocourt, L., Fabre, J. M., Maurel, P., and Vilarem, M. J. (2000) Dexamethasone induces pregnane X receptor and retinoid X receptor-alpha expression in human hepatocytes: synergistic increase of CYP3A4 induction by pregnane X receptor activators. *Mol. Pharmacol.* **58**, 361–372.

126. Huss, J. M. and Kasper, C. B. (1998) Nuclear receptor involvement in the regulation of rat cytochrome P450 3A23 expression. *J. Biol. Chem.* **273**, 16155–16162.

127. Pascussi, J. M., Gerbal-Chaloin, S., Fabre, J. M., Maurel, P., and Vilarem, M. J. (2000) Dexamethasone enhances constitutive androstane receptor expression in human hepatocytes: consequences on cytochrome P450 gene regulation. *Mol. Pharmacol.* **58**, 1441–1450.

128. Sugatani, J., Yamakawa, K., Tonda, E., Nishitani, S., Yoshinari, K., Degawa, M., Abe, I., Noguchi, H., and Miwa, M. (2004) The induction of human UDP-glucuronosyltransferase 1A1 mediated through a distal enhancer module by flavonoids and xenobiotics. *Biochem. Pharmacol.* **67**, 989–1000.

129. Sugatani, J., Nishitani, S., Yamakawa, K., Yoshinari, K., Sueyoshi, T., Negishi, M., and Miwa, M. (2005) Transcriptional regulation of human UGT1A1 gene expression: activated glucocorticoid receptor enhances constitutive androstane receptor/pregnane X receptor-mediated UDP-glucuronosyltransferase 1A1 regulation with glucocorticoid receptor-interacting protein 1. *Mol. Pharmacol.* **67**, 845–855.

130. Nebert, D. W., Jaiswal, A. K., Meyer, U. A., and Gonzalez, F. J. (1987) Human P-450 genes: evolution, regulation and possible role in carcinogenesis. *Biochem. Soc. Trans.* **15**, 586–589.

131. Nebert, D. W., Roe, A. L., Dieter, M. Z., Solis, W. A., Yang, Y., and Dalton, T. P. (2000) Role of the aromatic hydrocarbon receptor and [Ah] gene battery in the oxidative stress response, cell cycle control, and apoptosis. *Biochem. Pharmacol.* **59**, 65–85.

132. Kewley, R. J., Whitelaw, M. L., and Chapman-Smith, A. (2004) The mammalian basic helix–loop–helix/PAS family of transcriptional regulators. *Int. J. Biochem. Cell Biol.* **36**, 189–204.

133. Kazlauskas, A., Poellinger, L., and Pongratz, I. (2000) The immunophilin-like protein XAP2 regulates ubiquitination and subcellular localization of the dioxin receptor. *J. Biol. Chem.* **275**, 41317–41724.

134. Gu, Y. Z., Hogenesch, J. B., and Bradfield, C. A. (2000) The PAS superfamily: sensors of environmental and developmental signals. *Annu. Rev. Pharmacol. Toxicol.* **40**, 519–561.

135. Hankinson, O. (1995) The aryl hydrocarbon receptor complex. *Annu. Rev. Pharmacol. Toxicol.* **35**, 307–340.

136. Whitlock, J. P., Jr. (1999) Induction of cytochrome P4501A1. *Annu. Rev. Pharmacol. Toxicol.* **39**, 103–125.

137. Yueh, M. F., Bonzo, J. A., and Tukey, R. H. (2005) The role of Ah receptor in induction of human UDP-glucuronosyltransferase 1A1. *Methods Enzymol.* **400**, 75–91.

138. Birnbaum, L. S. and Tuomisto, J. (2000) Non-carcinogenic effects of TCDD in animals. *Food Addit. Contam.* **17**, 275–288.

139. Ciolino, H. P., Daschner, P. J., and Yeh, G. C. (1999) Dietary flavonols quercetin and kaempferol are ligands of the aryl hydrocarbon receptor that affect CYP1A1 transcription differentially. *Biochem. J.* **340** (Pt. 3), 715–722.

140. Uda, Y., Price, K. R., Williamson, G., and Rhodes, M. J. (1997) Induction of the anti-carcinogenic marker enzyme, quinone reductase, in murine hepatoma cells in vitro by flavonoids. *Cancer Lett.* **120**, 213–216.

141. Zhao, J. and Agarwal, R. (1999) Tissue distribution of silibinin, the major active constituent of silymarin, in mice and its association with enhancement of phase II enzymes: implications in cancer chemoprevention. *Carcinogenesis* **20**, 2101–2108.

142. Moon, Y. J., Wang, X., and Morris, M. E. (2006) Dietary flavonoids: effects on xenobiotic and carcinogen metabolism. *Toxicol. In Vitro* **20**, 187–210.

143. Eisenbrand, G. and Tang, W. (1993) Food-borne heterocyclic amines. Chemistry, formation, occurrence and biological activities. A literature review. *Toxicology* **84**, 1–82.

144. Lijinsky, W. and Shubik, P. (1964) Benzo(a)pyrene and other polynuclear hydrocarbons in charcoal-broiled meat. *Science* **145**, 53–55.

145. Sugimura, T. (2000) Nutrition and dietary carcinogens. *Carcinogenesis* **21**, 387–395.

146. Yueh, M. F., Huang, Y. H., Hiller, A., Chen, S., Nguyen, N., and Tukey, R. H. (2003) Involvement of the xenobiotic response element (XRE) in Ah receptor-mediated induction of human UDP-glucuronosyltransferase 1A1. *J. Biol. Chem.* **278**, 15001–15006.

147. Sugatani, J., Sueyoshi, T., Negishi, M., and Miwa, M. (2005) Regulation of the human UGT1A1 gene by nuclear receptors constitutive active/androstane receptor, pregnane X receptor, and glucocorticoid receptor. *Methods. Enzymol.* **400**, 92–104.

148. Bonzo, J. A., Belanger, A., and Tukey, R. H. (2007) The role of chrysin and the Ah receptor in induction of the human UGT1A1 gene in vitro and in transgenic UGT1 mice. *Hepatology* **45**, 349–360.

149. Metz, R. P. and Ritter, J. K. (1998) Transcriptional activation of the UDP-glucuronosyltransferase 1A7 gene in rat liver by aryl hydrocarbon receptor ligands and oltipraz. *J. Biol. Chem.* **273**, 5607–5614.

150. Sohn, O. S., Surace, A., Fiala, E. S., Richie, J. P., Jr., Colosimo, S., Zang, E., and Weisburger, J. H. (1994) Effects of green and black tea on hepatic xenobiotic metabolizing systems in the male F344 rat. *Xenobiotica* **24**, 119–127.

151. Bu-Abbas, A., Clifford, M. N., Ioannides, C., and Walker, R. (1995) Stimulation of rat hepatic UDP-glucuronosyl transferase activity following treatment with green tea. *Food Chem. Toxicol.* **33**, 27–30.

152. Emi, Y., Ikushiro, S., and Kato, Y. (2007) Thyroxine-metabolizing rat uridine diphosphate-glucuronosyltransferase 1A7 is regulated by thyroid hormone receptor. *Endocrinology* **148**, 6124–6133.

153. Kohn, M. C. (2000) Effects of TCDD on thyroid hormone homeostasis in the rat. *Drug Chem. Toxicol.* **23**, 259–277.

154. Wade, M. G., Parent, S., Finnson, K. W., Foster, W., Younglai, E., McMahon, A., Cyr, D. G., and Hughes, C. (2002) Thyroid toxicity due to subchronic exposure to a complex mixture of 16 organochlorines, lead, and cadmium. *Toxicol. Sci.* **67**, 207–218.

155. Pesatori, A. C., Consonni, D., Bachetti, S., Zocchetti, C., Bonzini, M., Baccarelli, A., and Bertazzi, P. A. (2003) Short- and long-term morbidity and mortality in the population exposed to dioxin after the "Seveso accident." *Ind. Health* **41**, 127–138.

156. Johnson, E., Shorter, C., Bestervelt, L., Patterson, D., Needham, L., Piper, W., Lucier, G., and Nolan, C. (2001) Serum hormone levels in humans with low serum concentrations of 2,3,7,8-TCDD. *Toxicol. Ind. Health* **17**, 105–112.

157. Safe, S. and McDougal, A. (2002) Mechanism of action and development of selective aryl hydrocarbon receptor modulators for treatment of hormone-dependent cancers (Review). *Int. J. Oncol.* **20**, 1123–1128.

158. Safe, S. H. and Zacharewski, T. (1997) Organochlorine exposure and risk for breast cancer. *Prog. Clin. Biol. Res.* **396**, 133–145.

159. Lu, F., Zahid, M., Saeed, M., Cavalieri, E. L., and Rogan, E. G. (2007) Estrogen metabolism and formation of estrogen-DNA adducts in estradiol-treated MCF-10F cells. The effects of 2,3,7,8-tetrachlorodibenzo-p-dioxin induction and catechol-O-methyltransferase inhibition. *J. Steroid Biochem. Mol. Biol.* **105**, 150–158.

160. Tanaka, J., Yonemoto, J., Zaha, H., Kiyama, R., and Sone, H. (2007) Estrogen-responsive genes newly found to be modified by TCDD exposure in human cell lines and mouse systems. *Mol. Cell Endocrinol.* **272**, 38–49.

161. Maliakal, P. P., Coville, P. F., and Wanwimolruk, S. (2001) Tea consumption modulates hepatic drug metabolizing enzymes in Wistar rats. *J. Pharm. Pharmacol.* **53**, 569–577.

162. Morel, F., Fardel, O., Meyer, D. J., Langouet, S., Gilmore, K. S., Meunier, B., Tu, C. P., Kensler, T. W., Ketterer, B., and Guillouzo, A. (1993) Preferential increase of glutathione S-transferase class alpha transcripts in cultured human hepatocytes by phenobarbital, 3-methylcholanthrene, and dithiolethiones. *Cancer Res.* **53**, 231–234.

163. Schrenk, D., Stuven, T., Gohl, G., Viebahn, R., and Bock, K. W. (1995) Induction of CYP1A and glutathione S-transferase activities by 2,3,7,8-tetrachlorodibenzo-p-dioxin in human hepatocyte cultures. *Carcinogenesis* **16**, 943–946.

164. Gao, X., Dinkova-Kostova, A. T., and Talalay, P. (2001) Powerful and prolonged protection of human retinal pigment epithelial cells, keratinocytes, and mouse leukemia cells against oxidative damage: the indirect antioxidant effects of sulforaphane. *Proc. Natl Acad. Sci. USA* **98**, 15221–15226.

165. Ramos-Gomez, M., Kwak, M. K., Dolan, P. M., Itoh, K., Yamamoto, M., Talalay, P., and Kensler, T. W. (2001) Sensitivity to carcinogenesis is increased and chemoprotective efficacy of enzyme inducers is lost in nrf2 transcription factor-deficient mice. *Proc. Natl Acad. Sci. USA* **98**, 3410–3415.

166. Thimmulappa, R. K., Mai, K. H., Srisuma, S., Kensler, T. W., Yamamoto, M., and Biswal, S. (2002) Identification of Nrf2-regulated genes induced by the chemopreventive agent sulforaphane by oligonucleotide microarray. *Cancer Res.* **62**, 5196–5203.

167. Nguyen, T., Sherratt, P. J., and Pickett, C. B. (2003) Regulatory mechanisms controlling gene expression mediated by the antioxidant response element. *Annu. Rev. Pharmacol. Toxicol.* **43**, 233–260.

168. Ma, Q., Kinneer, K., Bi, Y., Chan, J. Y., and Kan, Y. W. (2004) Induction of murine NAD(P)H:quinone oxidoreductase by 2,3,7,8-tetrachlorodibenzo-p-dioxin requires the CNC (cap 'n' collar) basic leucine zipper transcription factor Nrf2 (nuclear factor erythroid 2-related factor 2): cross-interaction between AhR (aryl hydrocarbon receptor) and Nrf2 signal transduction. *Biochem. J.* **377**, 205–213.

169. Kohle, C. and Bock, K. W. (2006) Activation of coupled Ah receptor and Nrf2 gene batteries by dietary phytochemicals in relation to chemoprevention. *Biochem. Pharmacol.* **72**, 795–805.

170. Miao, W., Hu, L., Scrivens, P. J., and Batist, G. (2005) Transcriptional regulation of NF-E2 p45-related factor (NRF2) expression by the aryl hydrocarbon receptor-xenobiotic response element signaling pathway: direct cross-talk between phase I and II drug-metabolizing enzymes. *J. Biol. Chem.* **280**, 20340–20348.

171. Radjendirane, V. and Jaiswal, A. K. (1999) Antioxidant response element-mediated 2,3,7,8-tetrachlorodibenzo-p-dioxin (TCDD) induction of human NAD(P)H:quinone oxidoreductase 1 gene expression. *Biochem. Pharmacol.* **58**, 1649–1655.

172. Ross, D. (2004) Quinone reductases multitasking in the metabolic world. *Drug Metab. Rev.* **36**, 639–654.

173. Friling, R. S., Bensimon, A., Tichauer, Y., and Daniel, V. (1990) Xenobiotic-inducible expression of murine glutathione S-transferase Ya subunit gene is controlled by an electrophile-responsive element. *Proc. Natl Acad. Sci. USA* **87**, 6258–6262.

174. Rushmore, T. H., Morton, M. R., and Pickett, C. B. (1991) The antioxidant responsive element. Activation by oxidative stress and identification of the DNA consensus sequence required for functional activity. *J. Biol. Chem.* **266**, 11632–11639.

175. Wasserman, W. W. and Fahl, W. E. (1997) Functional antioxidant responsive elements. *Proc. Natl Acad. Sci. USA* **94**, 5361–5366.

176. Itoh, K., Wakabayashi, N., Katoh, Y., Ishii, T., Igarashi, K., Engel, J. D., and Yamamoto, M. (1999) Keap1 represses nuclear activation of antioxidant responsive elements by Nrf2 through binding to the amino-terminal Neh2 domain. *Genes Dev.* **13**, 76–86.

177. Kang, M. I., Kobayashi, A., Wakabayashi, N., Kim, S. G., and Yamamoto, M. (2004) Scaffolding of Keap1 to the actin cytoskeleton controls the function of Nrf2 as key regulator of cytoprotective phase 2 genes. *Proc. Natl Acad. Sci. USA* **101**, 2046–2051.

178. Venugopal, R. and Jaiswal, A. K. (1996) Nrf1 and Nrf2 positively and c-Fos and Fra1 negatively regulate the human antioxidant response element-mediated expression of NAD(P)H:quinone oxidoreductase1 gene. *Proc. Natl Acad. Sci. USA* **93**, 14960–14965.

179. Itoh, K., Chiba, T., Takahashi, S., Ishii, T., Igarashi, K., Katoh, Y., Oyake, T., Hayashi, N., Satoh, K., Hatayama, I., et al. (1997) An Nrf2/small Maf heterodimer mediates the induction of phase II detoxifying enzyme genes through antioxidant response elements. *Biochem. Biophys. Res. Commun.* **236**, 313–322.

180. Jeyapaul, J. and Jaiswal, A. K. (2000) Nrf2 and c-Jun regulation of antioxidant response element (ARE)-mediated expression and induction of gamma-glutamylcysteine synthetase heavy subunit gene. *Biochem. Pharmacol.* **59**, 1433–1439.

181. Zhu, M. and Fahl, W. E. (2001) Functional characterization of transcription regulators that interact with the electrophile response element. *Biochem. Biophys. Res. Commun.* **289**, 212–219.

182. Chan, J. Y. and Kwong, M. (2000) Impaired expression of glutathione synthetic enzyme genes in mice with targeted deletion of the Nrf2 basic-leucine zipper protein. *Biochim. Biophys. Acta* **1517**, 19–26.

183. Chanas, S. A., Jiang, Q., McMahon, M., McWalter, G. K., McLellan, L. I., Elcombe, C. R., Henderson, C. J., Wolf, C. R., Moffat, G. J., Itoh, K., et al. (2002) Loss of the Nrf2 transcription factor causes a marked reduction in constitutive and inducible expression of the glutathione S-transferase Gsta1, Gsta2, Gstm1, Gstm2, Gstm3 and Gstm4 genes in the livers of male and female mice. *Biochem. J.* **365**, 405–416.

184. McMahon, M., Itoh, K., Yamamoto, M., Chanas, S. A., Henderson, C. J., McLellan, L. I., Wolf, C. R., Cavin, C., and Hayes, J. D. (2001) The Cap'n'Collar basic leucine zipper transcription factor Nrf2 (NF-E2 p45-related factor 2) controls both constitutive and inducible expression of intestinal detoxification and glutathione biosynthetic enzymes. *Cancer Res.* **61**, 3299–3307.

185. Bingham, S. and Riboli, E. (2004) Diet and cancer—the European prospective investigation into cancer and nutrition. *Nat. Rev. Cancer* **4**, 206–215.

186. Wakabayashi, N., Itoh, K., Wakabayashi, J., Motohashi, H., Noda, S., Takahashi, S., Imakado, S., Kotsuji, T., Otsuka, F., Roop, D. R., et al. (2003) Keap1-null mutation leads to postnatal lethality due to constitutive Nrf2 activation. *Nat. Genet.* **35**, 238–245.

187. Okawa, H., Motohashi, H., Kobayashi, A., Aburatani, H., Kensler, T. W., and Yamamoto, M. (2006) Hepatocyte-specific deletion of the keap1 gene activates Nrf2 and confers potent resistance against acute drug toxicity. *Biochem. Biophys. Res. Commun.* **339**, 79–88.

188. Uchida, K. (2003) 4-Hydroxy-2-nonenal: a product and mediator of oxidative stress. *Prog. Lipid Res.* **42**, 318–343.

189. Enomoto, A., Itoh, K., Nagayoshi, E., Haruta, J., Kimura, T., O'Connor, T., Harada, T., and Yamamoto, M. (2001) High sensitivity of Nrf2 knockout mice to acetaminophen hepatotoxicity associated with decreased expression of ARE-regulated drug metabolizing enzymes and antioxidant genes. *Toxicol. Sci.* **59**, 169–177.

190. Chan, K. and Kan, Y. W. (1999) Nrf2 is essential for protection against acute pulmonary injury in mice. *Proc. Natl Acad. Sci. USA* **96**, 12731–12736.

191. Munzel, P. A., Schmohl, S., Buckler, F., Jaehrling, J., Raschko, F. T., Kohle, C., and Bock, K. W. (2003) Contribution of the Ah receptor to the phenolic antioxidant-mediated expression of human and rat UDP-glucuronosyltransferase UGT1A6 in Caco-2 and rat hepatoma 5L cells. *Biochem. Pharmacol.* **66**, 841–847.

192. Chan, K., Han, X. D., and Kan, Y. W. (2001) An important function of Nrf2 in combating oxidative stress: detoxification of acetaminophen. *Proc. Natl Acad. Sci. USA* **98**, 4611–4616.

193. Shelby, M. K. and Klaassen, C. D. (2006) Induction of rat UDP-glucuronosyltransferases in liver and duodenum by microsomal enzyme inducers that activate various transcriptional pathways. *Drug Metab. Dispos.* **34**, 1772–1778.

194. Yueh, M. F. and Tukey, R. H. (2007) Nrf2-Keap1 signaling pathway regulates human UGT1A1 expression in vitro and in transgenic UGT1 mice. *J. Biol. Chem.* **282**, 8749–8758.

195. Ma, Q., Battelli, L., and Hubbs, A. F. (2006) Multiorgan autoimmune inflammation, enhanced lymphoproliferation, and impaired homeostasis of reactive oxygen species in mice lacking the antioxidant-activated transcription factor Nrf2. *Am. J. Pathol.* **168**, 1960–1974.

196. Corton, J. C., Anderson, S. P., and Stauber, A. (2000) Central role of peroxisome proliferator-activated receptors in the actions of peroxisome proliferators. *Annu. Rev. Pharmacol. Toxicol.* **40**, 491–518.

197. Lemberger, T., Desvergne, B., and Wahli, W. (1996) Peroxisome proliferator-activated receptors: a nuclear receptor signaling pathway in lipid physiology. *Annu. Rev. Cell Dev. Biol.* **12**, 335–363.

198. Kota, B. P., Huang, T. H., and Roufogalis, B. D. (2005) An overview on biological mechanisms of PPARs. *Pharmacol. Res.* **51**, 85–94.

199. Braissant, O., Foufelle, F., Scotto, C., Dauca, M., and Wahli, W. (1996) Differential expression of peroxisome proliferator-activated receptors (PPARs): tissue distribution of PPAR-alpha, -beta, and -gamma in the adult rat. *Endocrinology* **137**, 354–366.

200. Gearing, K. L., Gottlicher, M., Teboul, M., Widmark, E., and Gustafsson, J. A. (1993) Interaction of the peroxisome-proliferator-activated receptor and retinoid X receptor. *Proc. Natl Acad. Sci. USA* **90**, 1440–1444.

201. Barbier, O., Fontaine, C., Fruchart, J. C., and Staels, B. (2004) Genomic and non-genomic interactions of PPARalpha with xenobiotic-metabolizing enzymes. *Trends Endocrinol. Metab.* **15**, 324–330.

202. Barbier, O., Villeneuve, L., Bocher, V., Fontaine, C., Torra, I. P., Duhem, C., Kosykh, V., Fruchart, J. C., Guillemette, C., and Staels, B. (2003) The UDP-glucuronosyltransferase 1A9 enzyme is a peroxisome proliferator-activated receptor alpha and gamma target gene. *J. Biol. Chem.* **278**, 13975–13983.

203. Trottier, J., Verreault, M., Grepper, S., Monté, D., Bélanger, J., Kaeding, J., Caron, P., Inaba, T., and Barbier, O. (2006) Human UDP-glucuronosyltransferase (UGT)1A3 enzyme conjugates chenodeoxycholic in the liver. *Hepatology* **44**, 1158–1170.

204. Senekeo-Effenberger, K., Chen, S., Brace-Sinnokrak, E., Bonzo, J. A., Yueh, M. F., Argikar, U., Kaeding, J., Trottier, J., Remmel, R. P., Ritter, J. K., et al. (2007) Expression of the human UGT1 locus in transgenic mice by 4-chloro-6-(2,3-xylidino)-2-pyrimidinylthioacetic acid (WY-14643) and implications on drug metabolism through peroxisome proliferator-activated receptor alpha activation. *Drug Metab. Dispos.* **35**, 419–427.

205. Barbier, O., Torra, I. P., Duguay, Y., Blanquart, C., Fruchart, J. C., Glineur, C., and Staels, B. (2002) Pleiotropic actions of peroxisome proliferator-activated receptors in lipid metabolism and atherosclerosis. *Arterioscler. Thromb. Vasc. Biol.* **22**, 717–726.

206. Turgeon, D., Chouinard, S., Belanger, P., Picard, S., Labbe, J. F., Borgeat, P., and Belanger, A. (2003) Glucuronidation of arachidonic and linoleic acid metabolites by human UDP-glucuronosyltransferases. *J. Lipid Res.* **44**, 1182–1191.

207. Barbier, O., Duran-Sandoval, D., Pineda-Torra, I., Kosykh, V., Fruchart, J. C., and Staels, B. (2003) Peroxisome proliferator-activated receptor alpha induces hepatic expression of the human bile acid glucuronidating UDP-glucuronosyltransferase 2B4 enzyme. *J. Biol. Chem.* **278**, 32852–32860.

208. Chapman, M. J. (2003) Fibrates in 2003: therapeutic action in atherogenic dyslipidaemia and future perspectives. *Atherosclerosis* **171**, 1–13.

209. Ogilvie, B. W., Zhang, D., Li, W., Rodrigues, A. D., Gipson, A. E., Holsapple, J., Toren, P., and Parkinson, A. (2006) Glucuronidation converts gemfibrozil to a potent, metabolism-dependent inhibitor of CYP2C8: implications for drug–drug interactions. *Drug Metab. Dispos.* **34**, 191–197.

210. Sakai, M., Matsushima-Hibiya, Y., Nishizawa, M., and Nishi, S. (1995) Suppression of rat glutathione transferase P expression by peroxisome proliferators: interaction between Jun and peroxisome proliferator-activated receptor alpha. *Cancer Res.* **55**, 5370–5376.

211. Fan, L. Q., You, L., Brown-Borg, H., Brown, S., Edwards, R. J., and Corton, J. C. (2004) Regulation of phase I and phase II steroid metabolism enzymes by PPAR alpha activators. *Toxicology* **204**, 109–121.

212. Jiang, Y. J., Kim, P., Elias, P. M., and Feingold, K. R. (2005) LXR and PPAR activators stimulate cholesterol sulfotransferase type 2 isoform 1b in human keratinocytes. *J. Lipid Res.* **46**, 2657–2666.

213. Dixit, S. G., Tirona, R. G., and Kim, R. B. (2005) Beyond CAR and PXR. *Curr. Drug Metab.* **6**, 385–397.

214. Sporn, M. B., Suh, N., and Mangelsdorf, D. J. (2001) Prospects for prevention and treatment of cancer with selective PPARgamma modulators (SPARMs). *Trends Mol. Med.* **7**, 395–400.

215. Barbier, O., Girard, H., Inoue, Y., Duez, H., Villeneuve, L., Kamiya, A., Fruchart, J. C., Guillemette, C., Gonzalez, F. J., and Staels, B. (2005) Hepatic expression of the UGT1A9 gene is governed by hepatocyte nuclear factor 4alpha. *Mol. Pharmacol.* **67**, 241–249.

216. Ericsson, H., Hamren, B., Bergstrand, S., Elebring, M., Fryklund, L., Heijer, M., and Ohman, K. P. (2004) Pharmacokinetics and metabolism of tesaglitazar, a novel dual-acting peroxisome proliferator-activated receptor alpha/gamma agonist, after a single oral and intravenous dose in humans. *Drug Metab. Dispos.* **32**, 923–929.

217. Paumi, C. M., Smitherman, P. K., Townsend, A. J., and Morrow, C. S. (2004) Glutathione S-transferases (GSTs) inhibit transcriptional activation by the peroxisomal proliferator-activated receptor gamma (PPAR gamma) ligand, 15-deoxy-delta 12,14prostaglandin J2 (15-d-PGJ2). *Biochemistry* **43**, 2345–2352.

218. Kliewer, S. A. and Willson, T. M. (2002) Regulation of xenobiotic and bile acid metabolism by the nuclear pregnane X receptor. *J. Lipid Res.* **43**, 359–364.

219. Barbier, O., Torra, I. P., Sirvent, A., Claudel, T., Blanquart, C., Duran-Sandoval, D., Kuipers, F., Kosykh, V., Fruchart, J. C., and Staels, B. (2003) FXR induces the UGT2B4 enzyme in hepatocytes: a potential mechanism of negative feedback control of FXR activity. *Gastroenterology* **124**, 1926–1940.

220. Lu, Y., Heydel, J. M., Li, X., Bratton, S., Lindblom, T., and Radominska-Pandya, A. (2005) Lithocholic acid decreases expression of UGT2B7 in Caco-2 cells: a potential role for a negative farnesoid X receptor response element. *Drug Metab. Dispos.* **33**, 937–946.

221. Kitada, H., Miyata, M., Nakamura, T., Tozawa, A., Honma, W., Shimada, M., Nagata, K., Sinal, C. J., Guo, G. L., Gonzalez, F. J., et al. (2003) Protective role of hydroxysteroid sulfotransferase in lithocholic acid-induced liver toxicity. *J. Biol. Chem.* **278**, 17838–17844.

222. Uppal, H., Saini, S. P., Moschetta, A., Mu, Y., Zhou, J., Gong, H., Zhai, Y., Ren, S., Michalopoulos, G. K., Mangelsdorf, D. J., et al. (2007) Activation of LXRs prevents bile acid toxicity and cholestasis in female mice. *Hepatology* **45**, 422–432.

223. Verreault, M., Senekeo-Effenberger, K., Trottier, J., Bonzo, J. A., Bélanger, J., Kaeding, J., Staels, B., Caron, P., Tukey, R. H., and Barbier, O. (2006) The liver X-receptor alpha controls hepatic expression of the human bile acid-glucuronidating UGT1A3 enzyme in human cells and transgenic mice. *Hepatology* **44**, 368–378.

224. Gong, H., Guo, P., Zhai, Y., Zhou, J., Uppal, H., Jarzynka, M. J., Song, W. C., Cheng, S. Y., and Xie, W. (2007) Estrogen deprivation and inhibition of breast cancer growth in vivo through activation of the orphan nuclear receptor liver X receptor. *Mol. Endocrinol.* **21**, 1781–1790.

225. Kaeding, J., Bouchaert, E., Bélanger, J., Caron, P., Chouinard, S., Verreault, M., Larouche, O., Pelletier, G., Staels, B., Bélanger, A., et al. (2008) Activators of the farnesoid X receptor negatively regulate androgen glucuronidation in human prostate cancer LNCAP cells. *Biochem. J.* **410**, 245–253.

226. Frain, M., Swart, G., Monaci, P., Nicosia, A., Stampfli, S., Frank, R., and Cortese, R. (1989) The liver-specific transcription factor LF-B1 contains a highly diverged homeobox DNA binding domain. *Cell* **59**, 145–157.

227. Bartkowski, S., Zapp, D., Weber, H., Eberle, G., Zoidl, C., Senkel, S., Klein-Hitpass, L., and Ryffel, G. U. (1993) Developmental regulation and tissue distribution of the liver transcription factor LFB1 (HNF1) in Xenopus laevis. *Mol. Cell Biol.* **13**, 421–431.

228. Yamagata, K., Furuta, H., Oda, N., Kaisaki, P. J., Menzel, S., Cox, N. J., Fajans, S. S., Signorini, S., Stoffel, M., and Bell, G. I. (1996) Mutations in the hepatocyte nuclear factor-4alpha gene in maturity-onset diabetes of the young (MODY1). *Nature* **384**, 458–460.

229. Yamagata, K., Oda, N., Kaisaki, P. J., Menzel, S., Furuta, H., Vaxillaire, M., Southam, L., Cox, R. D., Lathrop, G. M., Boriraj, V. V., et al. (1996) Mutations in the hepatocyte nuclear factor-1alpha gene in maturity-onset diabetes of the young (MODY3). *Nature* **384**, 455–458.

230. Kanou, M., Usui, T., Ueyama, H., Sato, H., Ohkubo, I., and Mizutani, T. (2004) Stimulation of transcriptional expression of human UDP-glucuronosyltransferase 1A1 by dexamethasone. *Mol. Biol. Rep.* **31**, 151–158.

231. Auyeung, D. J., Kessler, F. K., and Ritter, J. K. (2001) An alternative promoter contributes to tissue- and inducer-specific expression of the rat UDP-glucuronosyltransferase 1A6 gene. *Toxicol. Appl. Pharmacol.* **174**, 60–68.

232. Gregory, P. A., Lewinsky, R. H., Gardner-Stephen, D. A., and Mackenzie, P. I. (2004) Regulation of UDP glucuronosyltransferases in the gastrointestinal tract. *Toxicol. Appl. Pharmacol.* **199**, 354–363.

233. Metz, R. P., Auyeung, D. J., Kessler, F. K., and Ritter, J. K. (2000) Involvement of hepatocyte nuclear factor 1 in the regulation of the UDP-glucuronosyltransferase 1A7 (UGT1A7) gene in rat hepatocytes. *Mol. Pharmacol.* **58**, 319–327.

234. Hansen, A. J., Lee, Y. H., Gonzalez, F. J., and Mackenzie, P. I. (1997) HNF1 alpha activates the rat UDP glucuronosyltransferase UGT2B1 gene promoter. *DNA Cell Biol.* **16**, 207–214.

235. Hansen, A. J., Lee, Y. H., Sterneck, E., Gonzalez, F. J., and Mackenzie, P. I. (1998) C/EBPalpha is a regulator of the UDP glucuronosyltransferase UGT2B1 gene. *Mol. Pharmacol.* **53**, 1027–1033.

236. Ishii, Y., Hansen, A. J., and Mackenzie, P. I. (2000) Octamer transcription factor-1 enhances hepatic nuclear factor-1alpha-mediated activation of the human UDP glucuronosyltransferase 2B7 promoter. *Mol. Pharmacol.* **57**, 940–947.

237. Chen, W. S., Manova, K., Weinstein, D. C., Duncan, S. A., Plump, A. S., Prezioso, V. R., Bachvarova, R. F., and Darnell, J. E., Jr. (1994) Disruption of the HNF-4 gene, expressed in visceral endoderm, leads to cell death in embryonic ectoderm and impaired gastrulation of mouse embryos. *Genes Dev.* **8**, 2466–2477.

238. Odom, D. T., Zizlsperger, N., Gordon, D. B., Bell, G. W., Rinaldi, N. J., Murray, H. L., Volkert, T. L., Schreiber, J., Rolfe, P. A., Gifford, D. K., et al. (2004) Control of pancreas and liver gene expression by HNF transcription factors. *Science* **303**, 1378–1381.

239. Kamiyama, Y., Matsubara, T., Yoshinari, K., Nagata, K., Kamimura, H., and Yamazoe, Y. (2007) Role of human hepatocyte nuclear factor 4alpha in the expression of drug-metabolizing enzymes and transporters in human hepatocytes assessed by use of small interfering RNA. *Drug Metab. Pharmacokinet.* **22**, 287–298.

240. Kamiya, A., Inoue, Y., and Gonzalez, F. J. (2003) Role of the hepatocyte nuclear factor 4alpha in control of the pregnane X receptor during fetal liver development. *Hepatology* **37**, 1375–1384.

241. Li, J., Ning, G., and Duncan, S. A. (2000) Mammalian hepatocyte differentiation requires the transcription factor HNF-4alpha. *Genes Dev.* **14**, 464–474.

242. Bhalla, S., Ozalp, C., Fang, S., Xiang, L., and Kemper, J. K. (2004) Ligand-activated pregnane X receptor interferes with HNF-4 signaling by targeting a common coactivator

PGC-1alpha. Functional implications in hepatic cholesterol and glucose metabolism. *J. Biol. Chem.* **279**, 45139–45147.

243. Li, T. and Chiang, J. Y. (2005) Mechanism of rifampicin and pregnane X receptor inhibition of human cholesterol 7 alpha-hydroxylase gene transcription. *Am. J. Physiol. Gastrointest. Liver Physiol.* **288**, G74-G84.

244. Miao, J., Fang, S., Bae, Y., and Kemper, J. K. (2006) Functional inhibitory cross-talk between constitutive androstane receptor and hepatic nuclear factor-4 in hepatic lipid/glucose metabolism is mediated by competition for binding to the DR1 motif and to the common coactivators, GRIP-1 and PGC-1alpha. *J. Biol. Chem.* **281**, 14537–14546.

245. Fang, H. L., Strom, S. C., Ellis, E., Duanmu, Z., Fu, J., Duniec-Dmuchowski, Z., Falany, C. N., Falany, J. L., Kocarek, T. A., and Runge-Morris, M. (2007) Positive and negative regulation of human hepatic hydroxysteroid sulfotransferase (SULT2A1) gene transcription by rifampicin: roles of hepatocyte nuclear factor 4alpha and pregnane X receptor. *J. Pharmacol. Exp. Ther.* **323**, 586–598.

246. Liu, L. and Klaassen, C. D. (1996) Regulation of hepatic sulfotransferases by steroidal chemicals in rats. *Drug Metab. Dispos.* **24**, 854–858.

247. Runge-Morris, M., Rose, K., and Kocarek, T. A. (1996) Regulation of rat hepatic sulfotransferase gene expression by glucocorticoid hormones. *Drug Metab. Dispos.* **24**, 1095–1101.

248. Runge-Morris, M., Wu, W., and Kocarek, T. A. (1999) Regulation of rat hepatic hydroxysteroid sulfotransferase (SULT2-40/41) gene expression by glucocorticoids: evidence for a dual mechanism of transcriptional control. *Mol. Pharmacol.* **56**, 1198–1206.

249. Duanmu, Z., Kocarek, T. A., and Runge-Morris, M. (2001) Transcriptional regulation of rat hepatic aryl sulfotransferase (SULT1A1) gene expression by glucocorticoids. *Drug Metab. Dispos.* **29**, 1130–1135.

250. Chouinard, S., Pelletier, G., Bélanger, A., and Barbier, O. (2006) Isoform-specific regulation of uridine diphosphate-glucuronosyltransferase 2B enzymes in the human prostate: differential consequences for androgen and bioactive lipid inactivation. *Endocrinology* **147**, 5431–5442.

251. Chouinard, S., Barbier, O., and Bélanger, A. (2007) UDP-glucuronosyltransferase 2B15 (UGT2B15) and UGT2B17 enzymes are major determinants of the androgen response in prostate cancer LNCaP cells. *J. Biol. Chem.* **282**, 33466–33474.

252. Harrington, W. R., Sengupta, S., and Katzenellenbogen, B. S. (2006) Estrogen regulation of the glucuronidation enzyme UGT2B15 in estrogen receptor-positive breast cancer cells. *Endocrinology* **147**, 3843–3850.

253. Starlard-Davenport, A., Lyn-Cook, B., and Radominska-Pandya, A. (2008) Novel identification of UDP-glucuronosyltransferase 1A10 as an estrogen-regulated target gene. *Steroids* **73**, 139–147.

254. Adams, J. B., Vrahimis, R., and Phillips, N. (1992) Regulation of estrogen sulfotransferase by estrogen in MCF-7 human mammary cancer cells. *Breast Cancer Res. Treat.* **22**, 157–161.

255. Seely, J., Amigh, K. S., Suzuki, T., Mayhew, B., Sasano, H., Giguere, V., Laganiere, J., Carr, B. R., and Rainey, W. E. (2005) Transcriptional regulation of dehydroepiandrosterone sulfotransferase (SULT2A1) by estrogen-related receptor alpha. *Endocrinology* **146**, 3605–3613.

256. Labrie, F., Luu-The, V., Bélanger, A., Lin, S. X., Simard, J., Pelletier, G., and Labrie, C. (2005) Is dehydroepiandrosterone a hormone? *J. Endocrinol.* **187**, 169–196.

257. Lu, C. and Li, A. P. (2001) Species comparison in P450 induction: effects of dexamethasone, omeprazole, and rifampin on P450 isoforms 1A and 3A in primary cultured hepatocytes from man, Sprague–Dawley rat, minipig, and beagle dog. *Chem. Biol. Interact.* **134**, 271–281.

4

NUCLEAR RECEPTOR-MEDIATED REGULATION OF DRUG TRANSPORTERS

OLIVER BURK

Department of Pharmacogenomics, Dr Margarete Fischer-Bosch Institute of Clinical Pharmacology, Stuttgart, Germany

4.1 INTRODUCTION

Drug disposition depends not only on the metabolism of a compound but also on active uptake and efflux by specific transporter proteins. Transporters are limiting intestinal drug absorption, mediating drug uptake into hepatocytes and elimination into bile, and participate in excretion and reabsorption processes in kidney. Microsomal enzyme inducers, which are well known to increase the expression of drug-metabolizing enzymes in liver and intestine, also induce the respective expression of drug transporters, thus resulting in limited bioavailability, enhanced hepatic uptake, and biliary excretion. The coordinated induction of drug metabolism and drug absorption and elimination efficiently protects the body from potentially harmful chemicals. In this chapter, our current knowledge of drug transporter induction by microsomal enzyme inducers is discussed. The main focus is on induction mediated by the nuclear receptors pregnane X receptor (PXR) and constitutive androstane receptor (CAR), which are activated by a large number of chemicals, including microsomal enzyme inducers of phenobarbital and glucocorticoids type.

Nuclear Receptors in Drug Metabolism   Edited by Wen Xie
Copyright © 2009 John Wiley & Sons, Inc.

4.2 DRUG TRANSPORTERS

Transporters of several gene families mediate the cellular uptake and export of chemicals. This section focuses on xenobiotic transporters of the ATP binding cassette (ABC), organic anion transporting polypeptide (*SLCO*/OATP), and the organic ion transporter (SLC22) families, which accept many commonly used drugs as substrates. ABC transporters mediate the ATP-dependent efflux from the inside of cells to the outside, whereas OATP and SLC22 transporter mediate the uptake into the cell (Figure 4.1). These transporters are highly expressed in liver, intestine, and kidney, organs that determine the absorption and elimination of chemicals. Additionally, they are expressed in blood–organ barriers, e.g., the blood–brain barrier, which further guard especially sensitive organs. Thus, they protect the body from the toxic actions

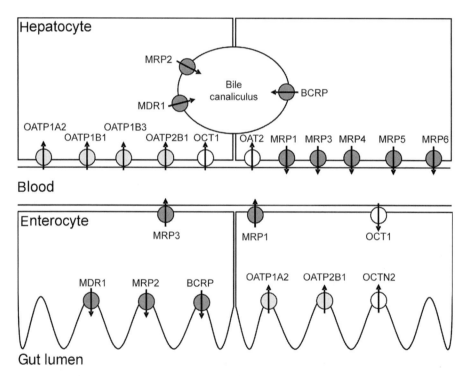

FIGURE 4.1 Localization of drug transporters in human liver and intestine. ABC efflux transporters MDR1/P-gp, MRP2, and BCRP are localized in the apical membrane of enterocytes and in the canalicular membrane of hepatocytes and thus limit drug absorption in the intestine and mediate elimination into bile. MRP1 and MRP3–6 are localized to the sinusoidal membrane of hepatocytes and transport chemicals into blood. Similarly, intestinal MRP1 and MRP3 in the basolateral membrane may play a role in the transport of chemicals from the gut lumen into blood. The functional role of apical OATP expression in enterocytes has not yet been defined; however, in liver the basolateral OATPs are believed to mediate the uptake of chemicals, together with OCT1 and OAT2. Apical OCTN2 in enterocytes is essential for the uptake of carnitine and may play an additional role in drug uptake.

of xenobiotics, often in concert with drug-metabolizing enzymes, which sometimes first have to convert the xenobiotic into a metabolite, which is suitable for transport.

4.2.1   ABC Efflux Transporters

Members of the ABC family, which are involved in the cellular export of drugs, comprise multidrug resistance (MDR) 1/P-glycoprotein, the multidrug resistance associated proteins (MRP), and breast cancer resistance protein (BCRP).

The human genome contains a single *MDR1* (*ABCB1*) gene, whereas two orthologs of human *MDR1* exist in rodents, which are denoted as *Mdr1a* (*Abcb1a*) and *Mdr1b* (*Abcb1b*) [1]. MDR1/P-glycoprotein (P-gp) plays an important role in the efflux of hydrophobic chemicals and peptides from the inside to the outside of cells. Initially it was discovered as the protein which bestows resistance against cytostatic drugs to cancer cells. Later on, it was shown that MDR1/P-gp is also expressed in nonmalignant cells of various organs. It is expressed in the brush border membrane of mature enterocytes in the intestine [2], where it limits absorption of chemicals from the gut lumen [3–5]. It was further shown that intestinal MDR1/P-gp also mediates drug elimination from the systemic circulation by secretion into the lumen of the gut [5]. In the liver, MDR1/P-gp is expressed in the canalicular membrane of hepatocytes [2], where it participates in biliary excretion [5, 6]. The proximal renal tubule cells of the kidney express MDR1/P-gp in their apical membranes [2], consistent with a role for the transporter in renal elimination [7]. Expression of MDR1/P-gp in the luminal membrane of endothelial cells of brain capillaries [8] limits the uptake of drugs into the CNS [9–11]. In general, localization of MDR1/P-gp expression argues for a physiological role of the transporter as a defense mechanism against the uptake of potentially harmful substances. MDR1/P-gp is likely to functionally interact with CYP3A drug-metabolizing enzymes, as both are coexpressed in intestine, liver, and kidney, show a broad overlap in their respective substrate specificities, and are often coinduced by the same microsomal enzyme MDR1/P-gp inducers. Especially in the intestine, the combined activities of CYP3A and MDR/P-gp constitute a functional barrier that protects the body from absorption of harmful compounds [12]. MDR1/P-gp transports an extremely broad range of hydrophobic chemicals, including many cytostatic anticancer drugs, HIV-protease inhibitors, antibiotics, cardiac drugs (digoxin, talinolol), calcium channel blockers, and immunosuppressive agents [13, 14].

MRPs belong to the ABCC subfamily, which consists of 13 members in humans, including MRP1–9 (ABCC1–6, ABCC10–12) [15]. MRPs are efflux transporters for structurally diverse amphipathic chemicals and organic anions. MRP2 is the only member of that subfamily that is consistently localized in the apical membrane of enterocytes, hepatocytes, and renal proximal tubule cells and thus participates in the secretion of organic anions back into the lumen of the gut, as well as into bile and urine [16–19]. MRP2 transports numerous anionic compounds, including drugs and conjugated drug metabolites. In liver, MRP2 also plays a role in bile acid secretion and bilirubin excretion. The importance of MRP2 for these processes is exemplified by the conjugated hyperbilirubinemia seen in patients with the Dubin–Johnson syndrome, who do not express functional MRP2 protein due to mutations in the *MRP2* gene [20]. The coexpression and coregulation of phase II drug metabolizing enzymes, which catalyze drug conjugation, and MRP2, which exports conjugated drugs, in intestine, liver, and kidney indicate a possible synergism in detoxification and excretion [13, 21]. In contrast to MRP2, MRP1 and MRP3–6 are localized in the basolateral (sinusoidal) membrane of hepatocytes, with MRP1 and MRP3 also being expressed in the respective membrane of enterocytes [22]. Therefore, these MRPs, especially MRP1 and MRP3, participate in the efflux of chemicals into the blood for subsequent renal elimination. If normal biliary elimination of toxic metabolites is impaired, as, for example, experimentally executed in animals by bile duct ligation or in Dubin–Johnson patients, secretion into blood and further renal elimination provide an important secondary mechanism of detoxification, which is adaptively up-regulated by an increase in hepatic MRP1 and MRP3 expression [23]. MRP1 is transporting cytotoxic drugs, as doxorubicin and vincristine, as well as conjugated metabolites such as bilirubin glucuronides [13, 24]. MRP3 transports a wide range of bile salts and some anticancer drugs, e.g., etoposide, vincristine, and methotrexate [13]. In humans, MRP4 is expressed at moderate levels in kidney with lower expression in liver and intestine [25]. It has been shown to mediate resistance to purine analogs that are used in cancer chemotherapy such as 6-mercaptopurine and thioguanine. Additionally, it transports nucleoside-based antiviral drugs and methotrexate [13]. MRP5 is widely expressed, including liver, colon, and kidney [13]. Similar to MRP4, it transports 6-mercaptopurine and thioguanine [13]. MRP6 is moderately expressed in liver and kidney, but it also shows low expression in other tissues including intestine [25]. It transports cytotoxic drugs that are also substrates for MRP1–3 [13].

ABCG2 (BCRP) is the most highly expressed ABC transporter in human small intestine [25, 26]; however also showing expression in liver, kidney, placenta, and in capillary endothelial epithelium [25, 27]. Expression of BCRP has been localized to apical membranes, the brush border membrane of enterocytes, and the canalicular membrane of hepatocytes [27]. The transporter confers resistance to many anticancer drugs such as mitoxantrone, anthracyclins, campothecin, and topotecan [23, 24], and participates in the efflux of folate and antifolates such as methotrexate [28], and of sulfated conjugates of steroids and xenobiotics [29]. Given the expression pattern of BCRP and its overlap in substrate specificity with MDR1/P-gp, it was hypothesized that BCRP could play a role similar to MDR1/P-gp in regulating drug absorption and disposition [13].

4.2.2 *SLCO*/OATP Transporters

SLCO/OATP transporters represent the major drug uptake system in liver. They transport a wide range of endogenous compounds (e.g., bile acids, thyroid hormones, prostaglandins, and conjugated steroids) and xenobiotics. Many clinically used drugs are substrates of OATP transporters, as digoxin, HMG-CoA reductase inhibitors (statins), methotrexate, antibiotics, and nonsteroidal anti-inflammatory drugs [30]. The OATP family includes 11 members in humans and 14 in rodents. Due to the low sequence homology, orthologs cannot be defined easily, especially in the *SLCO1A* and SLCO1B subfamilies. The single human gene *SLCO1A2* has five putative rodent orthologs, namely *Slco1a1*, *1a3*, *1a4*, *1a5*, and *1a6*, whereas only one rodent ortholog *Slco1b2* exists for the two human genes, *SLCO1B1* and *1B3* [31]. Rodent Oatp1a1 (formerly Oatp1) and 1a4 (formerly Oatp2) are predominantly expressed in liver, where the transporters are localized in the basolateral membrane of hepatocytes [32]. Both multispecific transporters show a considerable overlap in substrates, transporting bile salts, steroid conjugates, thyroid hormones, and drugs such as ouabain [30]. Oatp1a4 was further shown to represent a high affinity digoxin transporter in rodents [33]. Oatp1b2 (formerly Oatp4) is almost exclusively expressed in rodent liver [34] and localized in the basolateral membrane of hepatocytes [35]. Human OATP1A2 (formerly OATP-A) was originally cloned from liver [36]; however, it is widely expressed in extrahepatic tissues including brain, breast, and small intestine [37, 38]. In the latter tissue, it is colocalized with MDR1/P-gp in the apical brush border membrane of enterocytes and may thus play a role in the intestinal uptake of drugs such as fexofenadine [38]. It is most highly related to rat Oatp1a4, based on amino acid sequence homology [39]. OATP1B1 and OATP1B3 (formerly OATP-C/OATP2 and OATP8) are regarded to be exclusively expressed in human liver and localized in the basolateral membrane of hepatocytes [40, 41]. However, recently mRNA expression of their genes was also demonstrated in the small intestine [38]. On the amino acid level, both transporters exhibit around 65% sequence homology to rat Oatp1b2 [39]. OATP2B1, the former OATP-B, is most highly expressed in liver, in the basolateral membrane of hepatocytes, however also in other tissues, including small intestine and brain [39]. In small intestine, it is localized in the apical membrane of enterocytes and may participate in the absorption of anionic compounds from the lumen of the gut [42]. Compared to the other three human hepatic OATPs, it shows more limited substrate specificity [39].

4.2.3 SLC22/Organic Ion Transporters

The SLC22 transporter family can be divided into three subgroups, the organic cation transporters (OCTs), the organic cation transporters novel (OCTN), and the organic anion transporters (OATs). OCT1 (*SLC22A1*) is the most highly expressed drug transporter in liver, even exceeding MRP2 levels [25]. It is further expressed in small intestine and kidney. In these tissues, OCT1 is localized in the basolateral membrane of hepatocytes, enterocytes, and proximal tubule cells. OCT1 participates in the hepatic uptake of organic cations and transports many drugs such as metformin, endogenous compounds such as serotonin and prostaglandins, and model cations such as tetraethylammonium (TEA) [43]. OCT2 (*SLC22A2*) is mainly expressed in the kidney and there it is localized in the basolateral membrane of proximal tubule cells. Regarding substrates, it shows a broad overlap with OCT1 and OCT3 (*SLC22A3*), which is expressed in many tissues, including liver and kidney [43].

OCTN1 and OCTN2 (*SLC22A4*, *SLC22A5*) are expressed in human kidney, with OCTN2 also showing high expression in the small intestine [25]. In kidney both transporters are localized in the apical membrane of proximal tubules, whereas in intestine OCTN2 is localized in the brush border membrane of enterocytes. There it is essential for the intestinal uptake of carnitine [44]. Drugs that are transported by OCTN2 include β-lactam antibiotics, verapamil, and spironolactone [45, 46].

The members of the third subgroup OAT, such as OAT1–3 (*SLC22A6–8*), are highly expressed in kidney, and localized in the basolateral membrane of proximal tubules. Only OAT2 shows a significant expression in liver. There it is localized in the basolateral membrane of hepatocytes [43].

4.3   INDUCTION OF DRUG TRANSPORTERS BY ACTIVATION OF PXR AND CAR

The nuclear receptors pregnane X receptor (PXR, *NR1I2*) and constitutive androstane receptor (CAR, *NR1I3*) mediate the induction of gene expression by the glucocorticoid- and phenobarbital-type microsomal enzyme inducers, respectively. Compounds were assigned to the phenobarbital-type class of inducers, if they predominantly induce cytochrome P450 (*CYP*) *2B* genes, whereas glucocorticoid-type inducers result in strong induction of *CYP3A* gene expression.

Glucocorticoid-type microsomal enzyme inducers are agonist ligands of PXR. Ligand activation of PXR is demonstrated by induction of the activity of PXR-responsive *CYP3A* promoter/reporter gene constructs or by the induction of interactions with nuclear receptor coactivators. That way it was shown that pregnenolone-16α-carbonitrile (PCN), dexamethasone, RU486, 2-acetylaminofluorene (2-AAF), rifampin, St. John's wort extract (SJW)/hyperforin, phenobarbital, carbamazepine, avasimibe, and artemisinin are ligands of PXR [47–54], whereof PCN and dexamethasone are specific for rodent PXR and rifampin, phenobarbital and artemisinin specific for the human receptor [48, 54]. For only some of them, ligand binding was directly shown. The pronounced species specificity in binding of some ligands results from the divergent ligand binding domains of the rodent and human PXR orthologs, respectively, which only display about 75% amino acid identity [51]. PXR is highly expressed in liver and intestine and at much lower levels in some other tissues including kidney, brain, and breast [47, 51, 55–57]. In the small intestine, its expression is limited to the epithelial cells of the mucosa [55], which are also the site of efflux transporter expression.

By far most phenobarbital-type microsomal enzyme inducers activate CAR by just inducing the translocation of the receptor from the cytosol into the nucleus, without ligand binding. This mechanism has been demonstrated for CAR activators phenobarbital, diallyl sulfide (DAS), *trans*-stilbene oxide (TSO), bilirubin, and carbamazepine [58–62]. Nuclear localization is sufficient for the receptor to activate the expression of its target genes, as demonstrated by transient transfections of CAR expression plasmids into immortalized cell lines, there resulting in constitutive nuclear localization of CAR [58]. Only a few compounds have been shown to act as CAR agonist ligands. These comprise 1,4-bis-[2-(3,5-dichloropyridyloxy)]benzene (TCPOBOP), a mouse CAR ligand [48], 6-(4-chlorophenyl)imidazo[2,1-b]thiazole-5-carbaldehyde *O*-(3,4-dichlorobenzyl)oxime (CITCO), a ligand of the human receptor [63], and artemisinin, a ligand of both mouse and human CAR [54]. Nuclear translocation, which is induced by the ligands [58, 63], most likely differs from the mechanism exerted by nonligand phenobarbital-type inducers, as suggested by the observation that CITCO is not activating AMP-activated protein kinase, activation of which is a crucial step in the induction by phenobarbital [64]. As ligand binding induces conformational changes in CAR, it may be anticipated that induction by CAR ligands and nonligand phenobarbital-type inducers differs from each other, at least partially, with respect of the genes regulated. Similarly to PXR, CAR is most highly expressed in liver and intestine [65, 66], where it is localized to hepatocytes and the epithelial cells of the intestinal mucosa [67]. Low level CAR expression was further demonstrated in kidney [65, 68].

Some compounds, preferentially activating CAR such as phenobarbital and carbamazepine [62], are also ligands of PXR. However, both receptors not only share activators/ligands but also regulate overlapping sets of target genes [48, 69]. They mutually regulate even their respective prototypical target genes *CYP2B* and *CYP3A* [70]. Due to this cross talk and share of xenobiotic inducers, PXR- and CAR-dependent induction of transporters will here be discussed together.

The most common approaches to analyze the induction by PXR and CAR include the treatment of animals *in vivo* or of primary cells and cell lines *ex vivo* or *in vitro* with prototypical inducers. The major objection to these experiments is that they do not provide direct proof of the involvement of the nuclear receptor in question. Using a set of different activators for CAR and PXR will strengthen the conclusions [71]; however, as it was shown that, for example, the latter one regulates gene expression in a ligand and promoter-selective fashion [72, 73], it cannot be expected that all compounds of a given set will necessarily exert the same effects. PXR- or CAR-dependent induction can be proved by comparing the response in wild-type and respective knock-out mice. The second concern refers to the problems associated with *in vivo* treatment, which is dependent not only on the chemical but also on the fate of the chemical in the body. Thus, results may indirectly depend on the chemical or the nuclear receptor. Keeping that in mind, induction of ABC and OATP transporters by PXR and CAR has been compiled in Tables 4.1 and 4.2, respectively.

4.3.1   Induction in Liver and Intestine

In all species, which have been analyzed so far, PXR and CAR are most highly expressed in liver and intestine. These are also the main organs responsible for drug absorption and disposition, due to their abundant expression of drug-metabolizing enzymes and transporters. A plethora of data exists especially regarding the induction of transporters in liver. The results are not always consistent, which may reflect differences in species or strains, inducing compounds, and treatment regimens.

4.3.1.1 ABC Family

4.3.1.1.1 MDR1 (ABCB1)/*P-gp.* Few studies addressed the induction of rat *Mdr1* genes on the mRNA level, only where the two rodent orthologs of human *MDR1* can be differentiated. Rats, which were treated with PCN, spironolactone, or dexamethasone, did not show any induction of *Mdr1a* or *Mdr1b* in liver or intestine. In contrast, DAS

TABLE 4.1   Induction of ABC Drug Transporters by PXR and CAR

Transporter	Species	Inducer[a]	Organ/ Tissue/Cell[b]	Effect	NR[c]	Reference
Mdr1a	Rat	DAS	Liver	mRNA	*CAR*	[71]
Mdr1b		TSO	Liver	mRNA	*CAR*	[74]
Mdr1a, 1b		SJW	Intestine	Protein	*PXR*	[75]
		DEX	Liver	Protein	*PXR*	[56, 76]
			bbb	Protein		[56]
		PCN	Liver	Protein	*PXR*	[56]
			bbb	Protein		[56]
Mdr1a	Mouse	PCN	Liver	mRNA	**PXR**	[77, 78]
			Intestine	mRNA	**PXR**	[77, 79]
		RU486	Liver	mRNA	**PXR**	[78]
		TCPOBOP	Intestine	mRNA	**CAR**	[77]
Mdr1b		PCN	Liver, intestine	mRNA	**PXR**	[77]
Mdr1a, 1b		Rifampin SJW	Intestine	Protein	*PXR*	[80]
Mdr1	Pig	Rifampin	LLC-PK1 (k)	mRNA, protein	*PXR*	[81]
MDR1	Human	Rifampin	Fa2N-4 (l)	mRNA	*PXR*	[82]
			Hepatocytes	mRNA, protein		[53, 54, 63, 77, 83–86]
			LS174T (i)	mRNA		[87]
			LS180 (i)	mRNA, protein		[88, 89]
			Intestine	mRNA, protein		[90–93]
			PBMC	mRNA, protein		[94, 95]
			MG63 (os)	mRNA		[96]
		SJW/ hyperforin	Hepatocytes	mRNA	*PXR*	[84]
			Intestine	Protein		[75]
			PBMC	Protein		[97]
		Avasimibe	Hepatocytes	mRNA, protein	*PXR*	[53]
		PB	Fa2N-4 (l)	mRNA	*CAR/PXR*	[82]
			Hepatocytes	mRNA, protein		[53, 77, 85]
			LS180 (i)	Protein		[88]

TABLE 4.1   (*Continued*)

Transporter	Species	Inducer[a]	Organ/ Tissue/Cell[b]	Effect	NR[c]	Reference
		CBZ	LS174T (i)	mRNA	*CAR/PXR*	[87]
			Intestine	mRNA		[98]
			PBMC	mRNA, protein		[95]
		Artemisinin	Hepatocytes	mRNA	*PXR, CAR*	[54]
			LS174T (i)	mRNA		[54]
		CITCO	Hepatocytes	mRNA	*CAR*	[63]
Mrp2	Rat	PCN	FAO (l)	mRNA	*PXR*	[99]
			Hepatocytes	mRNA		[99]
			Liver	Protein		[100]
			bbb	Protein		[56]
		DEX	Hepatocytes	mRNA, protein	*PXR*	[99, 101]
			Liver	Protein		[100]
			bbb	Protein		[56]
		PB	H4IIE (l)	mRNA, protein	*CAR*	[101]
			Hepatocytes	mRNA		[99]
Mrp2	Mouse	PCN	Hepatocytes	mRNA	**PXR**	[99]
			Liver	mRNA, protein	**PXR**	[78]
			Intestine	mRNA	**PXR**	[77, 79]
		DEX	Hepatocytes	mRNA	**PXR**	[99]
		RU486	Liver	mRNA	**PXR**	[78, 102]
		2-AAF	Liver	mRNA	**PXR**	[50]
		VP-hPXR	Liver	mRNA	**PXR**	[103]
		PB	Liver	Protein	*CAR*	[104]
			Hepatocytes	mRNA		[99]
		TCPOBOP	Liver	mRNA	**CAR**	[61, 77]
			Liver	mRNA, protein	*CAR*	[105, 104]
		VP-mCAR	Liver	mRNA	**CAR**	[103]
		TCPOBOP	Kidney	mRNA	*CAR*	[104]
Mrp2	Pig	Rifampin	LLC-PK1 (k)	mRNA, protein	*PXR*	[81]
MRP2	Human	Rifampin	HepG2 (l)	mRNA, protein	*PXR*	[101, 106]
			Hepatocytes	mRNA, protein		[85, 86, 99, 107]
			Intestine	mRNA, protein		[19, 92, 93]

(*Continued*)

TABLE 4.1   (*Continued*)

Transporter	Species	Inducer[a]	Organ/Tissue/Cell[b]	Effect	NR[c]	Reference
		Hyperforin	Hepatocytes	mRNA	*PXR*	[99]
		CBZ	Liver	mRNA	*CAR/PXR*	[108]
			Intestine	mRNA, protein		[98]
		PB	HepG2 (l)	mRNA, protein	*CAR/PXR*	[101]
			Hepatocytes	mRNA		[85]
Mrp3	Rat	DAS	Liver	mRNA	*CAR*	[109]
Mrp3	Mouse	PCN	Liver	mRNA, protein	*PXR*	[105, 104]
			Liver	mRNA	**PXR**	[78, 79, 110]
		RU486	Liver	mRNA	**PXR**	[78, 102]
		TCPOBOP	Liver	mRNA	**CAR**	[77]
			Liver	mRNA, protein	*CAR*	[105, 104, 110]
MRP3	Human	Rifampin	HepG2 (l)	mRNA	*PXR*	[101]
			Hepatocytes	mRNA		[85]
			HuH7 (l)	mRNA		[102]
		PB	HepG2 (l)	mRNA	*CAR/PXR*	[111]
Mrp4	Rat	DAS	Liver	mRNA	*CAR*	[112]
		TSO	Liver	mRNA	*CAR*	[74]
Mrp4	Mouse	PB	Liver	mRNA, protein	**CAR**	[113]
			Liver	mRNA, protein	*CAR*	[104]
		TCPOBOP	Liver	mRNA, protein	**CAR**	[113]
			Liver	mRNA, protein	*CAR*	[105, 104]
		TSO	Liver	Protein	*CAR*	[60]
			Kidney	mRNA		[104]
MRP4	Human	PB	HepG2 (l)	mRNA	*CAR/PXR*	[113]
			Hepatocytes	mRNA		[113]
Mrp5	Mouse	TCPOBOP DAS	Liver	mRNA	*CAR*	[105]

TABLE 4.1   (*Continued*)

Transporter	Species	Inducer[a]	Organ/ Tissue/Cell[b]	Effect	NR[c]	Reference
MRP5	Human	Rifampin 2-AAF	HepG2 (l)	mRNA	*PXR*	[101]
Bcrp	Mouse	2-AAF	Liver	mRNA	**PXR**	[50]
BCRP	Human	Rifampin	Hepatocytes	mRNA	*PXR*	[85]
		PB	Hepatocytes	mRNA, protein	*CAR/PXR*	[85]
		CBZ	Liver	mRNA	*CAR/PXR*	[108]

[a]DAS, diallyl sulfide; TSO, *trans*-stilbene oxide; SJW, St. John's wort extract; DEX, dexamethasone; PCN, pregnenolone-16α-carbonitrile; TCPOBOP, 1,4-bis-[2-(3,5-dichloropyridyloxy)]benzene; PB, phenobarbital; CBZ, carbamazepine; CITCO, 6-(4-chlorophenyl)imidazo[2,1-b]thiazole-5-carbaldehyde *O*-(3,4-dichlorobenzyl)oxime; 2-AAF, 2-acetylaminofluorene.
[b]bbb, blood–brain barrier; PBMC, peripheral blood mononuclear cells; the tissue origin of cell lines is denoted as (k), kidney; (l), liver; (i), intestine; (os), osteosarcoma.
[c]Bold font: if induction has been shown to dependent on the respective nuclear receptor.

induced the expression of *Mdr1a* exclusively in liver [71]. DAS, which is regarded as an activator of CAR due to its capacity to induce *Cyp2b* genes, was only recently confirmed to be a true CAR activator [59]. Surprisingly, phenobarbital did not show any effect [71]. TSO, which was shown to promote nuclear localization of CAR [60], induces the expression of hepatic *Mdr1b* in a gender-dependent manner in Wistar–Kyoto rats, thereby suggesting regulation by CAR [74]. Males of these rats express much higher levels of CAR than females [121]. Induction of rat Mdr1 on the protein level was shown by several studies, using known ligands of PXR. SJW exclusively induced the expression of intestinal Mdr1/P-gp, whereas hepatic Mdr1/P-gp levels were not altered [75]. Furthermore, dexamethasone treatment was reported to induce Mdr1/P-gp expression in rat liver [76]. In contrast, treatment of primary rat hepatocytes with dexamethasone up to 100 μM did not alter Mdr1/P-gp expression [122]. Thus, whereas induction of rat Mdr1/P-gp protein in intestine by PXR ligands was consistently demonstrated, conflicting results exist regarding induction in liver.

Using PXR and CAR knock-out mice, it was unequivocally shown that both receptors are involved in the regulation of mouse *Mdr1a* gene expression in liver and small intestine. Treatment with PCN and RU486 induced the expression of Mdr1a mRNA in liver [77, 78] and small intestine [77, 79] of wild-type mice; however not in PXR knock-outs. A single study also demonstrated a PXR-dependent induction of *Mdr1b* in liver and intestine [77], which however was not seen in others [78, 79]. CAR-dependent induction of *Mdr1a* by TCPOBOP was only reported in the small intestine [77]. On the protein level, induction of mouse Mdr1/P-gp in the small intestine was achieved by SJW and rifampin [80]. In this study extremely high doses of rifampin, which is usually regarded as a human PXR-specific ligand, were used. At

TABLE 4.2 Induction of *SLCO*/OATP Transporters by PXR and CAR

Transporter	Species	Inducer[a]	Organ/Tissue/Cell[b]	Effect	NR[c]	Reference
SLCO1A2/OATP1A2	Human	CBZ	Liver	mRNA	*CAR/PXR*	[108]
		Rifampin	ZR-75-1 (bc) T47D (bc)	mRNA	*PXR*	[114]
Slco1a4/Oatp1a4	Rat	PCN	Liver	mRNA, protein	*PXR*	[115, 116]
		PB	Liver	Protein	*CAR*	[115, 116]
Slco1a4/Oatp1a4	Mouse	PCN	Liver	mRNA	**PXR**	[77, 79, 110, 117, 118]
			Liver	mRNA	*PXR*	[104, 119]
		2-AAF	Liver	mRNA	**PXR**	[50]
		LCA	Liver	mRNA	**PXR**	[118]
		VP-hPXR	Liver	mRNA	**PXR**	[103]
		PB	Liver	mRNA	*CAR*	[104, 110]
		TCPOBOP	Liver	mRNA	*CAR*	[104, 110]
		VP-mCAR	Liver	mRNA	**CAR**	[103]
SLCO1B1/OATP1B1	Human	Rifampin	HepG2 (l)	mRNA	*PXR*	[120]
			Hepatocytes	mRNA		[85]
		CBZ	Liver	mRNA	*CAR/PXR*	[108]
Slco1b2/Oatp1b2	Mouse	TCPOBOP, bilirubin	Liver	mRNA	**CAR**	[61]
		VP-mCAR	Liver	mRNA	**CAR**	[103]
SLCO1B3/OATP1B3	Human	Rifampin	Hepatocytes	mRNA	*PXR*	[86]
		CBZ	Liver	mRNA	*CAR/PXR*	[108]

[a]LCA, lithocholic acid; for further abbreviations, see Table 4.1.
[b]Origin of cell lines is denoted as (l), liver; (bc), breast cancer.
[c]See Table 4.1.

these high doses, which exceed more than 30-fold the dose used in humans, rifampin also seems to activate rodent PXR, as suggested by the induction of hepatic *Cyp3a* by high dose rifampin in rats [123]. However, hepatic Mdr1/P-gp was not induced [80].

The induction of *MDR1* by PXR and CAR activators in human liver has been analyzed in hepatoma cell lines and primary human hepatocytes. MDR1 mRNA was induced in the human hepatocyte-derived cell line Fa2N-4 by rifampin and phenobarbital (PB) treatment [82]. It has often been propagated that *MDR1* is not induced in human liver, on the basis of a report that failed to show consistent induction by rifampin and phenobarbital in primary human hepatocytes from two donors [124]. Since then, numerous reports have demonstrated inducibility of *MDR1* in primary cultures of human hepatocytes by treatment with the PXR ligands rifampin [53, 54, 63, 77, 83–86], hyperforin [84], avasimibe [53], artemisinin [54], which activates both PXR and CAR, phenobarbital [77, 85], and the CAR ligand CITCO [63]. As these studies usually have used hepatocyte cultures from a very limited number of donors, interindividual variability in induction is a major problem for the interpretation of the results. This is best illustrated by the fact that only one out of three cultures showed induction of MDR1 mRNA by rifampin and CITCO [63]. To address this interindividual variability in a large collection, we analyzed the induction of MDR1 in 26 human hepatocyte cultures, derived from different donors, which were treated with rifampin. The compound significantly induced MDR1 mRNA expression by 2.7 ± 1.5-fold ($p < 0.0001$), with individual cultures ranging from 0.7- to 8.1-fold induction. Three out of 26 cultures showed less than 50% induction (1.5-fold). Thus, studies, which use primary human hepatocytes, should be performed with an appropriate number of individual cultures to minimize the risk of false negative results.

In the human intestine, induction by PXR and CAR activators has been demonstrated in intestinal cell lines as well as *in vivo*. Treatment with numerous microsomal enzyme inducers, e.g., rifampin, clotrimazole, RU486, carbamazepine, nifedipine, artemisinin, all now known to activate PXR, induces the expression of *MDR1* in human intestinal LS180 and LS174T cells [54, 88–89]. Induction was only observed in cell lines with a high expression of PXR, e.g., LS174T, which express PXR up to 40% of the levels seen in liver [54], whereas *MDR1* was not induced by PXR activators in the intestinal Caco-2 cells, which express much lower levels of PXR [89]. *In vivo* induction of intestinal *MDR1* was demonstrated by comparing the expression in small intestinal biopsies before and after treatment of volunteers with the inducer. Using this experimental approach, it was shown that rifampin [90–93], SJW [75], and carbamazepine [98] induce the expression of MDR1/P-gp in human small intestine. MDR1 mRNA and P-gp protein were induced to a similar extent. Again, as with human hepatocyte cultures, pronounced variability in induction was observed, with some individuals not responding at all [75, 91].

4.3.1.1.2 MRP2 (ABCC2). The induction of rat hepatic *Mrp2* expression by PXR and CAR activators has been demonstrated by several studies using hepatoma cell lines, primary hepatocytes, and animals. The rat hepatoma FAO cell line showed an increase in Mrp2 mRNA expression by treatment with PCN [99]. Primary cultures of rat hepatocytes also demonstrated induction of *Mrp2* by treatment with the PXR ligands PCN and dexamethasone [99, 101]. Furthermore, Mrp2 protein expression was induced by PCN, dexamethasone, and spironolactone in rat liver [100]. The prototypical CAR activator phenobarbital induces *Mrp2* in the rat hepatoma H4IIE cell line [101] and Mrp2 mRNA in cultured rat primary hepatocytes [99], whereas induction of *Mrp2* in rat liver *in vivo* by activators of CAR has not consistently been observed. Treatment with phenobarbital, DAS, and TSO induced *Mrp2* expression in rat liver [112]; however, this is not confirmed by others, who do not observe induction of hepatic *Mrp2* by CAR activators phenobarbital and DAS [125, 109].

Using PXR and CAR knock-out mice, as well as transgenic mice expressing the activated VP-PXR and VP-CAR, it was clearly shown that mouse hepatic *Mrp2* is induced by activation of both receptors. Studies with PXR knock-out mice demonstrated that hepatic induction of *Mrp2* by PCN [78, 99], dexamethasone [99], RU486 [78, 102], and 2-AAF [50] depends on PXR. Transgenic mice, which specifically express the constitutively active VP-hPXR (a fusion of the activation domain of viral VP16 protein and human PXR) in liver, show an increased hepatic expression of Mrp2 mRNA [103]. PXR-dependent induction of Mrp2 mRNA by PCN was further observed in mouse intestine [77, 79]. Treatment with phenobarbital and TCPOBOP induced *Mrp2* in cultured mouse primary hepatocytes and in mouse liver [99, 105, 104]. Dependence on CAR was confirmed by induction of *Mrp2* mRNA in the liver of wild-type mice by TCPOBOP, which was absent in CAR knock-outs [61, 77]. Transgenic mice, which conditionally express the activated VP-mCAR, also showed induction of Mrp2 mRNA, if expression of the CAR transgene was induced [103].

PXR-mediated regulation of human hepatic and intestinal *MRP2* is suggested by rifampin-dependent induction in the HepG2 hepatoma cell line [101, 106], in cultured primary human hepatocytes [85, 86, 99, 107], and in duodenal biopsies of rifampin-treated volunteers *in vivo* [19, 92, 93]. Another PXR ligand, hyperforin, was also shown to induce MRP2 mRNA in primary human hepatocyte cultures [99]. Induction of MRP2 mRNA and protein in HepG2 cells and human hepatocytes by phenobarbital [85, 101], and an increased *MRP2* expression in the liver of patients with carbamazepine medication [108], as well as in the small intestine of volunteers treated with the same compound [98], further suggests a role for CAR in the regulation of *MRP2*.

4.3.1.1.3 MRP3 (ABCC3). Mrp3 mRNA expression is induced in rat liver by CAR activators phenobarbital and DAS [109]. Induction of hepatic *Mrp3* by phenobarbital was also demonstrated in mice [105, 104, 110]. However, it was clearly shown that this induction was independent of CAR [126, 127]. A role for PXR in induction by phenobarbital is ruled out by the facts that the compound is not a ligand of mouse PXR and that PXR knock-out mice show an even enhanced induction [110]. Similarly, induction of mouse and rat hepatic *Mrp3* by TSO [60, 112] proved to be independent of CAR [60, 74]. In conclusion, these results argue for a PXR/CAR-independent regulation of *Mrp3* by phenobarbital and TSO. The existence of such a pathway has first been suggested by the analysis of phenobarbital-dependent gene expression in wild-type and CAR knock-out mice [128]. In contrast to phenobarbital and TSO, induction of hepatic Mrp3 mRNA and protein by the CAR-ligand TCPOBOP [77, 105, 104, 110] was shown to depend on CAR [77], suggesting that CAR activation of *Mrp3* gene expression requires CAR to be activated by ligand binding. Besides CAR, PXR also plays a prominent role in *Mrp3* induction. PCN treatment of mice induces the expression of *Mrp3* in liver [105, 104]. This induction depends on PXR, as demonstrated by the increase of Mrp3 mRNA in PCN- and RU486-treated wild-type mice, which is not observed in PXR knock-outs [78, 79, 102, 110].

Rifampin induction of MRP3 mRNA in HepG2 and HuH7 human hepatoma cells [101, 102], as well as in cultured primary human hepatocytes [85], suggests PXR-dependent regulation of this gene also in humans. Additionally, it was shown that phenobarbital induces MRP3 mRNA expression in HepG2 cells [111].

4.3.1.1.4 MRP4 (ABCC4). Treatment of rats with CAR activators DAS and TSO results in increased hepatic Mrp4 mRNA expression [74, 112]. Similarly, CAR activators phenobarbital, TCPOBOP, and TSO induced *Mrp4* expression in the liver of mice [60, 105, 104, 113]. By using CAR knock-out mice and murine liver Hepa1c1c7 cells, which stably express the activated VP16-hCAR, it was clearly shown that induction by phenobarbital and TCPOBOP depends on CAR [113].

Phenobarbital also induces MRP4 mRNA expression in human HepG2 cells and cultured primary human hepatocytes [113]. HepG2 cells, which express the activated VP16-hCAR, show increased *MRP4* expression as compared to vector-infected cells [113]. In conclusion, these results strongly argue for a major role of CAR in the induction of hepatic *Mrp4*, which is conserved in evolution. However, it cannot be formally excluded that PXR may also play a role in the induction of human *MRP4* by phenobarbital. Treatment with a human CAR-specific ligand or gene-specific knock-out of PXR expression by siRNA may help to further clarify the mechanism of induction.

4.3.1.1.5 MRP5 (ABCC5). CAR activators TCPOBOP and DAS induce the expression of Mrp5 mRNA in the liver of treated mice. Treatment with phenobarbital and PXR ligands PCN, dexamethasone, and spironolactone does not induce or even inhibits *Mrp5* expression, respectively [105]. Human MRP5 mRNA is induced by 2-AAF and rifampin in HepG2 cells [101]. The respective roles of CAR and PXR in the regulation of *Mrp5* need to be elucidated in further studies.

4.3.1.1.6 BCRP (ABCG2). Mouse liver Bcrp mRNA is induced by treatment with 2-AAF in a PXR-dependent manner [50]. The induction of *BCRP* in cultured primary human hepatocytes by rifampin and phenobarbital [85] and the increased expression of BCRP mRNA in the liver of patients, who were treated with carbamazepine [108], further suggest PXR- and CAR-dependent regulation of *BCRP* in humans.

4.3.1.1.7 Further ABC Transporters. The induction of some other ABC transporters of the MRP subfamily has been reported in single studies. Mouse liver *Mrp1* is induced by TCPOBOP, depending on the expression of CAR [77]. However, this was not confirmed by further studies using CAR activators including TCPOBOP in rats and mice [109, 105]. Human hepatocytes treated with phenobarbital and rifampin also did not show an increased expression of MRP1, although in this study cultures from only two donors have been used [124]. Furthermore, the hepatic expression of *Mrp6* and *Mrp7* in mice was reported to be induced by CAR activators phenobarbital, TCPOBOP, and DAS [105]. Whether or not PXR and CAR regulate the expression of *Mrp1*, *Mrp6*, and *Mrp7* clearly requires further studies.

4.3.1.2 SLCO/OATP Family

4.3.1.2.1 Slco1a4/Oatp1a4 and SLCO1A2/OATP1A2. PXR- and CAR-dependent regulation of rodent *Slco1a4*/Oatp1a4 was first suggested by induction of rat hepatic Slco1a4 mRNA and Oatp1a4 protein by PCN and CAR activators phenobarbital and DAS [115, 116]. However, in contrast to DAS, phenobarbital induced Oatp1a4 protein expression without a concomitant induction of Slco1a4 mRNA [116], thereby suggesting that phenobarbital might induce Oatp1a4 expression independent of CAR. Transcriptional activation of *Slco1a4* by PCN was confirmed by increased hnRNA levels after treatment [116].

PCN also induced hepatic Slco1a4 mRNA in mice [104, 119]. Several studies using PXR knock-out mice established the essential role of PXR in the regulation of this transporter gene. Induction of Slco1a4 mRNA by PCN [77, 79, 110, 117, 118], lithocholic acid [118], and 2-AAF [50] proved to be annihilated in the liver of PXR knock-out mice. Increased expression of Slco1a4 mRNA in the livers of transgenic mice, which express the activated VP-hPXR, further confirmed the role of this nuclear receptor in the regulation of the gene [103]. Phenobarbital and TCPOBOP were shown to induce the expression of *Slco1a4* in mouse liver [104, 110]. A role for CAR in the regulation of *Slco1a4* was strongly suggested by the increase of mRNA expression in the livers of transgenic mice, which were induced to express the activated VP-mCAR [103]. However, another study did not report CAR-dependent induction of *Slco1a4* [77]. This might be explained by a negative role of nonliganded PXR in *Slco1a4* regulation. Such a role was suggested by the fact that phenobarbital is exclusively inducing the transporter in PXR knock-out mice [110].

The mRNA expression of *SLCO1A2*, the putative human ortholog of rodent *Slco1a4*, was reported to be induced in the liver of patients, who were treated with carbamazepine [108].

4.3.1.2.2 SLCO1B1/OATP1B1 and SLCO1B3/OATP1B3. Induction of SLCO1B1 mRNA by rifampin was shown in the human hepatoma HepG2 cells [120] and in five out of seven cultures of primary human hepatocytes [85]. The pronounced variability of the induction response between individuals most likely explains why a similar induction by rifampin was not observed in a second study, which used cultures from only two different donors [129]. Very recently, it was shown that rifampin treatment of primary human hepatocytes also results in the induction of SLCO1B3 mRNA expression [86]. Furthermore, *SLCO1B1* and *SLCO1B3* are expressed at increased levels in the liver of patients treated with carbamazepine [108]. Altogether these studies suggest a role for PXR, eventually also for CAR, in the regulation of human *SLCO1B* transporter genes, which has been confirmed for PXR by the identification of functional binding sites in the 5′ flanking region of *SLCO1B1* (see below).

4.3.1.2.3 Slco1b2/Oatp1b2. *Slco1b2* is the putative rodent ortholog of human *SLCO1B1* and *SLCO1B3*. Treatment of CAR-deficient mice with TCPOBOP did not result in induction of *Slco1b2*, whereas mRNA expression was clearly increased by TCPOBOP in the livers of wild-type mice [61]. A similar CAR-dependent induction of the gene was observed with bilirubin treatment of cultured hepatocytes from wild-type and CAR knock-out mice [61]. The role of CAR is further supported by the observation that transgenic mice, which express the activated VP-mCAR, show increased expression of *Slco1b2* in the liver [103].

4.3.1.3 Slc22/Organic Ion Transporter Family. Few studies have reported the induction of *Slc22a* genes by PXR activators. By using rat hepatoma RL-34 cells, cultured primary rat hepatocytes and animals, it was shown that PCN induced the expression of rat hepatic Oct1 (*Slc22a1*) mRNA. Induction of Oct1 transporter activity was further demonstrated *ex vivo* by increased uptake of the prototypical substrate MPP+ into hepatocytes, which were isolated from PCN-treated rats, compared to cells, which were isolated from control animals and *in vivo* by an increase in biliary excretion of TEA in PCN-treated rats [130].

A second Slc22a transporter may also be regulated by PXR, as human OCTN2 (*SLC22A5*) was shown to be induced by rifampin in cultured primary human hepatocytes [131].

4.3.2 Induction in Other Organs and Tissues

4.3.2.1 Kidney. *PXR* is expressed in the kidney of rodents [132], pigs [133], and humans [134]. Although its renal expression, especially in humans, is commonly regarded as being quite low compared to liver and intestine, it was recently shown that expression of the receptor is readily detectable on mRNA and protein level in human renal tubular epithelial cells [134]. Thus, renal induction by PXR ligands seems conceivable, provided that ligands can be present at sufficient doses in the systemic circulation. Rifampin treatment of the porcine renal LLC-PK1 cells, which resemble proximal tubular epithelial cells, induces the expression of *Mdr1* and *Mrp2* [81]. As rifampin is a ligand of pig PXR [135], these results may indicate renal transporter gene regulation by the receptor. In rodents, conflicting results regarding transporter induction by PXR and CAR in the kidney have been reported. Induction of *Mdr1a/b* and *Mrp1–3* by rodent PXR ligands and CAR activators has not been observed in rat and mouse kidney *in vivo* [71, 79, 109] and treatment with PXR ligands seems even to inhibit the renal expression of *Mrp4* in rats [136]. However, in another study, it was shown that Mrp2 and Mrp4 mRNA expressions are induced by TCPOBOP in mouse kidney [104]. The contradictory results may stem from the use of different mouse strains or CAR activators.

In contrast to liver, *Oct1* was not induced by PCN in the kidney of rats, whereas *Oct2* (*Slc22a2*) was induced [130].

4.3.2.2 Blood–Brain Barrier. *PXR* has been shown to be expressed in rat brain capillaries and treatment of isolated capillaries with PCN and dexamethasone *in* *vitro*, as well as exposing rats *in vivo* to these PXR ligands induced the expression and functional activity of Mdr1/P-gp and Mrp2 protein [56]. Furthermore, it was shown that rifampin and hyperforin induce *in vitro* and *in vivo* the expression and transport activity of Mdr1/P-gp in brain capillaries of transgenic mice, which express human PXR on a mouse PXR knock-out background [137]. The *hPXR* transgene was expressed in liver, intestine, and brain capillaries of these mice. The increase of Mdr1/P-gp activity by rifampin treatment tightens *in vivo* the blood–brain barrier to methadone, which is a substrate of the transporter [137].

4.3.2.3 Lymphocytes. Induction of human *MDR1* was observed in peripheral blood mononuclear cells (PBMC), i.e., lymphocytes, by treatment of volunteers with rifampin, SJW, and carbamazepine [94, 95, 97]. A substantial variability in induction was observed, with approximately 60% of individuals responding significantly to rifampin, in which response was not related to serum concentrations of the inducer [94]. Highly significant correlations between the mRNA expression levels of *PXR* and of the ABC transporter genes *MDR1*, *MRP1*, *MRP2*, and *BCRP* in human PBMC [138, 139] suggest that PXR may also be involved in the xenobiotic-independent regulation of these ABC transporters in PBMC, eventually via activation by endogenous ligands.

4.3.2.4 Osteosarcoma and Breast Cancer Cells. Induction of MDR1 mRNA by rifampin was observed in human MG63 osteosarcoma cells, which stably express a ligand-dependent VP16C-PXR fusion protein [96]. The stable cell line was generated to facilitate the search for PXR target genes in osteoblastic cells, which express only very low levels of endogenous *PXR* [140].

In breast cancer cells ZR-75-1 and T-47D, which also express only low levels of endogenous *PXR*, treatment with rifampin transiently induces the expression of SLCO1A2 mRNA in a dose- and time-dependent manner [114]. Induction by rifampin was shown to be independent of de novo protein biosynthesis. In contrast to *SLCO1A2*, *MDR1* was not induced [114]. The physiological relevance of these findings has to await further clarification.

4.4 INDUCTION OF DRUG TRANSPORTERS BY ACTIVATION OF PPARα

Peroxisome proliferator-activated receptor (PPAR) α (*NR1C1*) plays an essential role in fatty acid homeostasis, by controlling the expression of CYP4A and other enzymes involved in fatty acid metabolism [141]. Its endogenous ligands are fatty acids and fatty acid derivatives; however, it is also activated by xenobiotic ligands, which cause peroxisome proliferation. These peroxisome proliferators comprise hypolipidemic drugs, e.g., fibrates, and industrial chemicals, e.g., phthalates and perfluorinated fatty acids [142, 143]. Similarly to PXR and CAR, PPARα is highly expressed in liver. Ligand activation of the receptor induces not only the expression of CYP4A and other enzymes involved in fatty acid oxidation, but also the expression of drug transporters.

Long-term treatment of mice with the PPARα ligands clofibrate and ciprofibrate induces the expression of hepatic Mdr1a/1b mRNA and Mdr1/P-gp [105, 144, 145]. Using PPARα-deficient mice, it was further shown that this induction depends on PPARα [144, 145]. However, if cultured primary hepatocytes of wild-type and PPARα knock-out mice were treated *in vitro* with ciprofibrate for a shorter time, PPARα-dependent induction was not observed, thus arguing against a direct role of PPARα in *Mdr1a/1b* gene expression [144]. Possibly, activation of the receptor induces the expression of PXR, which in turn activates *Mdr1* gene expression. Such a mechanism is suggested by the induction of PXR mRNA in primary human hepatocytes by clofibrate and the identification of a functional PPARα response element in the human PXR promoter region [146].

The ABC transporters Mrp3, Mrp4, and Bcrp are induced by clofibrate in the livers of mice [105, 145] and it was shown that induction by clofibrate depends on PPARα [145].

Besides ABC transporters, PPARα agonists clofibrate and Wy14,643 further induce organic ion transporters of the Slc22 family. Both compounds induce Oct1 (*Slc22a1*) mRNA expression in rat H35 hepatoma cells and mouse liver. A similar induction was observed with PPARγ (*NR1C3*) agonists [147]. Induction by these compounds was shown to depend on a single PPAR response element at −2.4 kb in the 5′ flanking region of mouse Oct1, thus both PPARα and PPARγ transcriptionally activate the gene [147]. The induction further results in an increase of functional transporter expression, as the cellular uptake of a model Oct1 substrate is enhanced [147]. Additional Slc22a transporters induced by PPARα agonists comprise Octn1 (*Slc22a4*) and Octn2 (*Slc22a5*), which are induced in rat liver and hepatoma cells [148].

4.5 MOLECULAR MECHANISM OF PXR- AND CAR-DEPENDENT DRUG TRANSPORTER REGULATION

Studies with PXR- and CAR-deficient mice have unequivocally shown that *Mdr1a/1b*, *Mrp2–4*, *Bcrp*, *Slco1a4*, and *Slco1b2* are regulated by PXR and/or CAR. However, these studies do not prove that these transporters are directly regulated by the nuclear receptors, especially as the treatment of mice with inducers usually was performed for days. Typically, nuclear receptors regulate gene expression by binding to specific binding sites in promoter and enhancer elements of the target genes. PXR and CAR do so as heterodimers with retinoid X receptor (RXR) α (*NR2B1*), which bind to motifs, consisting of repeats of the general nuclear receptor half site with different orientation and spacing. Direct repeats (DR)1–4, everted repeats (ER)6, and inverted repeats (IR)0 have been identified as PXR/CAR binding sites in the regulatory regions of many genes encoding drug-metabolizing enzymes, which are induced by microsomal enzyme inducers (see Chapters 2 and 3). These binding sites have further been demonstrated to act as functional PXR/CAR response elements, which mediate induction by the receptors. In contrast to genes encoding drug-metabolizing enzymes, only a few transporter genes have been analyzed in such detail up to now. These comprise human *MDR1*, rat *Mrp2*, mouse *Mrp4*, rat *Slco1a4*, and human *SLCO1B1*, which will be discussed in the following.

4.5.1   MDR1 (ABCB1)

In a study analyzing the induction of human *MDR1* in intestinal cells, Geick et al. [87] identified a far distal enhancer region at −7.8 kb upstream of the transcriptional start site in the proximal promoter of *MDR1* as the single region, which mediated induction by rifampin. The 5′ flanking region between −10.2 and +0.26 kb of *MDR1*, with respect to the major transcriptional start site of the gene, was analyzed by transient transfections of promoter/reporter gene plasmids into intestinal LS174T cells, which were then treated with the human PXR ligand rifampin. The section between −8.0 and −7.7 kb mediated induction by rifampin and further conferred inducibility to the proximal *MDR1* promoter up to −1.8 kb, as well as to the heterologous thymidine kinase promoter, when cloned in front of these. It harbors a cluster of partially overlapping potential binding sites for PXR at −7.8 kb (Figure 4.2). Subsequent electrophoretic mobility shift assays demonstrated that all putative elements, with the exception of DR3, were specifically bound by PXR/RXRα heterodimers *in vitro*. Mutational analysis of the DR4(I) and ER6/DR4(III) binding sites, which were most strongly bound by PXR/RXRα, showed that only DR4(I) is essential for the induction of *MDR1* by rifampin in LS174T cells. Furthermore, a heterologous promoter/reporter gene, containing a dimer of DR4(I), was induced by rifampin, whereas a construct with a dimer of ER6/DR4(III) was not. Thus, the cluster of nuclear receptor binding sites at −7.8 kb constitutes an enhancer element, which mediates PXR-dependent induction of *MDR1*, by containing at least one functional PXR response element, the DR4(I) site.

In a subsequent study, it was further shown that *MDR1* is also regulated by CAR [149]. Electrophoretic mobility shift assays demonstrated specific binding of CAR/RXRα heterodimers to DR4(I) and somewhat weaker also to ER6/DR4(III). Additionally, CAR specifically bound to DR4(II) as a monomer. It was shown by gel shift analysis of the corresponding mutant binding sites that binding of CAR monomers required the intact 5′ half site of DR4(II) and additionally the AG dinucleotide 5′ in front of that. The binding site of CAR monomers in the *MDR1* −7.8-kb enhancer thus showed up to be the octamer sequence AGAGTTCA (Figure 4.2), which was previously demonstrated as the strongest binding site of CAR monomers [150]. As with PXR-mediated induction by rifampin, activation of *MDR1* by CAR depends on the −7.8-kb enhancer. In accordance with the *in vitro* binding studies, both the DR4(I) and the extended 5′ half site of DR4(II) were required for full activation of *MDR1* by CAR. The high affinity heterodimer binding site DR4(I) mediated 70% of activation by CAR, whereas the remaining 30% was mediated by binding of CAR monomers to the extended 5′ half site of DR4(II).

PXR and CAR cross talk at the DR4(I) element of the −7.8-kb enhancer of *MDR1*. Competition of both nuclear receptors for binding to DR4(I) has to be anticipated, as they bind to this element with comparable relative affinities [149]. However, cooperative regulation of *MDR1* expression by PXR and CAR is also conceivable, as CAR exhibits specific binding to the extended 5′ half site of DR4(II).

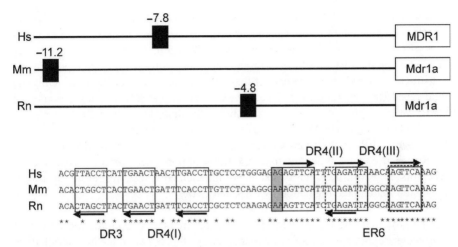

FIGURE 4.2 Phylogenetic conservation of the human −7.8-kb enhancer of *MDR1*. Sequences in the 5′ upstream regions of mouse (Mm) and rat (Rn) *Mdr1a* genes with sequence homology to the human (Hs) −7.8-kb enhancer of *MDR1* are depicted as black boxes in the upper part. The numbers in kb refer to the transcriptional start sites of the respective genes. The lower part shows the alignment of the respective enhancer sequences. Asterisks mark identical bases. The half sites of PXR and CAR binding motifs are boxed and their orientation is shown by arrows. The dinucleotide upstream of the 5′ half site of DR4(II), which is important for CAR monomer binding, is boxed in gray. Dashed boxes denote the half sites of the ER6 motif.

The rat and mouse *Mdr1a/1b* genes have also been shown to be induced by PXR- and CAR-dependent mechanisms (Table 4.1); however, the corresponding enhancer elements and binding sites of PXR and CAR have not yet been identified. A search for regions homologous to the human *MDR1* −7.8-kb enhancer in the genomic sequence databases of mouse and rat revealed that this enhancer is highly conserved in rodents. A region with 80% sequence identity to the human −7.8-kb enhancer is found at −11.2 and −4.8 kb of mouse and rat *Mdr1a*, respectively (Figure 4.2). The DR4(I), which is essential for the induction of human *MDR1* by xenobiotics, shows a single mismatch in rodents, whereas DR4(II) and DR4(III)/ER6 are even unchanged (Figure 4.2). However, a similar enhancer element is missing in the 5′ upstream region of *Mdr1b*. Thus, either the gene harbors different PXR/CAR response elements or the enhancer of *Mdr1a* acts as a long distance enhancer also for *Mdr1b*. The evolutionary conservation of the *MDR1* xenobiotic enhancer further points to the functional significance of the induction of MDR1 transporters, which will be discussed further below.

4.5.2 Mrp2 (Abcc2)

In the rat *Mrp2* gene, a single ER8 motif in the proximal promoter region at around −0.4 kb was identified, which mediates induction by PXR and CAR [99]. Electrophoretic mobility shift assays showed that PXR and CAR bound to this element as heterodimers with RXRα *in vitro*. This was the first report of an ER8 motif being a PXR and CAR binding site, which was later on confirmed by the systematic analyses of CAR and PXR binding specificities [120, 150]. Transcriptional activation, depending on the ER8 motif, was analyzed by transient transfection of reporter gene constructs, which harbored dimers of the wild-type or mutated ER8 motif. Only the wild-type ER8 construct was induced by PCN and activated by cotransfection of rat CAR. An *Mrp2* promoter construct, containing the first 1 kb of the 5′ flanking region, was similarly induced by PCN and activated by CAR. Altogether these results demonstrate that the ER8 in the rat proximal promoter of *Mrp2* acts as a functional PXR and CAR response element and mediates the transcriptional activation by these nuclear receptors; however, it cannot be excluded that elements, residing further upstream, may participate in the regulation by PXR and CAR.

In silico analysis of the human *MRP2* promoter does not reveal conservation of a similar proximal ER8 element. Furthermore, rifampin failed to induce the activity of a promoter/reporter gene construct, containing the first 5.2 kb of the human MRP2 5′ upstream region [106], thereby indicating that the elements mediating induction via PXR are not residing within this region. The response elements, which mediate the well-documented induction of human *MRP2* by activators of PXR and CAR, are still to be identified.

4.5.3   Mrp4 (Abcc4)

The 5′ upstream region between −10.3 and −4.8 kb of the mouse *Mrp4* gene was shown to mediate CAR-dependent induction by phenobarbital [113]. These 5.5 kb harbor numerous potential CAR binding sites, as analyzed *in silico* using the NUBIscan program for the identification of nuclear receptor binding sites [151]. Searching for DR1-5 and ER6 motifs, 13 putative sites were identified between −10.3 and −4.8 kb. An additional single putative ER6 motif was identified in the proximal promoter of *Mrp4*. The transcriptional activity of this 5′ flanking sequence, enriched in putative CAR sites, was analyzed in PXR and CAR knock-out mice, which were transfected *in vivo* with a reporter construct, which contains the region between −10.3 and −4.8 kb, cloned in front of the proximal *Mrp4* promoter. Induction by phenobarbital was only seen in the liver of PXR knock-out mice, thus demonstrating that CAR transcriptionally activates the *Mrp4* promoter in response to phenobarbital. In this study, no attempt was made to experimentally confirm CAR binding to any of the putative motifs, as much as no attempt was made to identify the response element(s), which mediate activation by CAR.

4.5.4   Slco1a4

The PXR-mediated induction of rat *Slco1a4* was shown to be dependent mainly on a distal cluster of PXR response elements at −8.0 kb of the 5′ upstream region of the gene [152]. Four potential PXR binding sites, arranged as DR3 motifs, were identified *in silico* in the 5′ flanking region of rat *Slco1a4*. The most proximal motif was found at −5.0 kb, whereas the other three clustered at about −8.0 kb. Electrophoretic mobility shift assays showed that three of the four putative sites were bound by PXR/RXRα heterodimers *in vitro*. The distal DR3-2 in the −8.0-kb region was bound most strongly, followed by DR3-4 in the same region and DR3-1 at −5.0 kb. Transient transfection of a reporter gene construct, containing the first 8.7 kb of the 5′ flanking region of rat *Slco1a4*, into rat H4IIE hepatoma cells demonstrated that cotransfected rat PXR transactivated the *Slco1a4* promoter in the presence of PCN. The analysis of deletion mutants in transient transfections showed that PXR-mediated induction by PCN mainly depends on the distal cluster of PXR binding sites at about −8.0 kb. A small region of 300 base pairs, including the cluster, was thereby shown to be sufficient for a strong induction by PCN. However, residual induction was even observed with a construct without any of these DR3 motifs, containing only the first 2.8 kb of the 5′ flanking region. As site-specific mutations have not been performed, the contribution of individual DR3 motifs in the distal −8.0-kb cluster to induction cannot be further dissected.

4.5.5 SLCO1B1

Ten putative PXR binding sites were identified *in silico* within the first 10 kb of the human *SLCO1B1* 5′ upstream sequence, by searching for DR3-5 and ER6-9 motifs. Electrophoretic mobility shift assays only confirmed the binding of PXR/RXRα heterodimers to the DR4-1 element at position −128/−111 and to the ER6-2 motif at −9957/−9940 [120]. *In vivo* functionality of these motifs was further analyzed by chromatin immunoprecipitation (ChIP) assays using human HepG2 cells. ChIP assays showed that PXR/RXRα heterodimers are bound to the two regions in HepG2 chromatin, which harbor the DR4-1 and ER6-2 motifs, respectively [120]. Chromatin binding was observed even in the absence of the PXR ligand rifampin and was not enhanced by ligand treatment. Although these two PXR binding sites were identified in the human *SLCO1B1* gene *in vitro* and *in vivo*, it was not analyzed whether binding of PXR to the sites also mediates the transcriptional activation of the *SLCO1B1* gene.

4.6 INDUCTION OF DRUG TRANSPORTER EXPRESSION AND DRUG DISPOSITION

In vitro, the extent of transporter induction is only moderate, compared to that of cytochrome P450 enzymes [85]. However, this is probably not an intrinsic feature of the regulation of transporters by PXR and CAR but rather a consequence of the differences in expression of transporter and cytochrome P450 genes during culture of hepatocytes. In contrast to cytochrome P450 genes, which are significantly (if not dramatically) down-regulated, expression of most transporters is unchanged or even enhanced by culturing human hepatocytes [85, 153]. Accordingly, the extent of induction of transporters and cytochrome P450 enzymes is similar *in vivo* [75, 90, 93] and thus induction of transporters may similarly modulate drug disposition.

Most drugs are not only transported but also metabolized, thus it is usually difficult to decide whether induction of transport plays a significant role in the increase in drug clearance. By using probe drugs such as digoxin, talinolol, or fexofenadine, which are transported as parent drugs with minimal metabolism [91, 154, 155], it was unequivocally shown that the induction of drug transport plays a significant role in drug disposition both in rodents and humans, thereby confirming earlier studies [156]. In rodents, Oatp1a4 mediates the uptake of digoxin into liver. Treatment of mice with PCN concomitantly induced hepatic *Oatp1a4* expression and digoxin uptake into liver [117]. In human volunteers, rifampin, SJW, and carbamazepine were shown to increase the expression of intestinal MDR1/P-gp and to decrease oral bioavailability and plasma levels of digoxin, talinolol, and fexofenadine [75, 90, 91, 98, 157, 158]. Rifampin also seems to induce hepatic efflux transporters, most likely MDR1/P-gp, as indicated by an increase of digoxin elimination into bile after rifampin treatment [5]. Carbamazepine was shown to additionally increase the renal clearance of talinolol, which may indicate induction of MDR1/P-gp expression in the tubular cells of the kidney [98]. Whereas effects on oral bioavailability, due to induction of intestinal MDR1/P-gp, were consistently observed, it was only reported by respective single studies that biliary and renal elimination are affected by treatment with inducers. Strikingly, only carbamazepine, which activates both PXR and CAR, and not the PXR ligand rifampin, showed an effect on the renal elimination of MDR1 probe drugs [5, 90, 91, 98], thereby indicating that renal CAR may be activated in clinical settings.

These studies further demonstrate that not only the induction of drug-metabolizing enzymes but also the induction of drug transporters is a mechanism underlying clinically relevant drug interactions. Accordingly, monitoring the drug plasma levels and increasing dosage, if necessary, has also been recommended for drugs, which are mainly transported and not metabolized such as digoxin, talinolol, and fexofenadine, if they are coadministered with rifampin [159].

4.7 CONCLUSIONS AND FUTURE PERSPECTIVES

As has been discussed above, activation of PXR and CAR by microsomal enzyme inducers results in increased expression of a battery of drug transporters including many ABC drug efflux transporters such as MDR1/P-gp, MRPs, and BCRP in intestine and liver as well as some hepatic uptake transporters such as Oatp1a4 in rodents and OATP1B1 in humans. Induction of some of these transporters has been shown to affect drug disposition with finally reducing the blood concentration of drugs. With regard to the protection from potentially harmful chemicals, the role of the induction of efflux transporters, especially in the intestine, is self-evident, whereas the induction of hepatic OATPs can only result in increased protection, if canalicular hepatic ABC transporters, which secrete the same chemicals or their metabolites into bile, are induced concomitantly. However, to protect the liver from the potential harmful action of chemicals, down-regulation of specific hepatic uptake transporters may also be appropriate. Actually, this was observed in mice. With the single exception of Oatp1a4, treatment of mice with PXR and CAR activators decreased the expression of most Oatps, which have been analyzed [119]. The mechanism of this down-regulation by PXR and CAR is not known and clearly should be addressed in the future.

Hepatic and intestinal induction of drug transporters is well documented: however, data are sparse in other tissues. It is especially challenging to analyze the induction of drug transporters in kidney and blood–brain barrier, as these tissues play important roles in the elimination of chemicals and in controlling the CNS accessibility of drugs, respectively.

Induction of only a few transporters has been analyzed on the molecular level. As it emerges that regulation by PXR and CAR is also influenced by the genomic context of their response elements, elucidation of the molecular mechanism by identifying the respective binding sites and response elements is crucial for a detailed understanding of induction. Furthermore, it will provide final proof of direct regulation by PXR and CAR, which is still missing for most transporters, despite the plethora of data obtained by treating animals, cells, and human volunteers with prototypical inducers.

REFERENCES

1. Hsu, S. I., Lothstein, L., and Horwitz, S. B. (1989) Differential overexpression of three mdr gene family members in multidrug-resistant J774. 2 mouse cells. Evidence that distinct P-glycoprotein precursors are encoded by unique mdr genes. *J. Biol. Chem.* **264**, 12053–12062.

2. Thiebaut, F., Tsuruo, T., Hamada, H., Gottesman, M. M., Pastan, I., and Willingham, M. C. (1987) Cellular localization of the multidrug-resistance gene product P-glycoprotein in normal human tissues. *Proc. Natl Acad. Sci. USA* **84**, 7735–7738.

3. Terao, T., Hisanga, E., Sai, Y., Tamai, I., and Tsuji, A. (1996) Active secretion of drugs from the small intestinal epithelium in rats by P-glycoprotein functioning as an absorption barrier. *J. Pharm. Pharmacol.* **48**, 1083–1089.

4. Stephens, R. H., O'Neill, C. A., Bennett, J., Humphrey, M., Henry, B., Rowland, M., and Warhurst, G. (2002) Resolution of P-glycoprotein and non-P-glycoprotein effects on drug permeability using intestinal tissues from mdr1a(–/–) mice. *Br. J. Pharmacol.* **135**, 2038–2046.

5. Drescher, S., Glaeser, H., Mürdter, T., Hitzl, M., Eichelbaum, M., and Fromm, M. F. (2003) P-glycoprotein-mediated intestinal and biliary digoxin transport in humans. *Clin. Pharmacol. Ther.* **73**, 223–231.

6. Annaert, P. P., Turncliff, R. Z., Booth, C. L., Thakker, D. R., and Brouwer, K. L. (2001) P-glycoprotein-mediated in vitro biliary excretion in sandwich-cultured rat hepatocytes. *Drug Metab. Dispos.* **29**, 1277–1283.

7. de Lannoy, I. A., Mandin, R. S., and Silverman, M. (1994) Renal secretion of vinblastine, vincristine and colchicine in vivo. *J. Pharmacol. Exp. Ther.* **268**, 388–395.

8. Cordon-Cardo, C., O'Brien, J. P., Casals, D., Rittman-Grauer, L., Biedler, J. L., Melamed, M. R., and Bertino, J. R. (1989) Multidrug-resistance gene (P-glycoprotein) is expressed by endothelial cells at blood-brain barrier sites. *Proc. Natl Acad. Sci. USA* **86**, 695–698.

9. Tatsuta, T., Naito, M., Oh-hara, T., Sugawara, I., and Tsuruo, T. (1992) Functional involvement of P-glycoprotein in blood-brain barrier. *J. Biol. Chem.* **267**, 20383–20391.

10. Schinkel, A. H., Smit, J. J., van Tellingen, O., Beijnen, J. H., Wagenaar, E., van Deemter, L., Mol, C. A., Van Der Valk, M. A., Robanus-Maandag, E. C., and te Riele, H. P.

(1994) Disruption of the mouse mdr1a P-glycoprotein gene leads to a deficiency in the blood-brain barrier and to increased sensitivity to drugs. *Cell* **77**, 491–502.

11. Tahara, H., Kusuhara, H., Fuse, E., and Sugiyama, Y. (2005) P-glycoprotein plays a major role in the efflux of fexofenadine in the small intestine and blood-brain barrier, but only a limited role in its biliary excretion. *Drug Metab. Dispos.* **33**, 963–968.

12. Zhang, Y. and Benet, L. Z. (2001) The gut as a barrier to drug absorption: combined role of cytochrome P450 3A and P-glycoprotein. *Clin. Pharmacokinet.* **40**, 159–168.

13. Chan, L. M., Lowes, S., and Hirst, B. H. (2004) The ABCs of drug transport in intestine and liver: efflux proteins limiting drug absorption and bioavailability. *Eur. J. Pharm. Sci.* **21**, 25–51.

14. Dietrich, C. G., Geier, A., and Oude Elferink, R. P. (2003) ABC of oral bioavailability: transporters as gatekeepers in the gut. *Gut* **52**, 1788–1795.

15. Dean, M., Rzhetsky, A., and Allikmets, R. (2001) The human ATP-binding cassette (ABC) transporter superfamily. *Genome Res.* **11**, 1156–1166.

16. Büchler, M., König, Brom, M., Kartenbeck, J., Spring, H., Horie, T., and Keppler, D. (1996) cDNA cloning of the hepatocyte canalicular isoform of the multidrug resistance protein, cMrp, reveals a novel conjugate export pump deficient in hyperbilirubinemic mutant rats. *J. Biol. Chem.* **271**, 15091–15098.

17. Schaub, T. P., Kartenbeck, J., König, J., Vogel, O., Witzgall, R., Kriz, W., and Keppler, D. (1997) Expression of the conjugate export pump encoded by the mrp2 gene in the apical membrane of kidney proximal tubules. *J. Am. Soc. Nephrol.* **8**, 1213–1221.

18. Schaub, T. P., Kartenbeck, J., König, J., Spring, H., Dorsam, J., Staehler, G., Storkel, S., Thon, W. F., and Keppler, D. (1999) Expression of the MRP2 gene-encoded conjugate export pump in human kidney proximal tubules and in renal cell carcinoma. *J. Am. Soc. Nephrol.* **10**, 1159–1169.

19. Fromm, M. F., Kauffmann, H. M., Fritz, P., Burk, O., Kroemer, H. K., Warzok, R. W., Eichelbaum, M., Siegmund, W., and Schrenk, D. (2000) The effect of rifampin treatment on intestinal expression of human MRP transporters. *Am. J. Pathol.* **157**, 1575–1580.

20. Paulusma, C. C., Kool, M., Bosma, P. J., Scheffer, G. L., ter Borg, F., Scheper, R. J., Tytgat, G. N., Borst, P., Baas, F., and Oude Elferink, R. P. (1997) A mutation in the human canalicular multispecific organic anion transporter gene causes the Dubin–Johnson syndrome. *Hepatology* **25**, 1539–1542.

21. Catania, V. A., Sanchez Pozzi, E. J., Luquita, M. G., Ruiz, M. L., Villanueva, S. S., Jones, B., and Mottino, A. D. (2004) Co-regulation of expression of phase II metabolizing enzymes and multidrug resistance-associated protein 2. *Ann. Hepatol.* **3**, 11–17.

22. Rost, D., Mahner, S., Sugiyama, Y., and Stremmel, W. (2002) Expression and localization of the multidrug resistance-associated protein 3 in rat small and large intestine. *Am. J. Physiol. Gastrointest. Liver Physiol.* **282**, G720-G726.

23. Leslie, E. M., Deeley, R. G., and Cole, S. P. (2005) Multidrug resistance proteins: role of P-glycoprotein, MRP1, MRP2, and BCRP (ABCG2) in tissue defense. *Toxicol. Appl. Pharmacol.* **204**, 216–237.

24. Haimeur, A., Conseil, G., Deeley, R. G., and Cole, S. P. (2004) The MRP-related and BCRP/ABCG2 multidrug resistance proteins: biology, substrate specificity and regulation. *Curr. Drug Metab.* **5**, 21–53.

25. Hilgendorf, C., Ahlin, G., Seithel, A., Artursson, P., Ungell, A. L., and Karlsson, J. (2007) Expression of thirty-six drug transporter genes in human intestine, liver, kidney, and organotypic cell lines. *Drug Metab. Dispos.* **35**, 1333–1340.

26. Taipalensuu, J., Törnblom, H., Lindberg, G., Einarsson, C., Sjöqvist, F., Melhus, H., Garberg, P., Sjöström, B., Lundgren, B., and Artursson, P. (2001) Correlation of gene expression of ten drug efflux proteins of the ATP-binding cassette transporter family in normal human jejunum and in human intestinal epithelial Caco-2 cells. *J. Pharmacol. Exp. Ther.* **299**, 164–170.

27. Maliepaard, M., Scheffer, G. L., Faneyte, I. F., van Gastelen, M. A., Pijnenborg, A. C., Schinkel, A. H., van De Vijver, M. J., Scheper, R. J., and Schellens, J. H. (2001) Subcellular localization and distribution of the breast cancer resistance protein transporter in normal human tissues. *Cancer Res.* **61**, 3458–3464.

28. Chen, R. Z., Robey, R. W., Belinsky, M. G., Shchaveleva, I., Ren, X. Q., Sugimoto, Y., Ross, D. D., Bates, S. E., and Kruh, G. D. (2003) Transport of methotrexate, methotrexate polyglutamates, and 17beta-estradiol 17-(beta-D-glucuronide) by ABCG2: effects of acquired mutations at R482 on methotrexate transport. *Cancer Res.* **63**, 4048–4054.

29. Suzuki, M., Suzuki, H., Sugimoto, Y., and Sugiyama, Y. (2003) ABCG2 transports sulfated conjugates of steroids and xenobiotics. *J. Biol. Chem.* **278**, 22644–22649.

30. Hagenbuch, B. and Meier, P. J. (2003) The superfamily of organic anion transporting polypeptides. *Biochim. Biophys. Acta* **1609**, 1–18.

31. Hagenbuch, B. and Meier, P. J. (2004) Organic anion transporting polypeptides of the OATP/SLC21 family: phylogenetic classification as OATP/SLCO superfamily, new nomenclature and molecular/functional properties. *Pflügers Arch.* **447**, 653–665.

32. Reichel, C., Gao, B., Van MOntfoort, J., Cattori, V., Rahner, C., Hagenbuch, B., Stieger, B., Kamisako, T., and Meier, P. J. (1999) Localization and function of the organic anion-transporting polypeptide Oatp2 in rat liver. *Gastroenterology* **117**, 688–695.

33. Noe, B., Hagenbuch, B., Stieger, B., and Meier, P. J. (1997) Isolation of a multispecific organic anion and cardiac glycoside transporter from rat brain. *Proc. Natl Acad. Sci. USA* **94**, 10346–10350.

34. Li, N., Hartley, D. P., Cherrington, N. J., and Klaassen, C. D. (2002) Tissue expression, ontogeny, and inducibility of rat organic anion transporting polypeptide 4. *J. Pharmacol. Exp. Ther.* **301**, 551–560.

35. Cattori, V., van Montfoort, J. E., Stieger, B., Landmann, L., Meijer, D. K., Winterhalter, K. H., Meier, P. J., and Hagenbuch, B. (2001) Localization of organic anion transporting polypeptide 4(Oatp4) in rat liver and comparison of its substrate specificity with Oatp1, Oatp2 and Oatp3. *Pflügers Arch.* **443**, 188–195.

36. Kullak-Ublick, G. A., Hagenbuch, B., Stieger, B., Schteingart, C. D., Hofmann, A. F., Wolkoff, A. W., and Meier, P. J. (1995) Molecular and functional characterization of an organic anion transporting polypeptide cloned from human liver. *Gastroenterology* **109**, 1274–1282.

37. Alcorn, J., Lu, X., Moscow, J. A., and McNamara, P. J. (2002) Transporter gene expression in lactating and nonlactating human mammary epithelial cells using real-time reverse transcription-polymerase chain reaction. *J. Pharmacol. Exp. Ther.* **303**, 487–496.

38. Glaeser, H., Bailey, D. G., Dresser, G. K., Gregor, J. C., Schwarz, U. I., McGrath, J. S., Jolicoeur, E., Lee, W., Leake, B. F., Tirona, R. G., et al. (2007) Intestinal drug transporter

expression and the impact of grapefruit juice in humans. *Clin. Pharmacol. Ther.* **81**, 362–370.

39. Kullak-Ublick, G. A., Ismair, M. G., Stieger, B., Landmann, L., Huber, R., Pizzagalli, F., Fattinger, K., Meier, P. J., and Hagenbuch, B. (2001) Organic anion-transporting polypeptide, B (OATP-B) and its functional comparison with three other OATPs of human liver. *Gastroenterology* **120**, 525–533.

40. König, J., Cui, Y., Nies, A. T., and Keppler, D. (2000) A novel human organic anion transporting polypeptide localized to the basolateral hepatocyte membrane. *Am. J. Physiol. Gastrointest. Liver Physiol.* **278**, G156–G164.

41. König, J., Cui, Y., Nies, A. T., and Keppler, D. (2000) Localization and genomic organization of a new hepatocellular organic anion transporting polypeptide. *J. Biol. Chem.* **275**, 23161–23168.

42. Kobayashi, D., Nozawa, T., Imai, K., Nezu, J. I., Tsuji, A., and Tamai, I. (2003) Involvement of human organic anion transporting polypeptide OATP-B (SLC21A9) in pH-dependent transport across intestinal apical membrane. *J. Pharmacol. Exp. Ther.* **306**, 703–708.

43. Koepsell, H. and Endou, H. (2004) The SLC22 drug transporter family. *Pflügers Arch.* **447**, 666–676.

44. Kato, Y., Suguira, M., Suguira, T., Wakayama, T., Kubo, Y., Kobayashi, D., Sai, Y., Tamai, I., Iseki, S., and Tsuji, A. (2006) Organic cation/carnitine transporter OCTN2 (Slc22a5) is responsible for carnitine transport across apical membranes of small intestinal epithelial cells in mouse. *Mol. Pharmacol.* **70**, 829–837.

45. Ganapathy, M. E., Huang, W., Rajan, D. P., Carter, A. L., Sugawara, M., Iseki, K., Leibach, F. H., and Ganapathy, V. (2000) Beta-lactam antibiotics as substrates for OCTN2, an organic cation/carnitine transporter. *J. Biol. Chem.* **275**, 1699–1707.

46. Grube M., Meyer zu Schwabedissen, H. E., Präger, D., Haney, J., Möritz, K. U., Meissner, K., Rosskopf, D., Eckel, L., Böhm, M., Jedlitschky, G., et al. (2006) Uptake of cardiovascular drugs into the human heart: expression, regulation, and function of the carnitine transporter OCTN2 (SLC22A5). *Circulation* **113**, 1114–1122.

47. Kliewer, S. A., Moore, J. T., Wade, L., Staudinger, J. L., Watson, M. A., Jones, S. A., McKee, D. D., Oliver, B. B., Willson, T. M., Zetterström, R. H., et al. (1998) An orphan nuclear receptor activated by pregnanes defines a novel steroid signaling pathway. *Cell* **92**, 73–82.

48. Moore, L. B., Parks, D. J., Jones, S. A., Bledsoe, R. K., Consler, T. G., Stimmel, J. B., Goodwin, B., Liddle, C., Blanchard, S. G., Willson, T. M., et al. (2000) Orphan nuclear receptors constitutive androstane receptor and pregnane X receptor share xenobiotic and steroid ligands. *J. Biol. Chem.* **275**, 15122–15127.

49. El-Sankary, W., Gibson, G. G., Ayrton, A., and Plant, N. (2001) Use of a reporter gene assay to predict and rank the potency and efficacy of CYP3A4 inducers. *Drug Metab. Dispos.* **29**, 1499–1504.

50. Anapolsky, A., Teng, S., Dixit, S., and Piquette-Miller, M. (2006) The role of pregnane X receptor in 2-acetylaminofluorene-mediated induction of drug transport and -metabolizing enzymes in mice. *Drug Metab. Dispos.* **34**, 405–409.

51. Lehmann, J. M., McKee, D. D., Watson, M. A., Willson, T. M., Moore, J. T., and Kliewer, S. A. (1998) The human orphan nuclear receptor PXR is activated by compounds that

regulate CYP3A4 gene expression and cause drug interactions. *J. Clin. Invest.* **102**, 1016–1023.

52. Moore, L. B., Goodwin, B., Jones, S. A., Wisely, G. B., Serabjit-Singh, C. J., Willson, T. M., Collins, J. L., and Kliewer, S. A. (2000) St John's wort induces hepatic drug metabolism through activation of the pregnane X receptor. *Proc. Natl Acad. Sci. USA* **97**, 7500–7502.

53. Sahi, J., Milad, M. A., Zheng, X., Rose, K. A., Wang, H., Stilgenbauer, L., Gilbert, D., Jolley, S., Stern, R. H., and Lecluyse, E. L. (2003) Avasimibe induces CYP3A4 and multiple drug resistance protein 1 gene expression through activation of the pregnane X receptor. *J. Pharmacol. Exp. Ther.* **306**, 1027–1034.

54. Burk, O., Arnold, K. A., Nussler, A. K., Schaeffeler, E., Efimova, E., Avery, B. A., Avery, M. A., Fromm M. F., and Eichelbaum, M. (2005) Antimalarial artemisinin drugs induce cytochrome P450 and MDR1 expression by activation of xenosensors pregnane X receptor and constitutive androstane receptor. *Mol. Pharmacol.* **67**, 1954–1964.

55. Bertilsson, G., Heidrich, J., Svensson, K., Asman, M., Jendeberg, L., Sydow-Bäckman, M., Ohlsson, R., Postlind, H., Blomquist, P., and Berkenstam, A. (1998) Identification of a human nuclear receptor defines a new signaling pathway for CYP3A induction. *Proc. Natl Acad. Sci. USA* **95**, 12208–12213.

56. Bauer, B., Hartz, A. M., Fricker, G., and Miller, D. S. (2004) Pregnane X receptor up-regulation of P-glycoprotein expression and transport function at the blood-brain-barrier. *Mol. Pharmacol.* **66**, 413–419.

57. Dotzlaw, H., Leygue, E., Watson, P., and Murphy, L. C. (1999) The human orphan receptor PXR messenger RNA is expressed in both normal and neoplastic breast tissue. *Clin. Cancer Res.* **5**, 2103–2107.

58. Kawamoto, T., Sueyoshi, T., Zelko, I., Moore, R., Washburn, K., and Negishi, M. (1999) Phenobarbital-responsive nuclear translocation of the receptor CAR in induction of the CYP2B gene. *Mol. Cell. Biol.* **19**, 6318–6322.

59. Fisher, C. D., Augustine, L. M., Maher, J. M., Nelson, D. M., Slitt, A. L., Klaassen, C. D., Lehman-McKeeman, L. D., and Cherrington, N. J. (2007) Induction of drug-metabolizing enzymes by garlic and allyl sulfide compounds via activation of constitutive androstane receptor and nuclear factor E2-related factor 2. *Drug Metab. Dispos.* **35**, 995–1000.

60. Slitt, A. L., Cherrington, N. J., Dieter, M. Z., Aleksunes, L. M., Scheffer, G. L., Huang, W., Moore, D. D., and Klaassen, C. D. (2006) Trans-stilbene oxide induces expression of genes involved in metabolism and transport in mouse liver via CAR and Nrf2 transcription factors. *Mol. Pharmacol.* **69**, 1554–1563.

61. Huang, W., Zhang, J., Chua, S. S., Qatanani, M., Han, Y., Granata, R., and Moore, D. D. (2003) Induction of bilirubin clearance by the constitutive androstane receptor. *Proc. Natl Acad. Sci. USA* **100**, 4156–4161.

62. Faucette, S. R., Zhang, T. C., Moore, R., Sueyoshi, T., Omiecinski, C. J., LeCluyse, E. L., Negishi, M., and Wang, H. (2007) Relative activation of human pregnane X receptor versus constitutive androstane receptor defines distinct classes of CYP2B6 and CYP3A4 inducers. *J. Pharmacol. Exp. Ther.* **320**, 72–80.

63. Maglich, J. M., Parks, D. J., Moore, L. B., Collins, J. L., Goodwin, B., Billin, A. N., Stoltz, C. A., Kliewer, S. A., Lambert, M. H., Willson, T. M., et al. (2003) Identification of a novel human constitutive androstane receptor (CAR) agonist and its use in the identification of CAR target genes. *J. Biol. Chem.* **278**, 17277–17283.

64. Rencurel, F., Foretz, M., Kaufmann, M. R., Stroka, D., Looser, R., Leclerc, I., da Silva Xavier, G., Rutter, G. A., Viollet, B., and Meyer, U. A. (2006) Stimulation of AMP-activated protein kinase is essential for the induction of drug metabolizing enzymes by phenobarbital in human and mouse liver. *Mol. Pharmacol.* **70**, 1925–1934.

65. Baes, M., Gulick, T., Choi, H. S., Martinoli, M. G., Simha, D., and Moore, D. D. (1994) A new orphan member of the nuclear hormone receptor superfamily that interacts with a subset of retinoic acid response elements. *Mol. Cell. Biol.* **14**, 1544–1552.

66. Choi, H. S., Chung, M., Tzameli, I., Simha, D., Lee, Y. K., Seol, W., and Moore, D. D. (1997) Differential transactivation by two isoforms of the orphan nuclear hormone receptor CAR. *J. Biol. Chem.* **272**, 23565–23571.

67. Wei, P., Zhang, J., Dowhan, D. H., Han, Y., and Moore, D. D. (2002) Specific and overlapping functions of the nuclear hormone receptors CAR and PXR in xenobiotic response. *Pharmacogenomics J.* **2**, 117–126.

68. Arnold, K. A., Eichelbaum, M., and Burk, O. (2004) Alternative splicing affects the function and tissue-specific expression of the human constitutive androstane receptor. *Nucl. Recept.* **2**, 1.

69. Rosenfeld, J., Vargas, R., Xie, W., and Evans, R. M. (2003) Genetic profiling defines the xenobiotic gene network controlled by the nuclear receptor pregnane X receptor. *Mol. Endocrinol.* **17**, 1268–1282.

70. Xie, W., Barwick, J. L., Simon, C. M., Pierce, A. M., Safe, S., Blumberg, B., Guzelian, P. S., and Evans, R. M. (2000) Reciprocal activation of xenobiotic response genes by nuclear receptors SXR/PXR and CAR. *Genes Dev.* **14**, 3014–3023.

71. Brady, J. M., Cherrington, N. J., Hartley, D. P., Buist, S. C., Li, N., and Klaassen, C. D. (2002) Tissue distribution and chemical induction of multiple drug resistance genes in rats. *Drug Metab. Dispos.* **30**, 838–844.

72. Song, X., Xie, M., Zhang, H., Li, Y., Sachdeva, K., and Yan, B. (2004) The pregnane X receptor binds to response elements in a genomic context-dependent manner and PXR activator rifampicin selectively alters the binding among target genes. *Drug Metab. Dispos.* **32**, 35–42.

73. Masuyama, H., Suwaki, N., Tateishi, Y., Nakatsukasa, H., Segawa, T., and Hiramatsu, Y. (2005) The pregnane X receptor regulates gene expression in a ligand- and promoter selective fashion. *Mol. Endocrinol.* **19**, 1170–1180.

74. Slitt, A. L., Cherrington, N. J., Fisher, C. D., Negishi, M., and Klaassen, C. D. (2006) Induction of genes for metabolism and transport by trans-stilbene oxide in livers of Sprague–Dawley and Wistar–Kyoto rats. *Drug Metab. Dispos.* **34**, 1190–1197.

75. Dürr, D., Stieger, B., Kullack-Ublick, G. A., Rentsch, K. M., Steinert, H. C., Meier, P. J., and Fattinger, K. (2000) St John's wort induces intestinal P-glycoprotein/MDR1 and intestinal and hepatic CYP3A4. *Clin. Pharmacol. Ther.* **68**, 598–604.

76. Demeule, M., Jodoin, J., Beaulieu, E., Brossard, M., and Beliveau, R. (1999) Dexamethasone modulation of multidrug transporters in normal tissues. *FEBS Lett.* **442**, 208–214.

77. Maglich, J. M., Stoltz, C. M., Goodwin, B., Hawkins-Brown, D., Moore, J. T., and Kliewer, S. A. (2002) Nuclear pregnane X receptor and constitutive androstane receptor regulate overlapping but distinct sets of genes involved in xenobiotic detoxification. *Mol. Pharmacol.* **62**, 638–646.

78. Teng, S. and Piquette-Miller, M. (2005) The involvement of the pregnane X receptor in hepatic gene regulation during inflammation in mice. *J. Pharmacol. Exp. Ther.* **312**, 841–848.

79. Cheng, X. and Klaassen, C. D. (2006) Regulation of mRNA expression of xenobiotic transporters by the pregnane X receptor in mouse liver, kidney, and intestine. *Drug Metab. Dispos.* **34**, 1863–1867.

80. Matheny, C. J., Ali, R. Y., Yang, X., and Pollack, G. M. (2004) Effect of prototypical inducing agents on P-glycoprotein and CYP3A expression in mouse tissues. *Drug Metab. Dispos.* **32**, 1008–1014.

81. Magnarin, M., Morelli, M., Rosati, A., Bartoli, F., Candussio, L., Giraldi, T., and Decorti, G. (2004) Induction of proteins involved in multidrug resistance (P-glycoprotein, MRP1, MRP2, LRP) and of CYP3A4 by rifampicin in LLC-PK1 cells. *Eur. J. Pharmacol.* **483**, 19–28.

82. Mills, J. B., Rose, K. A., Sadagopan, N., Sahi, J., and de Morais, S. M. (2004) Induction of drug metabolism enzymes and MDR1 using a novel human hepatocyte cell line. *J. Pharmacol. Exp. Ther.* **309**, 303–309.

83. Synold, T. W., Dussault, I., and Forman, B. M. (2001) The orphan nuclear receptor SXR coordinately regulates drug metabolism and efflux. *Nat. Med.* **7**, 584–590.

84. Watkins, R. E., Maglich, J. M., Moore, L. B., Wisely, G. B., Noble, S. M., Davis-Searles, P. R., Lambert, M. H., Kliewer, S. A., and Redinbo, M. R. (2003) 2.1 A crystal structure of human PXR in complex with St. John's wort compound hyperforin. *Biochemistry* **42**, 1430–1438.

85. Jigorel, E., Le Vee, M., Boursier-Neyret, C., Parmentier, Y., and Fardel, O. (2006) Differential regulation of sinusoidal and canalicular hepatic drug transporter expression by xenobiotics activating drug-sensing receptors in primary human hepatocytes. *Drug Metab. Dispos.* **34**, 1756–1763.

86. Dixit, V., Hariparsad, N., Li, F., Desai, P., Thummel, K. E., and Unadkat, J. D. (2007) Cytochrome P450 enzymes and transporters induced by anti-HIV protease inhibitors in human hepatocytes: implications for predicting clinical drug interactions. *Drug Metab. Dispos.* **35**, 1853–1859.

87. Geick, A., Eichelbaum, M., and Burk, O. (2001) Nuclear receptor response elements mediate induction of intestinal MDR1 by rifampin. *J. Biol. Chem.* **276**, 14581–14587.

88. Schuetz, E. G., Beck, W. T., and Schuetz, J. D. (1996) Modulators and substrates of P-glycoprotein and cytochrome P4503A coordinately up-regulate these proteins in human colon carcinoma cells. *Mol. Pharmacol.* **49**, 311–318.

89. Pfrunder, A., Gutmann, H., Beglinger, C., and Drewe, J. (2003) Gene expression of CYP3A4, ABC-transporters (MDR1 and MRP1-MRP5) and hPXR in three different human colon carcinoma cell lines. *J. Pharm. Pharmacol.* **55**, 59–66.

90. Greiner, B., Eichelbaum, M., Fritz, P., Kreichgauer, H. P., von Richter, O., Zundler, J., and Kroemer, H. K. (1999) The role of intestinal P-glycoprotein in the interaction of digoxin and rifampin. *J. Clin. Invest.* **104**, 147–153.

91. Westphal, K., Weinbrenner, A., Zschiesche, M., Franke, G., Knoke, M., Oertel, R., Fritz, P., von Richter, O., Warzok, R., Hachenberg, T., et al. (2000) Induction of P-glycoprotein by rifampin increases intestinal secretion of talinolol in human beings: a new type of drug/drug interaction. *Clin. Pharmacol. Ther.* **68**, 345–355.

92. Giessmann, T., Modess, C., Hecker, U., Zschiesche, M., Dazert, P., Kunert-Keil, C., Warzok, R., Engel, G., Weitschies, W., Cascorbi, I., et al. (2004) CYP2D6 genotype and induction of intestinal drug transporters by rifampin predict presystemic clearance of carvedilol in healthy subjects. *Clin. Pharmacol. Ther.* **75**, 213–222.

93. Oscarson, M., Burk, O., Winter, S., Schwab, M., Wolbold, R., Dippon, J., Eichelbaum, M., and Meyer, U. A. Effects of rifampicin on global gene expression in human small intestine. *Pharmacogenetics Genomics*, **17**, 907–918.

94. Ashgar, A., Gorski, J. C., Haehner-Daniels, B., and Hall, S. D. (2002) Induction of multidrug resistance-1 and cytochrome P450 mRNAs in human mononuclear cells by rifampin. *Drug Metab. Dispos.* **30**, 20–26.

95. Owen, A., Goldring, C., Morgan, P., Park, B. K., and Pirmohamed, M. (2006) Induction of P-glycoprotein in lymphocytes by carbamazepine and rifampicin: the role of nuclear hormone response elements. *Br. J. Clin. Pharmacol.* **62**, 237–242.

96. Ichikawa, T., Horie-Inoue, K., Ikeda, K., Blumberg, B., and Inoue, S. (2006) Steroid and xenobiotic receptor SXR mediates vitamin K2-activated transcription of extracellular matrix-related genes and collagen accumulation in osteoblastic cells. *J. Biol. Chem.* **281**, 16927–16934.

97. Hennessy, M., Kelleher, D., Spiers, J. P., Barry, M., Kavanagh, P., Back, D., Mulcahy, F., and Feely, J. (2002) St John's wort increases expression of P-glycoprotein: implications for drug interactions. *Br. J. Clin. Pharmacol.* **53**, 75–82.

98. Giessmann, T., May, K., Modess, C., Wegner, D., Hecker, U., Zschiesche, M., Dazert, P., Grube, M., Schroeder, E., Warzok, R., et al. (2004) Carbamazepine regulates intestinal P-glycoprotein and multidrug resistance protein MRP2 and influences disposition of talinolol in humans. *Clin. Pharmacol. Ther.* **76**, 192–200.

99. Kast, H. R., Goodwin, B., Tarr, P. T., Jones, S. A., Anisfeld, A. M., Stoltz, C. M., Tontonoz, P., Kliewer, S., Willson, T. M., and Edwards, P. A. (2002) Regulation of multidrug resistance-associated protein 2 (ABCC2) by the nuclear receptors pregnane X receptor, farnesoid X-activated receptor and constitutive androstane receptor. *J. Biol. Chem.* **277**, 2908–2915.

100. Johnson, D. R. and Klaassen, C. D. (2002) Regulation of rat multidrug resistance protein 2 by classes of prototypical microsomal enzyme inducers that activate distinct transcription pathways. *Toxicol. Sci.* **67**, 182–189.

101. Schrenk, D., Baus, P. R., Ermel, N., Klein, C., Vorderstemann, B., and Kauffmann, H. M. (2001) Up-regulation of transporters of the MRP family by drugs and toxins. *Toxicol. Lett.* **120**, 51–57.

102. Teng, S., Jekerle, V., and Piquette-Miller, M. (2003) Induction of ABCC3 (MRP3) by pregnane X receptor activators. *Drug Metab. Dispos.* **31**, 1296–1299.

103. Saini, S. P., Mu, Y., Gong, H., Toma, D., Uppal, H., Ren, S., Li, S., Poloyac, S. M., and Xie, W. (2005) Dual role of orphan nuclear receptor pregnane X receptor in bilirubin detoxification in mice. *Hepatology* **41**, 497–505.

104. Wagner, M., Halilbasic, E., Marschall, H. U., Zollner, G., Fickert, P., Langner, C., Zatloukal, K., Denk, H., and Trauner, M. (2005) CAR and PXR agonists stimulate hepatic bile acid and bilirubin detoxification and elimination pathways in mice. *Hepatology* **42**, 420–430.

105. Maher, J. M., Cheng, X., Slitt, A. L., Dieter, M. Z., and Klaassen, C. D. (2005) Induction of the multidrug resistance-associated protein family of transporters by chemical

activators of receptor-mediated pathways in mouse liver. *Drug Metab. Dispos.* **33**, 956–962.

106. Kauffmann, H. M., Pfannschmidt, S., Zöller, H., Benz, A., Vorderstemann, B., Webster, J. I., and Schrenk, D. (2002) Influence of redox-active compounds and PXR-activators on human MRP1 and MRP2 gene expression. *Toxicology* **171**, 137–146.

107. Dussault, I., Lin, M., Hollister, K., Wang, E. H., Synold, T. W., and Forman, B. M. (2001) Peptide mimetic HIV protease inhibitors are ligands for the orphan receptor SXR. *J. Biol. Chem.* **276**, 33309–33312.

108. Oscarson, M., Zanger, U. M., Rifki, O. F., Klein, K., Eichelbaum, M., and Meyer, U. A. (2006) Transcriptional profiling of genes induced in the livers of patients treated with carbamazepine. *Clin. Pharmacol. Ther.* **80**, 440–456.

109. Cherrington, N. J., Hartley, D. P., Li, N., Johnson, D. R., and Klaassen, C. D. (2002) Organ distribution of multidrug resistance proteins 1, 2, and 3 (Mrp 1, 2, and 3) mRNA and hepatic induction of Mrp3 by constitutive androstane receptor activators in rats. *J. Pharmacol. Exp. Ther.* **300**, 97–104.

110. Staudinger, J. L., Madan, A., Carol, K. M., and Parkinson, A. (2003) Regulation of drug transporter gene expression by nuclear receptors. *Drug Metab. Dispos.* **31**, 523–527.

111. Kiuchi, Y., Suzuki, H., Hirohashi, T., Tyson, C. A., and Sugiyama, Y. (1998) cDNA cloning and inducible expression of human multidrug resistance associated protein 3 (MRP3). *FEBS Lett.* **433**, 149–152.

112. Slitt, A. L., Cherrington, N. J., Maher, J. M., and Klaassen, C. D. (2003) Induction of multidrug resistance protein 3 in rat liver is associated with altered vectorial excretion of acetaminophen metabolites. *Drug Metab. Dispos.* **31**, 1176–1186.

113. Assem, M., Schuetz, E. G., Leggas, M., Sun, D., Yasuda, K., Reid, G., Zelcer, N., Adachi, M., Strom, S., Evans, R. M., et al. (2004) Interactions between hepatic Mrp4 and Sult2a as revealed by the constitutive androstane receptor and Mrp4 knockout mice. *J. Biol. Chem.* **279**, 22250–22257.

114. Miki, Y., Suzuki, T., Kitada, K., Yabuki, N., Shibuya, R., Moriya, T., Ishida, T., Ohuchi, N., Blumberg, B., and Sasano, H. (2006) Expression of the steroid and xenobiotic receptor and its possible target gene, organic anion transporting polypeptide-A, in human breast carcinoma. *Cancer Res.* **66**, 535–542.

115. Rausch-Derra, L. C., Hartley, D. P., Meier, P. J., and Klaassen, C. D. (2001) Differential effects of microsomal enzyme-inducing chemicals on the hepatic expression of rat organic anion transporters, Oatp1 and Oatp2. *Hepatology* **33**, 1469–1478.

116. Guo, G. L., Choudhuri, S., and Klaassen, C. D. (2002) Induction profile of rat organic anion transporting polypeptide 2 (oatp2) by prototypical drug-metabolizing enzyme inducers that activate gene expression through ligand-activated transcription factor pathways. *J. Pharmacol. Exp. Ther.* **300**, 206–212.

117. Staudinger J., Liu, Y., Madan, A., Habeebu, S., and Klaassen, C. D. (2001) Coordinate regulation of xenobiotic and bile acid homeostasis by pregnane X receptor. *Drug Metab. Dispos.* **29**, 1467–1472.

118. Staudinger, J. L., Goodwin, B., Jones, S. A., Hawkins-Brown, D., MacKenzie, K. I., LaTour, A., Liu, Y., Klaassen, C. D., Brown, K. K. Reinhard, J., et al. (2001) The nuclear receptor PXR is a lithocholic acid sensor that protects against liver toxicity. *Proc. Natl Acad. Sci. USA* **98**, 3369–3374.

119. Cheng, X., Maher, J., Dieter, M. Z., and Klaassen, C. D. (2005) Regulation of mouse organic-anion-transporting polypeptides (Oatps) in liver by prototypical microsomal enzyme inducers that activate distinct transcription factor pathways. *Drug Metab. Dispos.* **33**, 1276–1282.

120. Frank, C., Makkonen, H., Dunlop, T. W., Matilainen, M., Väisänen, S., and Carlberg, C. (2005) Identification of pregnane X receptor binding sites in the regulatory regions of genes involved in bile acid homeostasis. *J. Mol. Biol.* **346**, 505–519.

121. Yoshinari, K., Sueyoshi, T., Moore, R., and Negishi, M. (2001) Nuclear receptor CAR as a regulatory factor for the sexually dimorphic induction of CYP2B1 gene by phenobarbital in rat livers. *Mol. Pharmacol.* **59**, 278–284.

122. Turncliff, R. Z., Meier, P. J., and Brouwer, K. L. (2004) Effect of dexamethasone treatment on the expression and function of transport proteins in sandwich-cultured rat hepatocytes. *Drug Metab. Dispos.* **32**, 834–839.

123. Oesch, F., Arand, M., Benedetti, M. S., Castelli, M. G., and Dostert, P. (1996) Inducing properties of rifampicin and rifabutin for selected enzyme activities of the cytochrome P-450 and UDP-glucuronyltransferase superfamilies in female rat liver. *J. Antimicrob. Chemother.* **37**, 1111–1119.

124. Runge, D., Köhler, C., Kostrubsky, V. E., Jäger, D., Lehmann, T., Runge, D. M., May, U., Beer Stolz, D., Strom, S. C., Fleig, W. E., et al. (2000) Induction of Cytochrome P450 (CYP)1A1, CYP1A2, and CYP3A4 but not CYP2C9, CYP2C19, multidrug resistance (MDR-1) and multidrug resistance associated protein (MRP-1) by prototypical inducers in human hepatocytes. *Biochem. Biophys. Res. Commun.* **273**, 333–341.

125. Hagenbuch, N., Reichel, C., Stieger, B., Cattori, V., Fattinger, K. E., Landmann, L., Meier, P. J., and Kullak-Ublick, G. A. (2001) Effect of phenobarbital on the expression of bile salt and organic anion transporters of rat liver. *J. Hepatol.* **34**, 881–887.

126. Xiong, H., Yoshinari, K., Brouwer, K. L., and Negishi, M. (2002) Role of constitutive androstane receptor in the in vivo induction of Mrp3 and CYP2B1/2 by phenobarbital. *Drug Metab. Dispos.* **30**, 918–923.

127. Cherrington, N. J., Slitt, A. L., Maher, J. M., Zhang, X. X., Zhang, J., Huang, W., Wan, Y. J., Moore, D. D., and Klaassen, C. D. (2003) Induction of multidrug resistance protein 3 (MRP3) in vivo is independent of constitutive androstane receptor. *Drug Metab. Dispos.* **31**, 1315–1319.

128. Ueda, A., Hamadeh, H. K., Webb, H. K., Yamamoto, Y., Sueyoshi, T., Afshari, C. A., Lehmann, J. M., and Negishi, M. (2002) Diverse roles of nuclear orphan receptor CAR in regulating hepatic genes in response to phenobarbital. *Mol. Pharmacol.* **61**, 1–6.

129. Duret, C., Daujat-Chavanieu, M., Pascussi, J. M., Pichard-Garcia, L., Balaguer, P., Fabre, J. M., Vilarem, M. J., Maurel, P., and Gerbal-Chaloin, S. (2006) Ketoconazole and miconazole are antagonists of the human glucocorticoids receptor: consequences on the expression and function of the constitutive androstane receptor and the pregnane X receptor. *Mol. Pharmacol.* **70**, 329–339.

130. Maeda, T., Oyabu, M., Yotsumoto, T., Higashi, R., Nagata, K., Yamazue, Y., and Tamai, I. (2007) Effect of pregnane X receptor ligand on pharmacokinetics of substrates of organic cation transporter Oct1 in rats. *Drug Metab. Dispos.* **35**, 1580–1586.

131. Rae, J. M., Johnson, M. D., Lippman, M. E., and Flockhart, D. A. (2001) Rifampin is a selective, pleiotropic inducer of drug metabolism genes in human hepatocytes: studies with cDNA and oligonucleotide expression arrays. *J. Pharmacol. Exp. Ther.* **299**, 849–857.

132. Zhang, H., LeCluyse, E., Liu, L., Hu, M., Matoney, L., Zhu, W., and Yan, B. (1999) Rat pregnane X receptor: molecular cloning, tissue distribution, and xenobiotic regulation. *Arch. Biochem. Biophys.* **368**, 14–22.

133. Pollock, C. B., Rogatcheva, M. B., and Schook, L. B. (2007) Comparative genomics of xenobiotic metabolism: a porcine-human PXR gene comparison. *Mamm. Genome* **18**, 210–219.

134. Miki, Y., Suzuki, T., Tazawa, C., Blumberg, B., and Sasano, H. (2005) Steroid and xenobiotic receptor (SXR), cytochrome P450 3A4 and multidrug resistance gene 1 in human adult and fetal tissues. *Mol. Cell. Endocrinol.* **231**, 75–85.

135. Moore, L. B., Maglich, J. M., McKee, D. D., Wisely, B., Willson, T. M., Kliewer, S. A., Lambert, M. H., and Moore, J. T. (2002) Pregnane X receptor (PXR), constitutive androstane receptor (CAR), and benzoate X receptor (BXR) define three pharmacologically distinct classes of nuclear receptors. *Mol. Endocrinol.* **16**, 977–986.

136. Chen, C. and Klaassen, C. D. (2004) Rat multidrug resistance protein 4 (Mrp4, Abcc4): molecular cloning, organ distribution, postnatal renal expression, and chemical inducibility. *Biochem. Biophys. Res. Commun.* **317**, 46–53.

137. Bauer, B., Yang, X., Hartz, A. M., Olson, E. R., Zhao, R., Kalvass, J. C., Pollack, G. M., and Miller, D. S. (2006) In vivo activation of human pregnane X receptor tightens the blood-brain-barrier to methadone through P-glycoprotein up-regulation. *Mol. Pharmacol.* **70**, 1212–1219.

138. Owen, A., Chandler, B., Back, D. J., and Khoo, S. H. (2004) Expression of pregnane-X-receptor transcript in peripheral blood mononuclear cells and correlation with MDR1 mRNA. *Antivir. Ther.* **9**, 819–821.

139. Albermann, N., Schmitz-Winnenthal, F. H., Z'graggen, K., Volk, C., Hoffmann, M. M., Haefeli, W. E., and Weiss, J. (2005) Expression of the drug transporters MDR1/ABCB1, MRP1/ABCC1, MRP2/ABCC2, BCRP/ABCG2, and PXR in peripheral blood mononuclear cells and their relationship with the expression in intestine and liver. *Biochem. Pharmacol.* **70**, 949–958.

140. Tabb, M. M., Sun, A., Zhou, C., Grün, F., Errandi, J., Romero, K., Pham, H., Inoue, S., Mallick, S., Lin, M., et al. (2003) Vitamin K2 regulation of bone homeostasis is mediated by the steroid and xenobiotic receptor SXR. *J. Biol. Chem.* **278**, 43919–43927.

141. Desvergne, B. and Wahli, W. (1999) Peroxisome proliferator-activated receptors: nuclear control of metabolism. *Endocr. Rev.* **20**, 649–688.

142. Forman, B. M., Chen, J., and Evans, R. M. (1997) Hypolipidemic drugs, polyunsaturated fatty acids, and eicosanoids are ligands for peroxisome proliferator-activated receptors alpha and delta. *Proc. Natl Acad. Sci. USA* **94**, 4312–4317.

143. Peters, J. M., Hennuyer, N., Staels, B., Fruchart, J. C., Fievet, C., Gonzalez, F. J., and Auwerx, J. (1997) Alterations in lipoprotein metabolism in peroxisome proliferator-activated receptor alpha-deficient mice. *J. Biol. Chem.* **272**, 27307–27312.

144. Kok, T., Bloks, V. W., Wolters, H., Havinga, R., Jansen, P. L., Staels, B., and Kuipers, F. (2003) Peroxisome proliferator-activated receptor alpha (PPARalpha)-mediated regulation of multidrug resistance 2 (Mdr2) expression and function in mice. *Biochem. J.* **369**, 539–547.

145. Moffit, J. S., Aleksunes, L. M., Maher, J. M., Scheffer, G. L., Klaassen, C. D., and Manautou, J. E. (2006) Induction of hepatic transporters multidrug resistance-associated proteins (Mrp) 3 and 4 by clofibrate is regulated by peroxisome proliferator-activated receptor alpha. *J. Pharmacol. Exp. Ther.* **317**, 537–545.

146. Aouabdi, S., Gibson, G., and Plant, N. (2006) Transcriptional regulation of the PXR gene: identification and characterization of a functional peroxisome proliferator-activated receptor alpha binding site within the proximal promoter of PXR. *Drug Metab. Dispos.* **34**, 138–144.

147. Nie, W., Sweetser, S., Rinella, M., and Green, R. M. (2005) Transcriptional regulation of murine Slc22a1 by peroxisome proliferator agonist receptor-alpha and -gamma. *Am. J. Physiol. Gastrointest. Liver Physiol.* **288**, G207-G212.

148. Luci, S., Geissler, S., König, B., Koch, A., Stangl, G. I., Hirche, F., and Eder, K. (2006) PPARalpha agonists up-regulate organic cation transporters in rat liver cells. *Biochem. Biophys. Res. Commun.* **350**, 704–708.

149. Burk, O., Arnold, K. A., Geick, A., Tegude, H., and Eichelbaum, M. (2005) A role for constitutive androstane receptor in the regulation of human intestinal MDR1 expression. *Biol. Chem.* **386**, 503–513.

150. Frank, C., Gonzalez, M. M., Oinonen, C., Dunlop, T. W., and Carlberg, C. (2003) Characterization of DNA complexes formed by the nuclear receptor constitutive androstane receptor. *J. Biol. Chem.* **278**, 43299–43310.

151. Podvinec, M., Kaufmann, M. R., Handschin, C., and Meyer, U. A. (2002) NUBIScan, an in silico approach for prediction of nuclear receptor response elements. *Mol. Endocrinol.* **16**, 1269–1279.

152. Guo, G. L., Staudinger, J., Ogura, K., and Klaassen, C. D. (2002) Induction of rat organic anion transporting polypeptide 2 by pregnenolone-16α-carbonitrile is via interaction with pregnane X receptor. *Mol. Pharmacol.* **61**, 832–839.

153. Richert, L., Liguori, M. J., Abadie, C., Heyd, B., Mantion, G., Halkic, N., and Waring, J. F. (2006) Gene expression in human hepatocytes in suspension after isolation is similar to the liver of origin, is not affected by hepatocyte cold storage and cryopreservation, but is strongly changed after hepatocyte plating. *Drug Metab. Dispos.* **34**, 870–879.

154. Lacarelle, B., Rahmani, R., de Sousa, G., Durand, A., Placidi, M., and Cano, J. P. (1991) Metabolism of digoxin, digoxigenin digitoxosides and digoxigenin in human hepatocytes and liver microsomes. *Fundam. Clin. Pharmacol.* **5**, 567–582.

155. Dresser, G. K., Schwarz, U. I., Wilkinson, G. R., and Kim, R. B. (2003) Coordinate induction of both cytochrome P4503A and MDR1 by St. John's wort in healthy subjects. *Clin. Pharmacol. Ther.* **73**, 41–50.

156. Klaassen, C. D. and Slitt, A. L. (2005) Regulation of hepatic transporters by xenobiotic receptors. *Curr. Drug Metab.* **6**, 309–328.

157. Johne, A., Brockmöller, J., Bauer, S., Maurer, A., Langheinrich, M., and Roots, I. (1999) Pharmacokinetic interaction of digoxin with an herbal extract from St. John's wort (Hypericum perforatum). *Clin. Pharmacol. Ther.* **66**, 338–345.

158. Hamman, M. A., Bruce, M. A., Haehner-Daniels, B. D., and Hall, S. D. (2001) The effect of rifampin administration on the disposition of fexofenadine. *Clin. Pharmacol. Ther.* **69**, 114–121.

159. Niemi, M., Backman, J. T., Fromm, M. F., Neuvonen, P. J., and Kivistö, K. T. (2003) Pharmacokinetic interactions with rifampicin. *Clin. Pharmacokinet.* **42**, 819–850.

5

STRUCTURE AND FUNCTION OF PXR AND CAR

X. Edward Zhou and H. Eric Xu

Laboratory of Structural Sciences, Van Andel Research Institute, Grand Rapids, MI, USA

5.1 INTRODUCTION

The pregnane X receptor (PXR, NR1I2) and the constitutive androstane receptor (CAR, NR1I3) are DNA binding and ligand-regulated transcriptional factors that belong to the superfamily of nuclear receptors [1, 2]. Both receptors contain a highly conserved DNA binding domain and a moderately conserved ligand binding domain (LBD) that are connected by a flexible hinge region. The biological functions of PXR and CAR are mediated through a heterodimer complex with the retinoid X receptor (RXR), which also serves as an obligatory partner for many other nuclear receptors. Activation of either PXR or CAR initiates expression of cytochrome P450 enzymes and other proteins involved in metabolism and excretion of xenobiotic and endobiotic compounds; thus, PXR and CAR are key receptors that regulate the metabolism and detoxification of drugs and other chemicals in human body [3–7].

In common with most other nuclear receptors, agonist binding induces conformational changes in the receptor LBD that leads to receptor translocation to the nucleus, recruitment of coregulators, and initiation of gene expression through the assembly of transcriptional machinery [8]. In addition to its agonist-induced activity, CAR has high level constitutive activity that allows it to regulate the expression of target genes without direct binding of agonists. This constitutive activity of CAR can be repressed by an antagonist (also called an inverse agonist) such as androstanol, which turns off the constitutive transcription activity of CAR by disrupting the active

Nuclear Receptors in Drug Metabolism   Edited by Wen Xie
Copyright © 2009 John Wiley & Sons, Inc.

conformation of the receptor [9, 10]. In this chapter, we review structural studies of PXR and CAR, and our understanding of drug metabolism through the crystal structures of these two receptors.

5.2   STRUCTURE AND FUNCTION OF PXR

5.2.1   Crystal Structures of Human PXR LBD/Ligand Complexes

There are five published crystal structures of the human PXR LBD. The first two are an apo structure and a complex structure with the cholesterol-lowering drug 4-[2,2-bis(diethoxyphosphoryl)ethenyl]-2,6-ditert-butyl-phenol (SR12813), which were solved by the Redinbo and Kliewer groups [11]. Subsequent studies performed by the same groups determined the structures of human PXR LBD in complexes with hyperforin (the active agent of the antidepressant drug St. John's wort [12]), with the antibiotic rifampicin [13], T1317, a synthetic agonist for human PXR and liver X receptor (LXR) [14], and with SR12813 and a 25-residue coactivator motif (Figure 5.1) [15]. The overall features of PXR from these structures are similar, with the core domain of the PXR LBD comprising 10 α-helices (H1, H3, H3′, H4, H5, H7–H10, and H12) and 5 β-strands (S1, S1′, and S2–S4), instead of the typical 3 β-strands of most other nuclear receptors (Figures 5.1 and 5.3). Helix H12 is also termed the activation function 2 (AF-2) helix, because its conformation is the key determinant for the active status of the receptor (Figure 5.2c) [10]. In all PXR structures, the AF-2 helix adopts an active position by packing tightly against the core domain of the LBD (Figure 5.1c), suggesting that PXR is inclined to the activation conformation even in the absence of ligand. Ligand binding within the lower half of the LBD stabilizes the overall structure and thus the active conformation of the PXR, providing a mechanism for the receptor activation [16].

5.2.2   The Flexibility of the Ligand Binding Pocket of the PXR LBD

The PXR ligand binding pocket is formed by a scaffold formed by helices H3, H5, H7, and H10, and five β-strands. As in other nuclear receptors, the ligand binding pocket of PXR is within the bottom half of the LBD and is sandwiched between two layers of helices (Figures 5.1c and 5.2a). In terms of its general hydrophobic nature, the PXR pocket is similar to the ligand binding pockets of most nuclear receptor LBDs. However, the PXR pocket has several distinct features that facilitate PXR's function as a promiscuous drug-metabolizing nuclear receptor. The most critical feature of PXR LBD is the flexibility of the structural elements surrounding the ligand binding pocket. This flexibility allows PXR to bind promiscuously to distinct ligands of various sizes, from SR12813 (having a molecular weight of 507 Da), hyperforin (537 Da), to rifampicin (823 Da) (Figure 5.1b). This feature is primarily attributed to residues 177–321, which form the lower left part of the ligand binding pocket. This region, containing mainly loops—with an occasional short helix and two small β-strands in some determined structures—is very flexible and prone to conformational

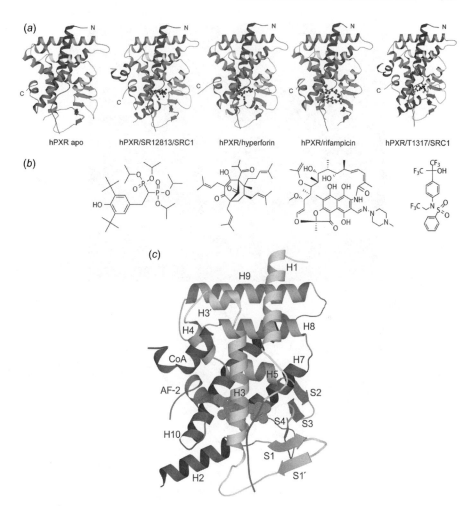

FIGURE 5.1 *Human PXR crystal structures.* (*a*) Diagrams of the crystal structures of the apo, ligand-bound, and coactivator-bound human PXR LBD. A rainbow ramp color code of blue to red is used to trace the amino acid chain from the N terminus to the C terminus. (*b*) Chemical structures of the PXR ligands. (*c*) The three-layered core structure of PXR LBD. The layers are shown by the brightness of the color: the first layer is the brightest and the third, the darkest. The helix AF-2 is colored red, and the coactivator motif is magenta. The ligand binding pocket is located at the bottom half of the LBD between the first and third layers. All helices, strands, and the coactivator motif are labeled, except for the helix H6, which is behind the core structure. See color insert for figure 5.1a and 5.1c.

changes, allowing the PXR pocket to expand from a volume of $1280 \, \text{Å}^3$ when binding to SR12813 to $1600 \, \text{Å}^3$ upon binding to rifampicin (Figure 5.2*a*) [10–12, 15].

The first part of this PXR flexible region (residues 177–228) is unique in the nuclear receptor superfamily. Within this region, residues 192–210 form one (in the coactivator-bound structures) or two (in hyperforin-bound structure) short α-helices

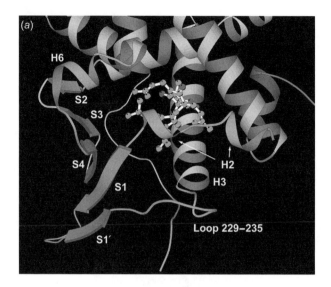

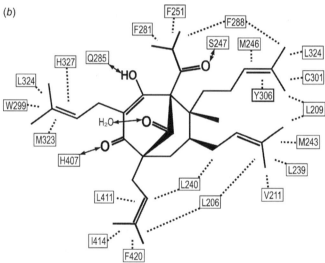

FIGURE 5.2 *PXR ligand binding and coactivator recruitment.* (*a*) A diagram of PXR ligand binding pocket bound to hyperforin. Cyan-colored regions are the "lid" of the pocket and its supporting regions; the orange region is the C-terminal AF-2 motif. The structural elements of the "lid" and the supporting regions are labeled. (*b*) Schematic representation of the ligand binding residues of PXR LBD in complex with hyperforin. A total of 23 residues are involved in binding to hyperforin, 6 of which are observed binding to ligands in all determined structures. (*c*) PXR LBD/coactivator interface. SRC-1 motif is colored magenta; helices H3, H3′, and H4, green; and helix AF-2, red. The charge clamp of the LBD is formed by residues K259 and E427, which form hydrogen bonds with the nitrogen of I689 and the carbonyl oxygen of L693 in the coactivator. See color insert for figure 5.2a and 5.2c.

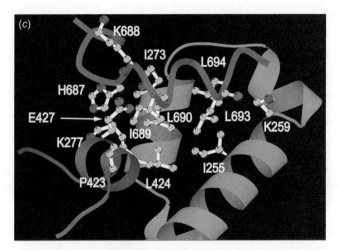

FIGURE 5.2   (*Continued*)

(H2), followed by a short loop and two β-strands (S1 and S1′, residues 212–228) that form a five-stranded sheet with strands S2, S3, and S4 in most determined PXR LBD structures (Figure 5.2*a*). Residues 177–191 are invisible in all crystal structures and are presumed to be disordered in a solvent-exposed environment. This disordered region, together with the subsequent helix H2, is the major basis for the promiscuous nature of the PXR pocket. Helix H2 and its flanking loops, at the bottom of the PXR LBD structures, are where helix H6 is positioned in other nuclear receptor LBDs. Generally, this region is flexible across the nuclear receptor family, suggesting that it may serve as a "lid" that governs entrance to the ligand binding pocket (Figure 5.2*a*). A similar helical structure and flexible region are also observed in the LBD structures of peroxisome proliferator-activated receptors (PPARs) [17, 18], indicating that there may be a common pathway of ligand entry and exit in these two nuclear receptors.

The second part of this flexible region (residues 229–235) is traced as an extended loop in the structures of the apo receptor and the complex bound to SR12813 or hyperforin, but it is disordered in the rifampicin complex structure. This region is located below the position of helix H2 and supports this helix as the lid of the ligand binding pocket (Figure 5.2*a*). The third part of the flexible region (residues 309–321) forms a loop in the apo and SR12813 complex structures but becomes a helix (H6) in the SR12813/SRC-1 (steroid receptor coactivator 1) and hyperforin complex structures. The different conformations of these structures further highlight the flexible nature of this region that forms the primary basis for promiscuity of the PXR ligand binding pocket (Figure 5.2*a*) [15].

Besides the secondary structural elements above, conformational changes of the ligand binding residues can in part contribute to the ligand promiscuity of PXR. Six residues (M243, S247, Q285, W299, H407, and F420) are shown to be involved in direct contacts with ligands in all determined human PXR structures (Figure 5.2*b*).

The side chains of several ligand binding residues (such as H407, M243, S247, and Q285) are in different conformations when the receptor binds to different ligands [11–15].

5.2.3   Coactivator Binding and Transcriptional Activity

The transcriptional activity of PXR is mediated through nuclear receptor coactivators, typified by the three members (SRC-1, SRC-2, and SRC-3) of the steroid receptor coactivator family [19, 16]. These coactivators are components of large protein complexes that contain chromatin-modifying enzymes to activate transcription. Agonist-bound nuclear receptors recruit coactivators, which in turn stabilize the receptor/ligand complex and potentiate transcriptional activation of the target genes in a ligand-dependent fashion. The recruitment of coactivators by nuclear receptors is primarily mediated through coactivator LXXLL motifs (where L is leucine and X is any amino acid) that bind to a charge-clamp pocket at the surface of the LBD. The sequence forming the charge-clamp pocket is most highly conserved among nuclear receptors. In PXR, this pocket is formed by the C-terminal AF-2 helix and helices H3, H3′, and H4, with residues I255, I273, and L424 making up the hydrophobic base of the pocket. The bound SRC-1 motif adopts a two-turn α-helical structure, with its leucine side chains docking against the hydrophobic base of the charge-clamp pocket (Figure 5.2c). Both ends of the α-helix are stabilized by charge interactions with K259 from helix H3 and E423 from the AF-2 helix [15]. This mode of coactivator binding is highly conserved in the nuclear receptor family and it is also observed in many other receptor/coactivator complexes.

5.2.4   Species Differences among LBD Sequences

While PXR has a remarkable ability to bind to diverse ligands, the receptors from different species show significant specificity in response to various agents. For example, human PXR can be activated by rifampicin, dexamethasone, and RU486, while mouse PXR can be activated by pregnenolone 16-α-carbonitrile and dexamethasone, but does not respond to rifampicin [20, 21]. Sequence alignment reveals significant sequence divergence in the PXR LBDs from different species (Figure 5.3), with only 70–80% identity among the mammalian PXRs. Currently, crystal structures are only available for the human PXR LBD/ligand complexes. On the basis of human structures and multiple sequence alignments, however, it can be shown that the divergence of the amino acid sequences in ligand binding pockets is crucial for the species specificity of PXR to ligands. Sequence alignment (Figure 5.3) indicates that among the six most frequent ligand binding residues in the human PXR LBD (M243, S247, Q285, W299, H407, and F420), only three residues (S247, W299, and F420) are conserved in mouse PXR and two (Q285 and W299) are conserved in a fish PXR. The low degree of conservation is the major reason for the difference in ligand binding specificity among PXR proteins from different species.

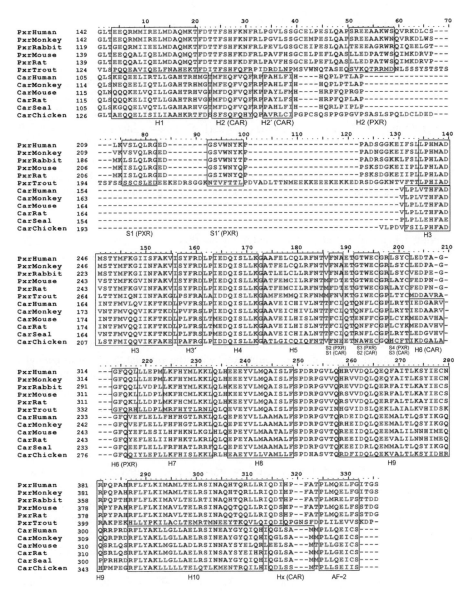

FIGURE 5.3 *Sequence alignment of PXR and CAR from different species.* Secondary structure elements are noted with H representing helices and S for strands. The color code for amino acid residues is as follows: red for acidic residues; blue, basic residues; green, hydrophobic residues; and grayish blue, polar residues. Three strands were observed in the CAR LBD (labeled as S1, S2, and S3) and five strands in the PXR LBD (labeled from S1 to S5). See color insert.

5.2.5   Heterodimerization with RXR

The functional unit of PXR is a heterodimer complex with RXR, which is also a common partner for many other nuclear receptors. PXR resides in cytoplasm in the absence of ligand [22]. Upon ligand binding, PXR is translocated into the nucleus, dimerizes with RXR, and the heterodimer complex binds to DNA response elements—a direct or inverted repeat of AG(G/T)TCA (DR-3, DR-4, or ER-6)—in the promoter region of its target genes [23, 24]. Neither a structure of a PXR/RXR heterodimer complex nor a structure of the PXR/RXR/DNA complex has been published. Thus, a multidomain structure of a PXR/RXR complex with or without DNA remains an important goal of crystallographic studies. However, on the basis of sequence conservation, the PXR/RXR heterodimer interface is mediated mainly through a coiled-coil structure of helix H10, as would be predicted from the heterodimer structures of RAR/RXR [25], PPAR/RXR [18], and LXR/RXR [26], as well as CAR/RXR (Figure 5.4*d*), which will be discussed in the next section.

5.3   STRUCTURE AND FUNCTION OF CAR

5.3.1   Crystal Structure of CAR

Currently there are four published CAR LBD structures, two human and two mouse. The two human structures are complexes with the agonist 6-(4-chlorophenyl)-imidazo[2,1-*b*]thiazole-5-carbaldehyde *O*-(3,4-dichlorobenzyl)oxime (CITCO) or 5β-pregnanedione [27], and the two mouse structures are complexes with a strong agonist, 1,4-bis[2-(3,5-dichloropyridyloxy)]benzene (TCPOBOP) [28], or an inverse agonist, androstanol [29] (Figures 5.4*a* and *b*). In all agonist-bound structures, the CAR LBD contains 11 α-helices, H1, H3, H3′, H4–H10, and AF-2 (which is the C-terminal activation function motif); two 3_{10} helices (H2 and H2′) located between helices H1 and H3; and three small β-strands (S1, S2, and S3) (Figures 5.3 and 5.4*c*). These helices are arranged into four helical layers: three layers of α-helices that compose the core LBD and an additional layer that consists of 3_{10} helices H2 and H2′. These two 3_{10} helices are unique to CAR among nuclear receptors (Figure 5.4*c*).

Importantly, in all active CAR structures, the residues between helices H10 and AF-2 adopt a short helix (Hx) structure instead of the extended loop normally observed in other nuclear receptors (Figure 5.4*c*). This short helix is tightly packed against helices H3 and H10 in the agonist-bound CAR structures. The rigid nature of helix Hx is shown to play a crucial role in maintaining the high level of constitutive activity of CAR by stabilizing the active conformation of the AF-2 helix. Mutations that destabilize Hx reduce AF-2, whereas mutations that stabilize Hx increase AF-2 [30, 28]. The androstanol-bound CAR LBD adopts the same conformation as agonist-bound CAR from residues 120 to 335. The major conformational difference is within the C-terminal region, from residue 335 to the C terminus. Androstanol-bound CAR has helix H11 (from residues 338 to 343), which is part of the continuous helix H10, in all agonist-bound CAR structures. Instead of a linker helix Hx in agonist-bound

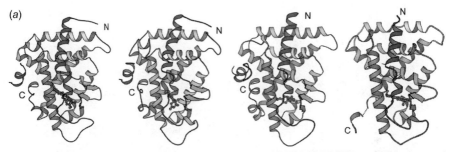

hCAR/CITCO/SRC1 hCAR/5β-pregnanedione/SRC1 mCAR/TCPOBOP/TIF2 mCAR/androstanol

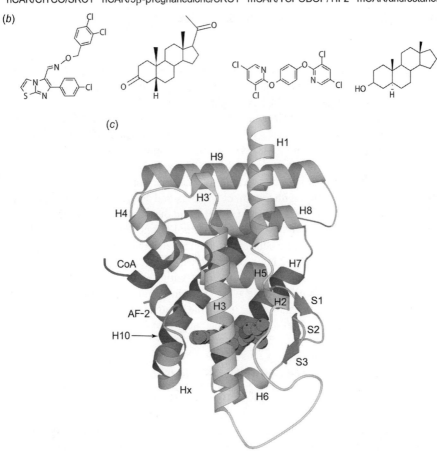

FIGURE 5.4 *Crystal structures of CAR.* (*a*) Diagrams of crystal structures of CAR in complexes with ligands or ligands and coactivator motifs. The color code is the same as used in Figure 5.1*a*. (*b*) Chemical structures of the respective CAR ligands for the crystal structures in part *a*. (*c*) The three-layered core structure of CAR LBD. The layers are shown by the same brightness levels as in Figure 5.1*c*. The helix AF-2 is colored red; the linker helix Hx, orange; and the coactivator motif, magenta. The ligand binding pocket is located at the bottom half of the LBD between the first and third layers. All helices, strands, and the coactivator motif are labeled. (*d*) Heterodimer of LBDs of CAR and RXR; a side view and a top view. See color insert for figure 5.4a, 5.4c and 5.4d.

(*d*)

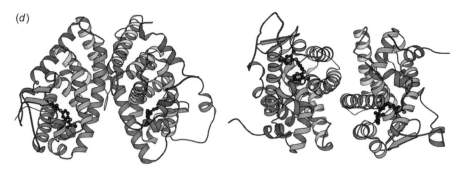

FIGURE 5.4   (*Continued*)

CAR, the region from residues 344 to 349 in androstanol-bound CAR structure is invisible. The helix AF-2, from residues 350 to 357, is positioned outside the core LBD in the androstanol-bound CAR structure (Figure 5.4*a*).

In addition, among all nuclear receptors, CAR has the shortest AF-2 helix (consisting of only eight amino acids in either mouse or human CAR; see Figure 5.3) and it is relatively rigid. In the TCPOBOP-bound mouse CAR structure, the AF-2 helix is packed tightly against the core LBD in an active conformation, which is directly stabilized by ligand binding. Additional stabilization of the AF-2 helix comes from a packing interaction with the linker helix Hx and a hydrogen bond between the C-terminal carboxylate group of the AF-2 helix and the side chain of K205 from helix H4. In contrast, the AF-2 helix from the androstanol-bound structure is extended away from the core domain into a solvent-exposed channel (Figure 5.6*b*). Electron density is observed only for one AF-2 helix from the two molecules in the asymmetric unit, as this AF-2 helix is stabilized by interactions with neighboring symmetry-related molecules. The conformation of the AF-2 helix in the androstanol-bound structure suggests that it is highly unstable and incapable of coactivator recruitment and transcriptional activation, in agreement with the antagonist properties of androstanol.

5.3.2   Ligand Binding and Activation of CAR

The human CAR ligand binding pocket is formed by helices H2–H7 and H10, plus β-strands S3 and S4; it has a volume of 675 Å^3, which is approximately half the size of the PXR pocket. There is little change in the CAR ligand binding pocket upon binding of 5β-pregnanedione or CITCO (Figure 5.5*a*), although 5β-pregnanedione is smaller than CITCO (and thus occupies less of the pocket). The binding mode of 5β-pregnanedione is clearly defined by electron density in the crystal structure, with hydrophobic interactions between the carbon backbone of the ligand and the side chains of the receptor residues of F161, M168, C202, L206, F217, Y224, F234, and L242, as well as hydrogen bonds between the C3 ketone of the ligand and the side chain of receptor residue H203, and between the C21 ketone and the side chain of receptor residue H160 (Figures 5.5*a* and *b*) [27].

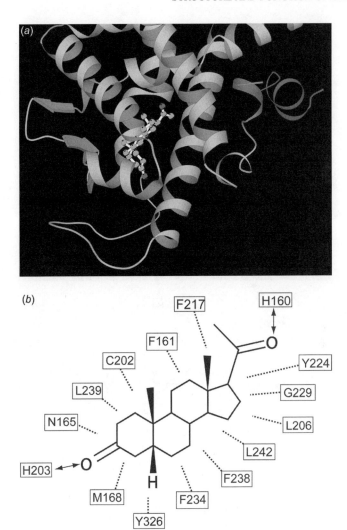

FIGURE 5.5 *Ligand binding of human and mouse CAR.* (*a*) A diagram of human CAR ligand binding pocket with bound 5β-pregnanedione. The orange region is the C-terminal helix AF-2 and magenta shows the coactivator motif. (*b*) Schematic representation of the ligand binding residues of human CAR LBD when binding 5β-pregnanedione. (*c*) A diagram of mouse CAR ligand binding pocket with bound TCPOBOP. Colors represent the same as in part *a*. (*d*) Schematic representation of the ligand binding residues of mouse CAR LBD when binding to TCPOBOP. The three aromatic rings of the ligand are labeled as A, B, and C from left to right. Shown are 24 receptor residues that directly bind to the ligand. See color insert for figure 5.5a and 5.5c.

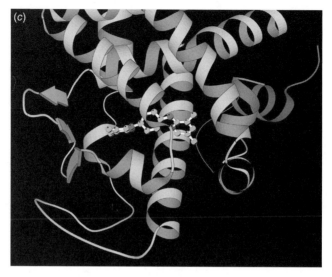

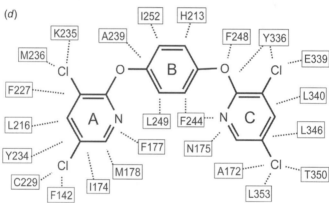

FIGURE 5.5 (*Continued*)

In contrast to the 5β-pregnanedione structure, the electron density of CITCO is less defined, especially the region of the oxime linker, which is invisible in the crystal structure of the human CAR/CITCO complex. The CITCO binding modes were determined by docking models using crystallographic data of the LBD [27]. Two ligand binding modes are identified, in which the ligand adopts U-shaped conformations that are consistent with the observable part of the electron density for the ligand in the crystal structure. On the basis of modeling studies, the LBD–ligand interaction is predominantly mediated through hydrophobic surfaces between the nonpolar residues of the LBD and the aromatic rings of the ligand. Additional interactions are mediated by the imidazothiazole heterocycle with hydrophilic residues (N165, C202, H203, and Y326) of CAR [27].

In the mouse CAR/TCPOBOP complex structure, the ligand binding pocket is shaped like a rectangular box with a volume of 525 Å^3. A total of 31 residues—from helices H5 to H10, and β-strands S2 and S3—contribute to the formation of the pocket (Figures 5.5c and d). The ligand TCPOBOP is a symmetric molecule with a twofold axis located across the center of phenyl ring B. In the crystal structure, the ligand was oriented with this symmetrical element almost perpendicular to the vertical axis of the LBD, with one of the two pyridine rings toward β-strands S2 and S3 and the other toward helices H10 and AF-2 (Figure 5.5c). The interactions between the ligand and the receptor are dominantly hydrophobic, particularly those between pyridine ring A of the ligand and the aromatic side chains of Y234 and F227, and pyridine ring C and the side chains of Y336 and F244 (Figure 5.5d). The strong sandwichlike hydrophobic packing between the pyridine rings and the aromatic side chains of the receptor explains the high affinity binding of the ligand to mouse CAR. In addition, the pyridine ring C packs closely toward the AF-2 helix, directly interacting with L353 of the AF-2 helix and L346 and T350 of the linker helix Hx. These interactions stabilize the AF-2 and the linker helices in the active conformation and serve as the key basis of TCPOBOP-mediated activation of mouse CAR [28].

In the crystal structure of the androstanol-bound mouse CAR complex, the ligand is located in the center of the binding pocket of the CAR LBD. As the volume of androstanol is only 270 Å^3—half of the volume of the CAR ligand binding pocket—the ligand appears to be "floating" within the pocket at almost the same distance from each pocket wall. The interactions between the ligand and the LBD are mainly between hydrophobic side chains of the LBD and the steroid rings of the ligand. Polar interactions between the ligand and the pocket are mediated by two hydrogen bonds between the hydroxyl group of the ligand and the side chains of LBD residues N175 and H213, the latter of which is bridged by a water molecule [29].

The most striking feature of the mouse CAR/androstanol structure is a kink between helices H10 and H11, which normally join as a continuous helix in all agonist-bound nuclear receptor LBD structures (Figure 5.6b). This kink is induced by the small size of androstanol; its binding leaves a void in the ligand binding pocket between the ligand and the normally straight helix H10. The kink allows this void to be filled by hydrophobic residues L340 and L343 of helix H11. The kink conformation is further stabilized by a hydrogen bond between E339 on the kink and Q245 at the loop between helices H6 and H7, pulling helix H11 in toward the ligand binding pocket and repositioning helix AF-2 away from the core LBD structure (see Figure 5.4c). The repositioning of helix AF-2 and the linker region disrupts the coactivator binding groove and switches mouse CAR from the active conformation to an inactive one [29].

Although the apo structure of CAR remains unsolved, the mechanism for the constitutive activity of this receptor can be largely discerned from the TCPOBOP-bound structure. Activation of nuclear receptors requires the AF-2 helix to form the charge-clamp pocket by packing closely against the core LBD; there are three distinct features in the TCPOBOP-bound structure that facilitate that packing. The first is the compact nature of the CAR LBD, with its ligand binding pocket half the size of the

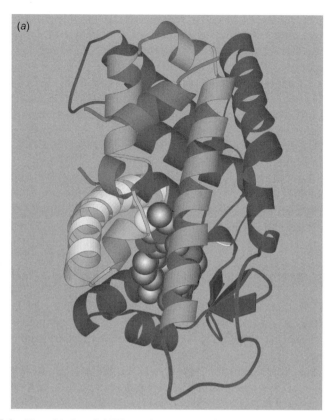

FIGURE 5.6 *Ligand-induced AF-2 conformational change and coactivator recruitment.* (*a*) CITCO-induced formation of the active conformation of helices H10 (yellow) and Hx (orange), and the coactivator binding groove framed by helices H3, H3′ (green), and AF-2 (red) with an SRC-1 motif bound to the coactivator binding groove of the human CAR LBD. (*b*) Antagonist-induced deactivation of the mouse CAR LBD. The helices H10, H11 (yellow), Hx (invisible), and AF-2 (red) adopt a conformation that disrupts the coactivator binding groove and deactivates the receptor. (*c*) The CAR LBD/coactivator interface with residues involved in the hydrophobic interactions. The coactivator motif is colored magenta; helices H3, H3′, and H4 are green; and helix AF-2 is red. The charge clamp is residues K187 and E355, which form hydrogen bonds with the nitrogen of L744 and carbonyl oxygen of L748 of the coactivator. The second charge clamp of the LBD is formed by R193 and E198, which are bound respectively to residues D750 and R746 of the coactivator motif; this clamp further stabilizes the coactivator binding. See color insert.

PXR pocket, thus providing overall stability of the LBD for packing of the AF-2 helix in the active conformation. The second feature is the unique linker helix Hx; its position greatly facilitates the AF-2 helix positioning. The role of helix Hx in CAR activation is highlighted by the mutation of L346 to phenylalanine, which dramatically increases CAR's ability to interact with coactivators [31, 28]. Residue L346 is located

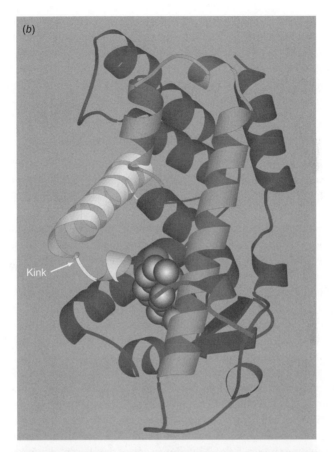

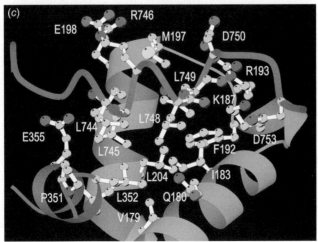

FIGURE 5.6   (*Continued*)

at the center of helix Hx, and it is predicted that the large side chain of phenylalanine in the mutant L346F will penetrate into the pocket, thus locking the linker helix into a position that stabilizes the AF-2 helix in the active conformation. The third feature is the short AF-2 helix with its C-terminal carboxylate group hydrogen bonded with K205 of helix H4 and S337 of helix H10. These three structural features provide an integrated basis for the constitutive activity of CAR [28].

5.3.3 Cofactor Recruitment and Transcription Activation

Transcription activation by CAR is finely regulated by interaction with coactivators or corepressors through its hydrophobic cofactor binding groove. Both mouse and human CAR recruit TIF2 (transcriptional intermediary factor 2 or SRC-2) as a coactivator, the binding of which is enhanced further by agonists or inhibited by antagonists [32]. The binding of CAR to corepressors, which is promoted by antagonists, results in deactivation of the basal activity of the receptor [32]. In the mouse CAR/TIF2 complex, the LXXLL motif of TIF2 adopts a two-turn α-helix that binds to the cofactor groove through hydrophobic interactions between the leucine residues in the TIF2 motif and hydrophobic side chains of the cofactor binding groove. Each end of the TIF2 helix is held by the charge clamp formed by the conserved glutamic acid residue E355 from the AF-2 helix and the lysine residue K187 at the end of helix H3. These two charge-clamp residues form hydrogen bonds, respectively, with the backbone nitrogen of L744 and the carbonyl oxygen of L748 of TIF2. Again, this mode of coactivator interaction is common among nuclear receptors, in agreement with a conserved mechanism of nuclear receptor activation.

Interestingly, CAR contains a second charge clamp formed by residues R193 from helix H3′ and E198 from helix H4. This second charge clamp forms additional interactions with residues D750 and R746 of the TIF2 motif, further stabilizing the coactivator binding (Figure 5.6c). Mutation in any of the charge-clamp residues results in a great loss of binding affinity for the TIF2 motif, implying that each charge-clamp interaction is important for coactivator binding and transcriptional activation of this receptor [28].

5.3.4 Species Differences in CAR Ligand Binding and Activation

It has been observed over the last decade that CAR from different species displays different ligand specificities. The most well documented example is TCPOBOP; a strong agonist that activates mouse CAR but has no activity toward human CAR. CITCO is another example: it serves as a modest agonist for human CAR but shows no activity toward mouse CAR [33, 34]. This species specificity predominantly results from the difference in the amino acid sequences of the LBDs, which contribute to their distinct structural features and distinct ligand binding properties. Sequence alignment shows that 10 residues out of the 31 that comprise ligand binding pocket are not conserved between the mouse and human receptors (see Figure 5.3).

Interestingly, a large number of the nonconserved residues are those that have smaller side chains in human CAR and larger side chains in mouse CAR, which makes the human CAR ligand binding pocket larger. The CITCO binding pocket of the human CAR LBD has a volume of 675 Å^3, 29% larger than the 525 Å^3 volume of the TCPOBOP binding pocket of mouse CAR. Additionally, the ligand binding specificity of CAR is constrained by the topology of the pocket. The mouse CAR pocket is a rectangular box well adapted to the binding of TCPOBOP, while the human CAR pocket is round and fitted for the bulky ligand CITCO. Furthermore, the specificity of TCPOBOP is shown to reside in a single residue difference between the human and mouse receptors [28]. In mouse CAR, a small residue T350 at this position leaves enough pocket space for the extended shape of the TCPOBOP, whereas a large residue, M340, at the same position of the human CAR sterically blocks the binding of TCPOBOP [35, 28].

5.3.5   Heterodimerization with RXR

The CAR/RXR dimer interface has been characterized in the human and mouse CAR structures. The main interactions of the dimerization are between the helices H10, H7, H9, and the N-terminal loop region of H9, from both CAR and RXR. The central part of the interface is a coiled-coil structure formed by hydrophobic interactions between N-terminal halves of H10 of both receptors. Surrounding this central interface is a network of hydrogen bonds formed between helix H7 and the loop region between helices H8 and H9 of the two LBDs. Similar to RXR heterodimers with other receptors (such as PPARγ, RAR, and LXR), the CAR/RXR dimer is nonperfectly symmetrical in relation to a C$_2$-axis along the dimer interface. Compared with the PPARγ/RXR dimer, the CAR/RXR dimer has a 10% larger interface, facilitating a more potent dimerization interaction, which may be important for CAR activation and signaling (see Figure 5.4d) [28, 27].

5.4   CONCLUDING REMARKS

Structural analyses of the nuclear receptors PXR and CAR have greatly contributed to our understanding of the biochemical mechanisms and ligand recognition that regulate gene expression for xenobiotic metabolism. The structural models for drug-induced transcription by PXR and CAR are important for studying drug metabolism in humans and for designing effective and safe drug therapies. As both PXR and CAR are promiscuous in ligand binding, further structural analyses of receptor–drug complexes will provide additional information and new aspects of the modes of drug-induced gene expression. Both PXR and CAR are DNA binding and ligand-regulated transcriptional factors, but structures of a multidomain PXR or CAR bound to DNA and coactivators remain elusive because these multidomain proteins are refractory to crystallization. Structure determination of multidomain and multiprotein complexes of nuclear receptors has become a grand challenge for structural biologists. Solving these structures will help to unveil the complicated layers of transcription regulation by these drug/xenobiotic receptors.

REFERENCES

1. Chawla, A., Repa, J. J., Evans, R. M., and Mangelsdorf, D. J. (2001) Nuclear receptors and lipid physiology: opening the X-files. *Science* **294** (5548), 1866–1870.

2. Robinson-Rechavi, M., Escriva Garcia, H., and Laudet, V. (2003) The nuclear receptor superfamily. *J. Cell Sci.* **116** (Pt. 4), 585–586.

3. Waxman, D. J. (1999) P450 gene induction by structurally diverse xenochemicals: central role of nuclear receptors CAR, PXR, and PPAR. *Arch. Biochem. Biophys.* **369** (1), 11–23.

4. Willson, T. M. and Kliewer, S. A. (2002) PXR, CAR and drug metabolism. *Nat. Rev. Drug Discov.* **1** (4), 259–266.

5. Handschin, C. and Meyer, U. A. (2003) Induction of drug metabolism: the role of nuclear receptors. *Pharmacol. Rev.* **55** (4), 649–673.

6. Tirona, R. G. and Kim, R. B. (2005) Nuclear receptors and drug disposition gene regulation. *J. Pharm. Sci.* **94** (6), 1169–1186.

7. Timsit, Y. E. and Negishi, M. (2007) CAR and PXR: the xenobiotic-sensing receptors. *Steroids* **72** (3), 231–246.

8. Evans, R. (2004) A transcriptional basis for physiology. *Nat. Med.* **10** (10), 1022–1026.

9. Tzameli, I. and Moore, D. D. (2001) Role reversal: new insights from new ligands for the xenobiotic receptor CAR. *Trends Endocrinol. Metab.* **12** (1), 7–10.

10. Qatanani, M. and Moore, D. D. (2005) CAR, the continuously advancing receptor, in drug metabolism and disease. *Curr. Drug Metab.* **6** (4), 329–339.

11. Watkins, R. E., Wisely, G. B., Moore, L. B., Collins, J. L., Lambert, M. H., Williams, S. P., Willson, T. M., Kliewer, S. A., and Redinbo, M. R. (2001) The human nuclear xenobiotic receptor PXR: structural determinants of directed promiscuity. *Science* **292** (5525), 2329–2333.

12. Watkins, R. E., Maglich, J. M., Moore, L. B., Wisely, G. B., Noble, S. M., Davis-Searles, P. R., Lambert, M. H., Kliewer, S. A., and Redinbo, M. R. (2003) 2.1 Å crystal structure of human PXR in complex with the St. John's wort compound hyperforin. *Biochemistry* **42** (6), 1430–1438.

13. Chrencik, J. E., Orans, J., Moore, L. B., Xue, Y., Peng, L., Collins, J. L., Wisely, G. B., Lambert, M. H., Kliewer, S. A., and Redinbo, M. R. (2005) Structural disorder in the complex of human pregnane X receptor and the macrolide antibiotic rifampicin. *Mol. Endocrinol.* **19** (5), 1125–1134.

14. Xue, Y., Chao, E., Zuercher, W. J., Willson, T. M., Collins, J. L., and Redinbo, M. R. (2007) Crystal structure of the PXR-T1317 complex provides a scaffold to examine the potential for receptor antagonism. *Bioorg. Med. Chem.* **15** (5), 2156–2166.

15. Watkins, R. E., Davis-Searles, P. R., Lambert, M. H., and Redinbo, M. R. (2003) Coactivator binding promotes the specific interaction between ligand and the pregnane X receptor. *J. Mol. Biol.* **331** (4), 815–828.

16. Orans, J., Teotico, D. G., and Redinbo, M. R. (2005) The nuclear xenobiotic receptor pregnane X receptor: recent insights and new challenges. *Mol. Endocrinol.* **19** (12), 2891–2900.

17. Nolte, R. T., Wisely, G. B., Westin, S., Cobb, J. E., Lambert, M. H., Kurokawa, R., Rosenfeld, M. G., Willson, T. M., Glass, C. K., and Milburn, M. V. (1998) Ligand binding and co-activator assembly of the peroxisome proliferator-activated receptor-gamma. *Nature* **395** (6698), 137–143.

18. Gampe, R. T., Jr., Montana, V. G., Lambert, M. H., Miller, A. B., Bledsoe, R. K., Milburn, M. V., Kliewer, S. A., Willson, T. M., and Xu, H. E. (2000) Asymmetry in the PPARgamma/RXRalpha crystal structure reveals the molecular basis of heterodimerization among nuclear receptors. *Mol. Cell* **5** (3), 545–555.

19. Masuyama, H., Suwaki, N., Tateishi, Y., Nakatsukasa, H., Segawa, T., and Hiramatsu, Y. (2005) The pregnane X receptor regulates gene expression in a ligand- and promoter-selective fashion. *Mol. Endocrinol.* **19** (5), 1170–1180.

20. Kocarek, T. A., Schuetz, E. G., Strom, S. C., Fisher, R. A., and Guzelian, P. S. (1995) Comparative analysis of cytochrome P4503A induction in primary cultures of rat, rabbit, and human hepatocytes. *Drug Metab. Dispos.* **23** (3), 415–421.

21. Lehmann, J. M., McKee, D. D., Watson, M. A., Willson, T. M., Moore, J. T., and Kliewer, S. A. (1998) The human orphan nuclear receptor PXR is activated by compounds that regulate CYP3A4 gene expression and cause drug interactions. *J. Clin. Invest.* **102** (5), 1016–1023.

22. Mangelsdorf, D. J. and Evans, R. M. (1995) The RXR heterodimers and orphan receptors. *Cell* **83** (6), 841–850.

23. Sueyoshi, T. and Negishi, M. (2001) Phenobarbital response elements of cytochrome P450 genes and nuclear receptors. *Annu. Rev. Pharmacol. Toxicol.* **41**, 123–143.

24. Kliewer, S. A., Goodwin, B., and Willson, T. M. (2002) The nuclear pregnane X receptor: a key regulator of xenobiotic metabolism. *Endocr. Rev.* **23** (5), 687–702.

25. Bourguet, W., Vivat, V., Wurtz, J. M., Chambon, P., Gronemeyer, H., and Moras, D. (2000) Crystal structure of a heterodimeric complex of RAR and RXR ligand-binding domains. *Mol. Cell* **5** (2), 289–298.

26. Svensson, S., Ostberg, T., Jacobsson, M., Norström, C., Stefansson, K., Hallén, D., Johansson, I. C., Zachrisson, K., Ogg, D., and Jendeberg, L. (2003) Crystal structure of the heterodimeric complex of LXRalpha and RXRbeta ligand-binding domains in a fully agonistic conformation. *EMBO J.* **22** (18), 4625–4633.

27. Xu, R. X., Lambert, M. H., Wisely, B. B., Warren, E. N., Weinert, E. E., Waitt, G. M., Williams, J. D., Collins, J. L., Moore, L. B., Willson, T. M., et al. (2004) A structural basis for constitutive activity in the human CAR/RXRalpha heterodimer. *Mol. Cell* **16** (6), 919–928.

28. Suino, K., Peng, L., Reynolds, R., Li, Y., Cha, J. Y., Repa, J. J., Kliewer, S. A., and Xu, H. E. (2004) The nuclear xenobiotic receptor CAR: structural determinants of constitutive activation and heterodimerization. *Mol. Cell* **16** (6), 893–905.

29. Shan, L., Vincent, J., Brunzelle, J. S., Dussault, I., Lin, M., Ianculescu, I., Sherman, M. A., Forman, B. M., and Fernandez, E. J. (2004) Structure of the murine constitutive androstane receptor complexed to androstanol: a molecular basis for inverse agonism. *Mol. Cell* **16** (6), 907–917.

30. Dussault, I., Lin, M., Hollister, K., Fan, M., Termini, J., Sherman, M. A., and Forman, B. M. (2002) A structural model of the constitutive androstane receptor defines novel interactions that mediate ligand-independent activity. *Mol. Cell. Biol.* **22** (15), 5270–5280.

31. Jyrkkärinne, J., Mäkinen, J., Gynther, J., Savolainen, H., Poso, A., and Honkakoski, P. (2003) Molecular determinants of steroid inhibition for the mouse constitutive androstane receptor. *J. Med. Chem.* **46** (22), 4687–4695.

32. Lempiäinen, H., Molnár, F., Macias, G. M., Peräkylä, M., and Carlberg, C. (2005) Antagonist- and inverse agonist-driven interactions of the vitamin D receptor and the

constitutive androstane receptor with corepressor protein. *Mol. Endocrinol.* **19** (9), 2258–2272.

33. Blizard, D., Sueyoshi, T., Negishi, M., Dehal, S. S., and Kupfer, D. (2001) Mechanism of induction of cytochrome p450 enzymes by the proestrogenic endocrine disruptor pesticide-methoxychlor: interactions of methoxychlor metabolites with the constitutive androstane receptor system. *Drug Metab. Dispos.* **29** (6), 781–785.

34. Kobayashi, K., Yamanaka, Y., Iwazaki, N., Nakajo, I., Hosokawa, M., Negishi, M., and Chiba, K. (2005) Identification of HMG-CoA reductase inhibitors as activators for human, mouse and rat constitutive androstane receptor. *Drug Metab. Dispos.* **33** (7), 924–929.

35. Ueda, A., Kakizaki, S., Negishi, M., and Sueyoshi, T. (2002) Residue threonine 350 confers steroid hormone responsiveness to the mouse nuclear orphan receptor CAR. *Mol. Pharmacol.* **61** (6), 1284–1288.

6

XENOBIOTIC RECEPTOR COFACTORS AND COREGULATORS

JOHN Y. L. CHIANG

Department of Microbiology, Immunology and Biochemistry, Northeastern Ohio Universities College of Medicine, Rootstown, OH, USA

6.1 REGULATION OF PXR AND CAR NUCLEAR TRANSLOCATION

6.1.1 Activation and Nuclear Translocation of CAR

The classical CAR activator phenobarbital (PB) does not bind to CAR. A highly potent "PB-like" activator TCPOBOP (1,4-bis [2-(3,5-dichloropyridyloxy)] benzene) binds the murine but not human CAR [1, 2]. RXR agonists called rexinoids including 9-*cis*-retinoic acid and synthetic compounds can directly activate some RXR heterodimers (i.e., PPAR/RXR and FXR/RXR, referred to as permissive agonists) but not the others (i.e., RAR/RXR and TR/RXR, nonpermissive) [3]. The effect of rexinoids on CAR transactivation is complex [3]. Rexinoids (9-*cis*-retinoic acid and all-*trans*-retinoic acid) block PB induction of the CAR target genes CYP2B1 and CYP2B2 [4], and decrease TCPOBOP-activated CAR induction of CYP2B10 in mouse primary hepatocytes [5].

Activation and nucleocytoplasmic translocation of CAR have been studied extensively. CAR is normally localized in the cytoplasm of liver cells. When treated with PB, CAR is translocated to the nucleus. This is the first step of activation of CAR in response to drug treatment. PB does not bind to the ligand binding domain (LBD) of human or mouse CAR. TCPOBOP binds mouse but not human CAR. In general, active transport of nuclear receptors from the cytoplasm to the nucleus requires a short sequence called the nuclear localization signal (NLS) that is exposed

Nuclear Receptors in Drug Metabolism　Edited by Wen Xie
Copyright © 2009 John Wiley & Sons, Inc.

when a ligand binds to a nuclear receptor. A typical NLS (RRARQARRR) is rich in basic amino acids and is located in the DNA binding domain (DBD). The NLS binds importins, which escort CAR into the nucleus. The human CAR does not have an NLS. A leucine-rich sequence (L/MXXLXXL) located within the C-terminal 30 amino acid region in the mouse and human CAR has been identified as a signal for xenobiotic responsive nuclear translocation [6]. This sequence is named the xeno-biochemical response signal (XRS). In mouse CAR, the NLS does not play a role in nuclear localization likely because mutations in the NLS weaken its nuclear import activity [7]. Thus, CAR may be translocated to the nucleus by both ligand-dependent and ligand-independent pathways [6, 8]. Other amino acid residues also have been identified to be required for nuclear translocation. The Ser202 located in the DBD has been shown to be essential for nuclear translocation of mouse CAR [9]. Mutation of Ser202 to Asp prevents CAR translocation into the nucleus of TCPOBOP-treated HepG2 cells. In HepG2 cells, S202 can be phosphorylated. Dephosphorylation of S202 is required in xenobiotic response and nuclear translocation of mouse CAR. In transformed cells and primary mouse hepatocytes, CAR is predominantly located in the nucleus even in the absence of an activator [10].

Two NLS have been identified in rat CAR [11]. NLS1 (amino acid residues 100–108) is rich in basic amino acid residues and is located in the hinge region between AF-1 and DBD. NLS2 is spread between amino acid residues 111 and 320 in the LBD. A cytoplasmic retention region (220–258) has been identified in rat CAR. In addition, a sequence similar to the XRS is also located in rat CAR. This XRS does not function as an NLS. A typical nuclear expert sequence (NES) is leucine rich and is located in the C-terminal region. The relative activity of NLS and NES may determine nucleocytoplasmic shuttling of human and rat CAR [12, 13]. A tetratricopeptide repeat protein called cytoplasmic CAR retention protein (CCRP) may interact with CAR and retain CAR in the cytoplasm of HepG2 cells [14]. CAR, CCRP, and heat shock protein 90 form a complex that recruits protein phosphatase A2 (PPA2) in response to PB [15]. A PPA2 inhibitor okadaic acid promotes accumulation of CAR in the nucleus, suggesting a phosphorylation-dependent nuclear translocation of CAR [8].

Swales and Negishi [16] have proposed the mechanisms for ligand-dependent and ligand-independent nuclear translocation for CAR. In this model, PB may bind to a membrane receptor, which activates a signal transduction pathway to regulate CAR activation. CAR forms a complex with HSP90 and a cochaperone CCRP, and subsequently recruits PP2A. This complex is translocated to the nucleus. In the nucleus, calmodulin-dependent kinase may be involved in activation since inhibitors of Ca^+–calmodulin-dependent kinase inactivate CAR activity. Nuclear receptor coactivators are then recruited to interact with the CAR/RXR heterodimer bound to the PB-responsive enhancer module (PBREM) in the promoter and transactivates the target genes.

6.1.2 Nuclear Translocation of PXR

Recent studies show that PXR is primarily located in the cytoplasm of untreated mouse liver [17]. Upon treatment with a ligand pregnenolone-16α-carbonitrile (PCN), mouse

PXR is translocated to the nucleus [7, 17]. Nuclear translocation of human PXR is regulated by a bipartite NLS (amino acid residues 66–92) located in the DBD [7]. This NLS binds importin α and importin β peptides and target PXR to the nuclear pore. The nuclear translocation of PXR is essential for regulation of its target gene CYP3A4. The XRS and the activation function 2 (AF-2) domain are also involved in nuclear translocation of PXR. Similar to CAR, PXR forms a complex with HSP90 and CCRP in the cytoplasm of HepG2 cells [17]. The bipartite NLS overlaps with the DBD in PXR and is required for nuclear translocation. The conserved basic amino acid lysine (K70) in the 5′ region (RRXXKR) of the NLS is substituted by a serine in human, rat, and mouse CAR. Thus, a single amino acid substitution of K70S in the CAR weakens its nuclear translocation activity. In cultured cells, overexpressed PXR and CAR are translocated into the nuclei by a ligand-independent mechanism. The NLS in the DBD and the XRS in the LBD may play differential roles in ligand-dependent nuclear-cytoplasmic shuffling of PXR.

6.2 NUCLEAR RECEPTOR COREGULATORS AND EPIGENETIC REGULATION OF GENE TRANSCRIPTION

The activated receptor sequentially recruits or exchanges coactivators with the core-pressor complex to the chromatins. The chromatin contains a nucleosome structure of 146 base pairs of DNA wrapped around the histone octamers containing two copies each of histone H2A, H2B, H3, and H4. Chromatin remodeling enzymes catalyze the covalent modification of histones to allow transcription factor binding. Histone acetyl-transferase (HAT) activity associated with coactivators such as CBP/P300 acetylates lysine residues (K) in H3 tails, and also transcription factors to activate gene transcription. On the other hand, histone deacetylases (HDACs) remove acetyl groups and inhibit gene transcription [18]. Protein arginine methyltransferases (PRMT1/5) and coactivator-associated arginine methyltransferase 1 (CARM-1) methylate arginine (R) or lysine residues in histone tails to either activate or inhibit gene transcription. Other ATP-dependent modulator protein complexes (Swi/Snf, and ATPases BRM and BRG-1) and ATP-independent mSin3A also are involved in remodeling of chromatin structure to inhibit RNA polymerase II activity [19]. Epigenetic "histone code" and coactivators and corepressors integrate extra- and intracellular signals to determine tissue, species, and gene-specific regulation of gene transcription by nuclear receptors and transcription factors [20–23].

6.2.1 Ligand-Independent and Ligand-Dependent Recruitment of Nuclear Receptor Coactivators

The p160 family of coactivators including steroid receptor coactivator 1 (SRC-1), TIF/GRIP (SRC-2), and ACTR/pCIP (SRC-3) [24] stimulates the transcriptional activity of nuclear receptors. The p160 family of coactivators has three LXXLL motifs in the receptor-interacting domain (RID) that interacts with the AF-2 domain of nuclear receptors. SRC-1 is a weak HAT that interacts with p300/CBP and recruits CARM-1.

PXR is a promiscuous xenobiotic receptor that has very broad ligand binding activity. Species-specific ligand binding and activation of PXR has been well documented [25]. Rifampicin binds and activates human but not mouse PXR, whereas PCN activates mouse but not human PXR. Crystal structure analysis of PXR LBD in complex with the cholesterol-lowering drug SR12813 and an SRC-1 peptide containing an LXXLL motif reveals that SRC-1 binds to the AF-2 helix on the surface of PXR and stabilizes the LBD and promotes specific interaction between ligand and PXR [26]. The binding of coactivator to the surface of PXR may limit its access to ligands, thus providing specificity to ligand binding. The crystal structure of the LBD of PXR complexed with hyperforin reveals an unusually large but flexible ligand binding cavity [27, 28]. Coactivator binding to the surface of PXR limits the access of ligand to the ligand binding cavity and may allow the specific interaction between ligand and PXR [26]. PXR interacts with peroxisome proliferators-activated receptor γ coactivator-1a (PGC-1α) in a ligand-dependent manner [29]. PGC-1α is a potent coactivator induced during fasting and plays a critical role in glucose, lipid, and energy metabolism [30–32].

The crystal structure of the CAR/RXR heterodimer reveals an unusually large dimerization surface and a small ligand binding pocket [33]. The binding of an inverse agonist androstenol in the ligand binding pocket stabilizes the AF-2 of both CAR and RXR in the active confirmation [34, 35]. The CAR/RXR complex interacts with SRC-1 via the LBD of CAR/RXR and the receptor interaction domains (RIDs) of the coactivator. Addition of a CAR-specific ligand, TCPOBOP, increases the affinity of SRC-1 binding to CAR. The RXR-specific ligand, 9-*cis*-retinoic acid, increases the affinity of RXR binding to a second SRC-1 site [36]. Addition of both receptor ligands increases affinity for one of the two SRC-1 sites. Endogenous ligands of CAR, androstanol and androstenol, inhibit the constitutive activity of CAR by dissociating SRC-1 from binding to CAR [37]. It has been reported that an SRC-2 coactivator, glucocorticoid receptor-interacting protein 1 (GRIP-1) interacts with CAR and mediates ligand-independent nuclear translocation and activation of CAR [38]. The binding of CAR to GRIP-1 is through the C-terminal region of the CAR, and is increased by a CAR agonist and decreased by a CAR antagonist.

Another study shows that ligand-independent activation of CAR requires subnuclear targeting by PGC-1α [39]. Both the N-terminal LXXLL motif and C-terminal serine/arginine domain (RS domain) of PGC-1α are required for activation of PXR and CAR. GST pull down assay reveals that CAR indirectly interacts with both the LXXLL motif and RS domain. The RS domain of PGC-1α is required for CAR localization at the nuclear speckles. Activating signal cointegrator-2 (ASC-2), also named as PPAR-interacting protein (PRIP), interacts with CAR in the presence of a ligand [40]. ASC-2 may be involved in acetaminophen-induced hepatotoxicity mediated by CAR since transgenic mice expressing a dominant negative fragment of ASC-2 encompassing the N-terminal LXXLL motif are resistant to acetaminophen-induced hepatotoxicity. CAR also interacts with PPAR binding protein (PBP) [41]. PBP is also known as thyroid hormone receptor associated protein 220 (TRAP 220), vitamin D receptor-interacting protein 205 (DRIP205) or mediator 1. PBP interacts with the C-terminal region of CAR, and is essential for PB-induced CAR nuclear translocation [42]. PBP may enhance either nuclear import or nuclear retention of

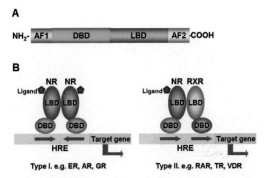

A

NH₂- AF1 | DBD | LBD | AF2 -COOH

B

Type I. e.g. ER, AR, GR

Type II. e.g. RAR, TR, VDR

FIGURE 2.1 Nuclear receptor domain structure and its signaling mode. (*a*) Modular structure of nuclear receptors. (*b*) Signaling mode of the type I and type II nuclear receptors.

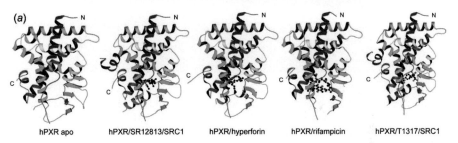

(*a*)

hPXR apo | hPXR/SR12813/SRC1 | hPXR/hyperforin | hPXR/rifampicin | hPXR/T1317/SRC1

FIGURE 5.1 *Human PXR crystal structures.* (*a*) Diagrams of the crystal structures of the apo, ligand-bound, and coactivator-bound human PXR LBD. A rainbow ramp color code of blue to red is used to trace the amino acid chain from the N terminus to the C terminus.

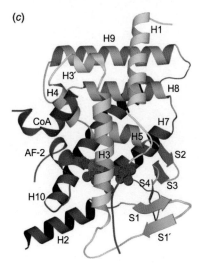

(*c*)

FIGURE 5.1 (*c*) The three-layered core structure of PXR LBD. The layers are shown by the brightness of the color: the first layer is the brightest and the third, the darkest. The helix AF-2 is colored red, and the coactivator motif is magenta. The ligand binding pocket is located at the bottom half of the LBD between the first and third layers. All helices, strands, and the coactivator motif are labeled, except for the helix H6, which is behind the core structure.

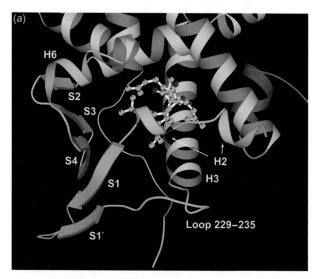

FIGURE 5.2 *PXR ligand binding and coactivator recruitment.* (*a*) A diagram of PXR ligand binding pocket bound to hyperforin. Cyan-colored regions are the "lid" of the pocket and its supporting regions; the orange region is the C-terminal AF-2 motif. The structural elements of the "lid" and the supporting regions are labeled.

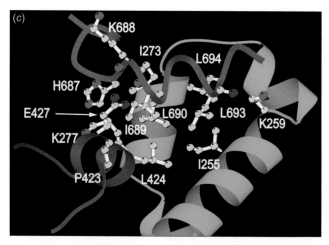

FIGURE 5.2 (*c*) PXR LBD/coactivator interface. SRC1 motif is colored magenta; helices H3, H3′, and H4, green; and helix AF-2, red. The charge clamp of the LBD is formed by residues K259 and E427, which form hydrogen bonds with the nitrogen of I689 and the carboxyl oxygen of L693 in the coactivator.

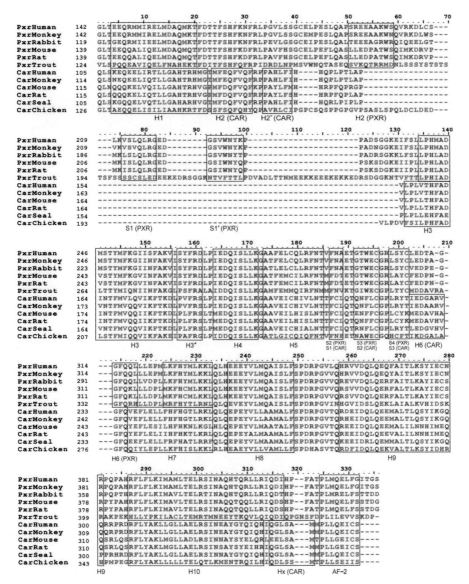

FIGURE 5.3 *Sequence alignment of PXR and CAR from different species.* Secondary structure elements are noted with H representing helices and S for strands. The color code for amino acid residues is as follows: red for acidic residues; blue, basic residues; green, hydrophobic residues; and grayish blue, polar residues. Three strands were observed in the CAR LBD (labeled as S1, S2, and S3) and five strands in the PXR LBD (labeled from S1 to S5).

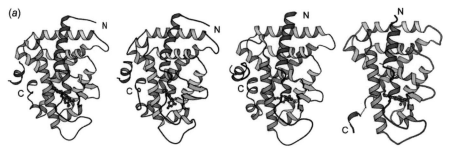

(a)

hCAR/CITCO/SRC1   hCAR/5β-pregnanedione/SRC1   mCAR/TCPOBOP/TIF2   mCAR/androstanol

FIGURE 5.4 *Crystal structures of CAR.* (*a*) Diagrams of crystal structures of CAR in complexes with ligands or ligands and coactivator motifs. The color code is the same as used in Figure 5.1*a*.

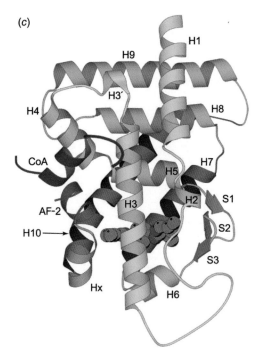

(c)

FIGURE 5.4 (*c*) The three-layered core structure of CAR LBD. The layers are shown by the same brightness levels as in Figure 5.1*c*. The helix AF-2 is colored red; the linker helix Hx, orange; and the coactivator motif, magenta. The ligand binding pocket is located at the bottom half of the LBD between the first and third layers. All helices, strands, and the coactivator motif are labeled.

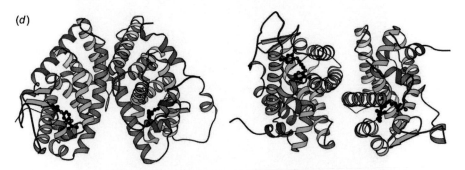

FIGURE 5.4 (*d*) Heterodimer of LBDs of CAR and RXR; a side view and a top view.

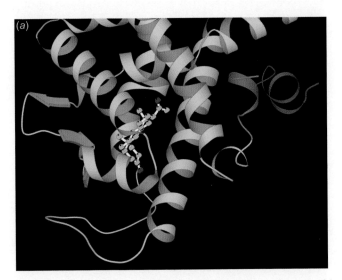

FIGURE 5.5 *Ligand binding of human and mouse CAR.* (*a*) A diagram of human CAR ligand binding pocket with bound 5β-pregnanedione. The orange region is the C-terminal helix AF-2 and magenta shows the coactivator motif.

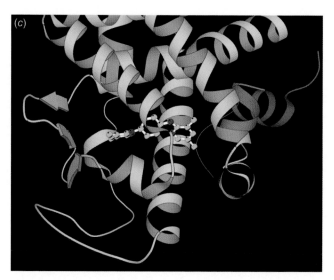

FIGURE 5.5 (*c*) A diagram of mouse CAR ligand binding pocket with bound TCPOBOP. Colors represent the same as in part *a*.

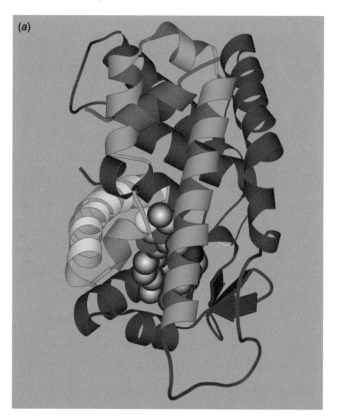

FIGURE 5.6 *Ligand-induced AF-2 conformational change and coactivator recruitment.* (*a*) CITCO-induced formation of the active conformation of helices H10 (yellow) and Hx (orange), and the coactivator binding groove framed by helices H3, H3′ (green), and AF-2 (red) with an SRC1 motif bound to the coactivator binding groove of the human CAR LBD.

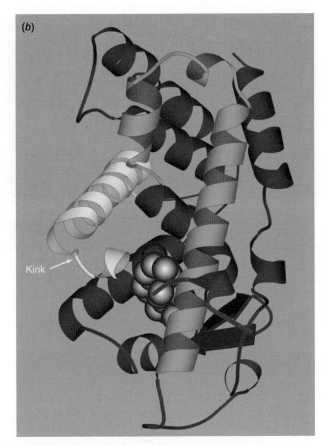

FIGURE 5.6 (*b*) Antagonist-induced deactivation of the mouse CAR LBD. The helices H10, H11 (yellow), Hx (invisible), and AF-2 (red) adopt a conformation that disrupts the coactivator binding groove and deactivates the receptor.

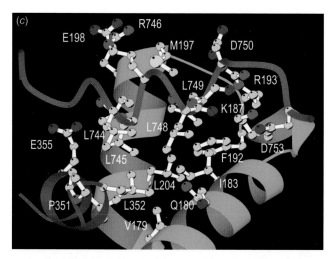

FIGURE 5.6 (*c*) The CAR LBD/coactivator interface with residues involved in the hydrophobic interactions. The coactivator motif is colored magenta; helices H3, H3′, and H4 are green; and helix AF-2 is red. The charge clamp is residues K187 and E355, which form hydrogen bonds with the nitrogen of L744 and carboxyl oxygen of L748 of the coactivator. The second charge clamp of the LBD is formed by R193 and E198, which are bound respectively to residues D750 and R746 of the coactivator motif; this clamp further stabilizes the coactivator binding.

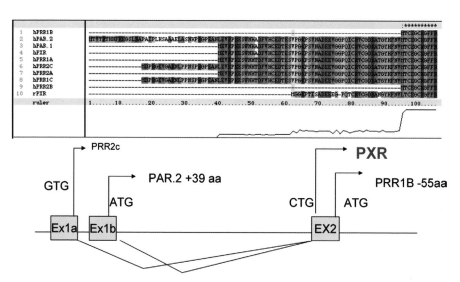

FIGURE 9.2 PXR has multiple isoforms with different amino terminal ends. (*a*) Amino acid alignment of PXR est isoforms. PXR is the most abundant isoform. (*b*) Generation of PXR, PAR.2, and PRR isoforms utilizing different initiation codons in exons 1a, 1b, and 2.

CAR in hepatocytes. However, PRIP is not required for nuclear translocation of CAR. Activation of CAR is known to increase acetaminophen hepatotoxicity by inducing acetaminophen-metabolizing enzymes in the liver [43]. Deficiency of PBP, but not of PRIP, abrogates acetaminophen hepatotoxicity [41].

6.2.2 Nuclear Receptor Corepressor in Xenobiotic Receptor Signaling

Nuclear receptor corepressors inhibit gene transcription by three possible mechanisms: competition for coactivators, inhibition of transcription factor binding to DNA, and active repression [22]. Corepressors recruit HDACs, Sim3A, Swi/Snf, and BRG/BRM1, and ubiquitous repressors NCoR (nuclear receptor corepressor) and SMRT (silencing mediator of retinoid and thyroid) to remodel chromatin and inhibit transactivating activity of nuclear receptors. In the absence of a ligand, PXR interacts with NCoR and inhibits target gene transcription. A potent PXR inducer paclitaxel (Taxol) activates PXR by reducing PXR–NCoR interaction [44]. SMRT forms a repressive complex with PXR in the absence of a ligand [44]. Rifampicin can dissociate SMRT from PXR by exchanging SMRT with a p160 family coactivator, receptor-associated coactivator 3 (RAC3) [45]. SMRT, but not NCoR, inhibits not only basal but also rifampicin-induced transcriptional activity of PXR [46]. Interestingly, rifampicin also increases the interaction of PXR with SMRT as well as SRC-1. All these studies suggest that dissociation of the corepressor from PXR plays an important role in PXR-mediated induction of PXR target genes. Ketoconazole, an antifungal drug, partially inhibits PXR-mediated transcription of the CYP3A4 gene [46]. Ketoconazole inhibits PXR interaction with SRC-1 and results in inhibition of Cyp3a11 and Mdr-1 gene expression in mice, and this effect seems to be specific for PXR and CAR [47]. Thus, differential interaction of coactivators and corepressors induced by various xenobiotics may alter CAR- and PXR-mediated transcription.

6.3 PXR AND CAR CROSS TALK WITH OTHER NUCLEAR RECEPTORS AND TRANSCRIPTION FACTORS

6.3.1 PXR and CAR Interact with Other Nuclear Receptors

PXR and CAR are able to interact with several nuclear receptors involved in lipid metabolism. Mammalian two-hybrid assay and immunoprecipitation assays show that PXR strongly interacts with hepatic nuclear factors 4α (HNF4α) and rifampicin is required [29]. HNF4α is an orphan receptor that plays a central role in maintaining lipid homeostasis. Liver-specific ablation of the HNF4α gene in mice causes lipid accumulation in hepatocytes [48]. HNF4α transactivation activity is strongly stimulated by PGC-1α.

Bile acids and metabolites are endogenous PXR ligands that activate PXR to metabolize bile acids [49]. This may be a mechanism for detoxifying bile acids in the liver and intestine [50]. Bile acids activate farnesoid X receptor (FXR), which induces expression of small heterodimer partner (SHP), an atypical orphan receptor that lacks the DBD. SHP then inhibits transactivation of CYP7A1 gene by orphan receptor, liver-related homologue-1 (LRH-1) [51]. SHP is known to interact with most if not all nuclear receptors in a ligand-dependent or ligand-independent manner. As expected, SHP interacts with PXR and inhibits PXR induction of the CYP3A family of drug-metabolizing cytochrome P450 enzymes in the liver and intestine *in vitro* [52]. It has been suggested that the bile acid-activated FXR/SHP pathway not only inhibits bile acid synthesis but also inhibits PXR induction of drug metabolisms mediated by CYP3A4 [52]. SHP also inhibits the transactivation of the CYP2B gene by CAR in cultured hepatoma cells [53]. In GST pull down assay, HDAC1, Sim3A, and SMRT, but not NCoR, interact with SHP. Interestingly, rifampicin-activated PXR strongly inhibits SHP gene expression in human primary hepatocytes suggesting that PXR inhibition of SHP to enhance PXR induction of CYP3A4 in hepatocytes [54]. It is unlikely that SHP inhibits CYP3A4 expression and drug metabolism *in vivo*.

PB or TCPOBOP treatment reduces the mRNA expression of the gluconeogenic gene, phosphoenoylpyruvate carboxylase (PEPCK) in mouse liver [55, 56]. Similar to PXR, CAR may inhibit CYP7A1 by competing with HNF4α for PGC-1α and GRIP [57]. CAR may also compete with HNF4α for binding to the HNF4α binding site in the CYP7A1 promoter [57]. Similarly, the same mechanisms may also inhibit carnitine palmitoyltransferase 1 (CPT1) and enoyl-CoA isomerase (ECI) gene transcription by CAR [55]. This may be a general mechanism for PXR and CAR inhibition of genes in glucose and lipid metabolism. CAR also interacts with estrogen receptor (ER) and inhibits ER signaling by squelching p160 coactivators [58].

6.3.2   PXR and CAR Interact with Other Transcription Factors

It is well documented that many nuclear receptors can interact with other transcription factors and either activate or inhibit gene transcription as a coactivator or corepressor [20]. PXR and CAR are able to interact with several transcription factors involved in lipid, glucose, and energy metabolism in the liver. It has been reported that the forkhead transcription factor O1 (FoxO1) interacts with CAR in the presence of TCPOBOP. FoxO1 is a coactivator of CAR that stimulates TCPOBOP-activated CAR induction of CYP3A4 and CYP2B6 gene transcription [59]. FoxO1 also interacts with PXR and stimulates PCN-activated PXR induction of CYP3A4. In contrast, PXR and CAR interact with FoxO1 and inhibit FoxO1 activation of its target genes PEPCK and G6Pase [59]. A recent study reports that PXR interacts with FoxA2 and inhibits FoxA2 target gene CPT1A involved in fatty acid β-oxidation and 3-hydroxy-3-methylglutaryl CoA synthase 2 (HMGCS2) in ketogenesis [60]. The DBD of FoxA2 interacts with the DBD of PXR and prevents FoxA2 binding to the CPT1A and HMGCS2 promoters.

CAR and PXR also interact with POU domain proteins such as Pit-1 and Oct-1 [61]. The interactions between Pit-1 and nuclear receptors are ligand dependent, and helix 12 in AF-2 domain of receptors is involved. In the absence of a receptor ligand, Pit-1 inhibits PXR and CAR heterodimer formation with RXR. The POU domain proteins repression of PXR and CAR may involve HDAC. Conversely, CAR and PXR repress Pit-1 by competing for coactivators [61]. POU domain proteins interact with many nuclear receptors. The physiological significance of POU domain proteins interaction with nuclear receptors is not clear.

6.4   PXR AND CAR REGULATION OF LIPID AND GLUCOSE HOMEOSTASIS

6.4.1   CAR and PXR Regulation of Fatty Acid Metabolism

Several recent studies have implicated CAR and PXR in regulating fatty acid metabolism in the liver (Figure 6.1) [55–57, 62]. PB treatment reduces CPTA1 and ECI gene expression in fatty acid β-oxidation in wild type but not CAR null mice, suggesting that CAR inhibits fatty acid β-oxidation [55, 56]. Inhibition of fatty acid oxidation increases accumulation of triglycerides in the liver. Interestingly, a recent study implicates CAR in pathogenesis of nonalcoholic steatohepatitis (NASH) [63]. A methionine and choline-deficient (MCD) diet has been used to induce NASH in mice. These investigators report that feeding the MCD diet for 16 weeks, the CAR+/+ mice developed fibrosis, increased expression of mRNA for liver fibrosis markers, collagen α1 and inhibitor of metalloproteinase-1. The MCD diet also causes nuclear translocation of CAR and increased expression of the CAR target genes CYP2B10 and CYP3A11 in CAR+/+ mice. In CAR−/− mice, these fibrosis phenotypes were less severe. This study suggests that CAR causes the worsening of the hepatic injury and fibrosis in the dietary model of NASH and CAR may play a critical role in the progression of steatosis to steatohepatitis.

PXR activates stearoyl CoA desaturases (SCD-1), the key enzyme in triglyceride synthesis (Figure 6.1) [60]. In transgenic mice overexpressing a constitutive PXR, SCD-1 gene expression is induced and lipid droplets accumulate in the liver of PXR transgenic mice [62]. In Scd-1 null mice, triglyceride levels are reduced. However, the mechanism of PXR induction of the SCD-1 gene is not clear. Insulin induces the SCD-1 gene via SREBP-1c (Figure 6.1). It was speculated that PXR might interact with SREBP and result in inhibiting SCD-1 gene transcription. In contrast, a recent report shows that PXR induction of SCD-1 is independent of the SREBP-1c pathway [62]. PXR also induces free fatty acid transporter CD36 and several lipogenic genes including fatty acid synthase (FAS) (Figure 6.1).

6.4.2   PXR and CAR Regulation of Bile Acid Metabolism

In the liver, cholesterol is catabolized to bile acids. CYP7A1 is the first and rate-limiting enzyme of the classical bile acid biosynthetic pathway. It is known that PCN and dexamethasone feeding to rats strongly inhibits bile acid synthesis and CYP7A1 activity [64]. This effect is not mediated by glucocorticoid receptor. PXR has recently been implicated in mediating PCN inhibition of CYP7A1 gene transcription. In Pxr knock-out mice, CYP7A1 mRNA levels increase, suggesting that PXR inhibits CYP7A1 gene expression (Figure 6.1) [50]. PCN induces the hepatic bile acid transporter organic anion transport peptide 2 (OATP2) mRNA expression in

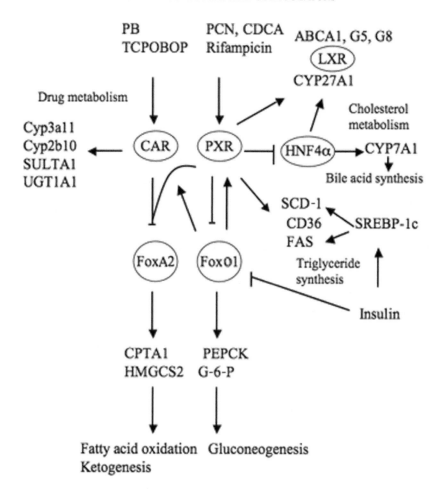

Energy metabolism

FIGURE 6.1   PXR and CAR cross talk with HNF4α and FoxO1 and regulate drug, lipid, glucose, and energy metabolism. PXR inhibits FoxO1 induction of PEPCK and G-6-P in gluconeogenesis. PXR agonists may reduce glucose and increase insulin sensitivity. CAR inhibits FoxA2 induction of carnitine palmitoyltransferase 1 (CPT1) and enoyl-CoA isomerase (ECI) in fatty acid β-oxidation and 3-hydroxy-3-methylglutaryl CoA synthase (HMGCS2) in ketogenesis. CAR and PXR may inhibit energy metabolism during starvation. PXR and CAR stimulate triglyceride synthesis by inducing stearoyl CoA desaturase-1 (SCD-1). PXR induces free fatty acid transporter (CD36) and fatty acid synthase (FAS). Thus, PXR and CAR may inhibit fatty acid oxidation but stimulate triglyceride synthesis and contribute to hepatic steatosis. FoxO1 stimulates PXR and CAR induction of genes involved in three phases of drug metabolism including CYP3A4, CYP2B6, SULTA1, and UGT1A1. PXR inhibits HNF4α transactivation of CYP7A1 in bile acid synthesis. In the intestine, PXR induces CYP27A1, which hydroxylases cholesterol to 27-hydroxycholesterol, an endogenous ligand of LXRα. LXRα induces cholesterol efflux transporters ABCA1, ABCG1, and ABCG8. Insulin signaling inhibits FoxO1 activity by phosphorylation and inhibits the gluconeogenic genes. Insulin induces SREBP-1c via activation of LXRα. SREBP plays a major role in fatty acids, triglycerides, and cholesterol synthesis.

wild-type mice. CAR and PXR induce phase II drug conjugation enzymes sulfotransferase (SULTA1) and UDP-glucuronosyltransferase (UGT1A1), and phase III drug transporters multidrug resistant protein (MDR)-related protein (MRP2/4) (Figure 6.1) [65]. It appears that PXR and CAR play important roles in bile acid metabolism. An intermediate in bile acid synthesis, 5β-cholestane-3α, 7α, 12α-triol is an endogenous ligand of mouse PXR [49, 66]. This triol is a substrate of both mouse CYP3A11 and CYP27A1 (Figure 6.1), the latter is a mitochondrial 25 and 27-hydroxylase involved in oxidative cleavage of steroid side chain. PXR induces CYP3A11 to hydroxylate triol to 5β-cholestane-3α, 7α, 12α, 25-tetraol and leads to synthesis of cholic acid in mouse liver. Trio does not activate PXR and does not induce CYP3A4 in human livers. In cerebrotendinous xanthomatosis (CTX) patients with mutations of the CYP27A1 gene, triol is accumulated at very high levels in serum.

Cholestasis is a chronic liver disease caused by disruption of bile flow and is also induced by drugs, pregnancy, hepatitis, and inflammatory agents [67–69]. PXR plays a critical role in detoxification of bile acids [70, 71]. Feeding lithocholic acid causes liver damage in Pxr null mice, but transgenic mice expressing a human PXR are protected from bile acid toxicity [50, 70]. PXR has been shown to inhibit CYP7A1 gene transcription [29]. ChIP assay has confirmed the presence of PXR, HNF4α, and PGC-1α in the human CYP7A1 chromatin containing both PXR and HNF4α binding sites. Rifampicin dissociates PGC-1α from CYP7A1 chromatin. It appears that PXR inhibits CYP7A1 gene transcription by two mechanisms: PXR competes with HNF4α for PGC-1α, and PXR interacts with HNF4α and prevents PGC-1α and HNF4α from transactivation the CYP7A1 gene [29, 72]. Thus, PXR plays a protective role against cholestatic liver injury in mice by inducing bile acid detoxification and inhibiting bile acid synthesis. All these studies suggest that PXR may coordinately regulate bile acid and drug metabolism [25, 73].

6.4.3 PXR Regulation of Cholesterol Metabolism

Oxysterols and cholesterol metabolites are able to activate PXR and induce CYP3A4 in rat and mouse hepatocytes [74, 75]. Interestingly, a potent liver orphan receptor α (LXRα) agonist T0901817 is also a PXR agonist that induces LXR-regulated genes involved in bile acid, fatty acid and triglyceride synthesis [76]. Several intermediates of the de novo cholesterol synthesis including squalene metabolites and 24(S), 25-epoxycholesterol are potent PXR ligands and inducers of CYP3A4 expression in primary hepatocytes. Pxr null mice fed a high cholesterol and cholic acid diet develop acute hepatorenal failure, suggesting that PXR may play a role in detoxification of cholesterol metabolites [77]. Rifampicin-activated PXR is able to induce CYP27A1 expression in intestine, but not in liver cells [78]. In the intestine, rifampicin treatment increases 27-hydroxycholesterol, which is a weak endogenous ligand of LXRα. LXRα induces cholesterol efflux by inducing ABCA1 and ABCG5/G8 transporters in liver and intestine (Figure 6.1). Induction of CYP3A4 is correlated to increased plasmid HDL and ApoA1 levels, and rifampicin and other PXR agonists increase ApoA1 expression and serum HDL-cholesterol levels in rodents [79]. Bile acids are known to reduce HDL cholesterol, plasma ApoA1, and hepatic ApoA1 mRNA expression via activation of FXR, which inhibits ApoA1 gene expression [80, 81]. The inhibitory effect of bile acids is more pronounced in Pxr null mice and is attenuated in human PXR transgenic mice [82].

6.4.4 CAR and PXR Regulation of Glucose and Energy Metabolism

Insulin signaling inhibits the gluconeogenic genes PEPCK and glucose-6-phosphatase (G6Pase) in the liver [83, 84] (Figure 6.1). The inhibitory effect of insulin on gluconeogenesis is mediated through FoxO1. FoxO1 and PGC-1α induce gluconeogenesis during fasting. Insulin signaling phosphorylates and inactivates FoxO1, and results in inhibiting gluconeogenesis. TCPOBOP or PB treatment reduces PEPCK mRNA levels in wild type but not CAR−/− mice. CAR and PXR inhibit FoxO1 activity, thus inhibiting the PEPCK and G6Pase genes in the presence of their agonists [59]. This may explain that diabetic patients treated with PB have reduced glucose level and increased insulin sensitivity. FoxO1 is a coactivator of CAR and PXR and activates drug metabolism induced by PXR and CAR. Thus, FoxO1 cross talks with xenobiotic receptors reciprocally coregulate target genes and thereby coordinately regulate both drug metabolism and glucose metabolism.

PXR cross talks with FoxA2 and inhibits CPT1a1 and HMGCS2 genes in fatty acid β-oxidation and ketogenesis, thus regulates energy metabolism (Figure 6.1) [56, 59]. PGC-1α interaction with CAR provides a link between xenobiotic response and nutritional state [39]. Fasting induces PGC-1α, which activates HNF4α and FoxO1, and stimulates gluconeogenesis and fatty acid oxidation for energy metabolism. CAR mRNA expression is increased in fasted animal and the effect of TCPOBOP is greater in fasted than fed animals [57]. Increased CAR expression during fasting may attenuate induction of gluconeogenesis and fatty acid oxidation. Therefore, PXR and CAR interaction with FoxA2 may cause drug-induced hypertriglyceridemia in fasting [60].

TCPOBOP treatment decreases serum thyroid hormone levels. CAR apparently is involved in thyroid hormone metabolism since CAR−/− mice treated with TCPOBOP failed to alter thyroid hormone levels [85]. During fasting, thyroid hormone levels decrease in a CAR-dependent manner because the CAR target genes SULTs and UGT1A1 regulate thyroid hormone levels. CAR target genes CYP2b10, oatp2, sult2a1, and ugt1a1 are induced during fasting by increasing cAMP levels [86]. PGC-1α is induced by cAMP response element binding (CREB) protein induced during fasting. Mice lacking CAR have decreased thyroid hormone levels and decreased resistance to weight loss during fasting and caloric restriction.

6.5 CONCLUSION

PXR and CAR are metabolic regulators that are activated by xenobiotics and endobiotics. PXR and CAR interact with nuclear receptor (HNF4α, SHP), transcription factors (FoxO1, FoxA2), and coactivators (PGC-1α, GRIP-1) and cross-regulate drug, glucose, lipids, and energy metabolism. Activation of PXR and CAR target genes detoxify drugs, bile acids, and cholesterol metabolites. PXR and CAR play a critical role in protecting liver against drug- and bile acid-induced cholestasis and liver injury. Drug activation of PXR and CAR is known to cause drug–drug interaction in humans. Activation of PXR and CAR also affects glucose and fatty liver metabolism. PXR and CAR inhibit gluconeogenesis, fatty acid oxidation, and ketogenesis, and thus decrease energy metabolism during fasting. PXR and CAR stimulate triglyceride synthesis and may contribute to hepatic steatosis and development of NASH in animal models. On the other hand, CAR and PXR activators may reduce serum glucose concentration and increase insulin sensitivity. CAR and PXR activators may be used for treatment of diabetes. However, development of drugs targeting PXR and CAR for treating liver diseases may be challenging since PXR and CAR activators may have both positive and negative effects on glucose and lipid metabolism.

REFERENCES

1. Tzameli, I., Pissios, P., Schuetz, E. G., and Moore, D. D. (2000) The xenobiotic compound 1,4-bis[2-(3,5-dichloropyridyloxy)]benzene is an agonist ligand for the nuclear receptor CAR. *Mol. Cell Biol.* **20**, 2951–2958.

2. Moore, L. B., Parks, D. J., Jones, S. A., Bledsoe, R. K., Consler, T. G., Stimmel, J. B., Goodwin, B., Liddle, C., Blanchard, S. G., Willson, T. M., et al. (2000) Orphan nuclear receptors constitutive androstane receptor and pregnane X receptor share xenobiotic and steroid ligands. *J. Biol. Chem.* **275**, 15122–15127.

3. Tzameli, I., Chua, S. S., Cheskis, B., and Moore, D. D. (2003) Complex effects of rexinoids on ligand dependent activation or inhibition of the xenobiotic receptor, CAR. *Nucl. Recept.* **1**, 2.

4. Yamada, H., Yamaguchi, T., and Oguri, K. (2000) Suppression of the expression of the CYP2B1/2 gene by retinoic acids. *Biochem. Biophys. Res. Commun.* **277**, 66–71.

5. Kakizaki, S., Karami, S., and Negishi, M. (2002) Retinoic acids repress constitutive active receptor-mediated induction by 1,4-bis[2-(3,5-dichloropyridyloxy)]benzene of the CYP2B10 gene in mouse primary hepatocytes. *Drug Metab. Dispos.* **30**, 208–211.

6. Zelko, I., Sueyoshi, T., Kawamoto, T., Moore, R., and Negishi, M. (2001) The peptide near the C terminus regulates receptor CAR nuclear translocation induced by xenochemicals in mouse liver. *Mol. Cell Biol.* **21**, 2838–2846.

7. Kawana, K., Ikuta, T., Kobayashi, Y., Gotoh, O., Takeda, K., and Kawajiri, K. (2003) Molecular mechanism of nuclear translocation of an orphan nuclear receptor, SXR. *Mol. Pharmacol.* **63**, 524–531.

8. Kawamoto, T., Sueyoshi, T., Zelko, I., Moore, R., Washburn, K., and Negishi, M. (1999) Phenobarbital-responsive nuclear translocation of the receptor CAR in induction of the CYP2B gene. *Mol. Cell Biol.* **19**, 6318–6322.

9. Hosseinpour, F., Moore, R., Negishi, M., and Sueyoshi, T. (2006) Serine 202 regulates the nuclear translocation of constitutive active/androstane receptor. *Mol. Pharmacol.* **69**, 1095–1102.

10. Swales, K., Kakizaki, S., Yamamoto, Y., Inoue, K., Kobayashi, K., and Negishi, M. (2005) Novel CAR-mediated mechanism for synergistic activation of two distinct elements within the human cytochrome P450 2B6 gene in HepG2 cells. *J. Biol. Chem.* **280**, 3458–3466.

11. Kanno, Y., Suzuki, M., Nakahama, T., and Inouye, Y. (2005) Characterization of nuclear localization signals and cytoplasmic retention region in the nuclear receptor CAR. *Biochim. Biophys. Acta* **1745**, 215–222.

12. Kanno, Y., Suzuki, M., Miyazaki, Y., Matsuzaki, M., Nakahama, T., Kurose, K., Sawada, J., and Inouye, Y. (2007) Difference in nucleocytoplasmic shuttling sequences of rat and human constitutive active/androstane receptor. *Biochim. Biophys. Acta* **1773**, 934–944.

13. Kumar, S., Saradhi, M., Chaturvedi, N. K., and Tyagi, R. K. (2006) Intracellular localization and nucleocytoplasmic trafficking of steroid receptors: an overview. *Mol. Cell Endocrinol.* **246**, 147–156.

14. Kobayashi, K., Sueyoshi, T., Inoue, K., Moore, R., and Negishi, M. (2003) Cytoplasmic accumulation of the nuclear receptor CAR by a tetratricopeptide repeat protein in HepG2 cells. *Mol. Pharmacol.* **64**, 1069–1075.

15. Yoshinari, K., Kobayashi, K., Moore, R., Kawamoto, T., and Negishi, M. (2003) Identification of the nuclear receptor CAR:HSP90 complex in mouse liver and recruitment of protein phosphatase 2A in response to phenobarbital. *FEBS Lett.* **548**, 17–20.

16. Swales, K. and Negishi, M. (2004) CAR, driving into the future. *Mol. Endocrinol.* **18**, 1589–1598.

17. Squires, E. J., Sueyoshi, T., and Negishi, M. (2004) Cytoplasmic localization of pregnane X receptor and ligand-dependent nuclear translocation in mouse liver. *J. Biol. Chem.* **279**, 49307–49314.

18. Jenuwein, T. and Allis, C. D. (2001) Translating the histone code. *Science* **293**, 1074–1080.

19. Kadonaga, J. T. (1998) Eukaryotic transcription: an interlaced network of transcription factors and chromatin-modifying machines. *Cell* **92**, 307–313.

20. Glass, C. K. and Rosenfeld, M. G. (2000) The coregulator exchange in transcriptional functions of nuclear receptors. *Genes Dev.* **14**, 121–141.

21. Perissi, V., Aggarwal, A., Glass, C. K., Rose, D. W., and Rosenfeld, M. G. (2004) A corepressor/coactivator exchange complex required for transcriptional activation by nuclear receptors and other regulated transcription factors. *Cell* **116**, 511–526.

22. Rosenfeld, M. G., Lunyak, V. V., and Glass, C. K. (2006) Sensors and signals: a coactivator/corepressor/epigenetic code for integrating signal-dependent programs of transcriptional response. *Genes Dev.* **20**, 1405–1428.

23. O'Malley, B. W. (2007) Coregulators: from whence came these "master genes". *Mol. Endocrinol.* **21**, 1009–1013.

24. McKenna, N. J., Lanz, R. B., and O'Malley, B. W. (1999) Nuclear receptor coregulators: cellular and molecular biology. *Endocr. Rev.* **20**, 321–344.

25. Kliewer, S. A. and Willson, T. M. (2002) Regulation of xenobiotic and bile acid metabolism by the nuclear pregnane X receptor. *J. Lipid Res.* **43**, 359–364.

26. Watkins, R. E., Davis-Searles, P. R., Lambert, M. H., and Redinbo, M. R. (2003) Coactivator binding promotes the specific interaction between ligand and the pregnane X receptor. *J. Mol. Biol.* **331**, 815–828.

27. Watkins, R. E., Wisely, G. B., Moore, L. B., Collins, J. L., Lambert, M. H., Williams, S. P., Willson, T. M., Kliewer, S. A., and Redinbo, M. R. (2001) The human nuclear xenobiotic receptor PXR: structural determinants of directed promiscuity. *Science* **292**, 2329–2333.

28. Watkins, R. E., Maglich, J. M., Moore, L. B., Wisely, G. B., Noble, S. M., Davis-Searles, P. R., Lambert, M. H., Kliewer, S. A., and Redinbo, M. R. (2003) 2.1 A crystal structure

of human PXR in complex with the St. John's wort compound hyperforin. *Biochemistry* **42**, 1430–1438.

29. Li, T. and Chiang, J. Y. (2005) Mechanism of rifampicin and pregnane X receptor inhibition of human cholesterol 7 alpha-hydroxylase gene transcription. *Am. J. Physiol. Gastrointest. Liver Physiol.* **288**, G74–G84.

30. Rhee, J., Inoue, Y., Yoon, J. C., Puigserver, P., Fan, M., Gonzalez, F. J., and Spiegelman, B. M. (2003) Regulation of hepatic fasting response by PPARgamma coactivator-1alpha (PGC-1): requirement for hepatocyte nuclear factor 4alpha in gluconeogenesis. *Proc. Natl Acad. Sci. USA* **100**, 4012–4017.

31. Zhang, Y., Castellani, L. W., Sinal, C. J., Gonzalez, F. J., and Edwards, P. A. (2004) Peroxisome proliferator-activated receptor-gamma coactivator 1alpha (PGC-1alpha) regulates triglyceride metabolism by activation of the nuclear receptor FXR. *Genes Dev.* **18**, 157–169.

32. Rhee, J., Ge, H., Yang, W., Fan, M., Handschin, C., Cooper, M., Lin, J., Li, C., and Spiegelman, B. M. (2006) Partnership of PGC-1alpha and HNF4alpha in the regulation of lipoprotein metabolism. *J. Biol. Chem.* **281**, 14683–14690.

33. Suino, K., Peng, L., Reynolds, R., Li, Y., Cha, J. Y., Repa, J. J., Kliewer, S. A., and Xu, H. E. (2004) The nuclear xenobiotic receptor CAR: structural determinants of constitutive activation and heterodimerization. *Mol. Cell* **16**, 893–905.

34. Shan, L., Vincent, J., Brunzelle, J. S., Dussault, I., Lin, M., Ianculescu, I., Sherman, M. A., Forman, B. M., and Fernandez, E. J. (2004) Structure of the murine constitutive androstane receptor complexed to androstenol: a molecular basis for inverse agonism. *Mol. Cell* **16**, 907–917.

35. Xu, R. X., Lambert, M. H., Wisely, B. B., Warren, E. N., Weinert, E. E., Waitt, G. M., Williams, J. D., Collins, J. L., Moore, L. B., Willson, T. M., et al. (2004) A structural basis for constitutive activity in the human CAR/RXRalpha heterodimer. *Mol. Cell* **16**, 919–928.

36. Wright, E., Vincent, J., and Fernandez, E. J. (2007) Thermodynamic characterization of the interaction between CAR-RXR and SRC-1 peptide by isothermal titration calorimetry. *Biochemistry* **46**, 862–870.

37. Forman, B. M., Tzameli, I., Choi, H. S., Chen, J., Simha, D., Seol, W., Evans, R. M., and Moore, D. D. (1998) Androstane metabolites bind to and deactivate the nuclear receptor CAR-beta. *Nature* **395**, 612–615.

38. Min, G., Kemper, J. K., and Kemper, B. (2002) Glucocorticoid receptor-interacting protein 1 mediates ligand-independent nuclear translocation and activation of constitutive androstane receptor in vivo. *J. Biol. Chem.* **277**, 26356–26363.

39. Shiraki, T., Sakai, N., Kanaya, E., and Jingami, H. (2003) Activation of orphan nuclear constitutive androstane receptor requires subnuclear targeting by peroxisome proliferator-activated receptor gamma coactivator-1 alpha. A possible link between xenobiotic response and nutritional state. *J. Biol. Chem.* **278**, 11344–11350.

40. Choi, E., Lee, S., Yeom, S. Y., Kim, G. H., Lee, J. W., and Kim, S. W. (2005) Characterization of activating signal cointegrator-2 as a novel transcriptional coactivator of the xenobiotic nuclear receptor constitutive androstane receptor. *Mol. Endocrinol.* **19**, 1711–1719.

41. Jia, Y., Guo, G. L., Surapureddi, S., Sarkar, J., Qi, C., Guo, D., Xia, J., Kashireddi, P., Yu, S., Cho, Y. W., et al. (2005) Transcription coactivator peroxisome proliferator-activated

receptor-binding protein/mediator 1 deficiency abrogates acetaminophen hepatotoxicity. *Proc. Natl Acad. Sci. USA* **102**, 12531–12536.

42. Guo, D., Sarkar, J., Ahmed, M. R., Viswakarma, N., Jia, Y., Yu, S., Sambasiva Rao, M., and Reddy, J. K. (2006) Peroxisome proliferator-activated receptor (PPAR)-binding protein (PBP) but not PPAR-interacting protein (PRIP) is required for nuclear translocation of constitutive androstane receptor in mouse liver. *Biochem. Biophys. Res. Commun.* **347**, 485–495.

43. Zhang, J., Huang, W., Qatanani, M., Evans, R. M., and Moore, D. D. (2004) The constitutive androstane receptor and pregnane X receptor function coordinately to prevent bile acid-induced hepatotoxicity. *J. Biol. Chem.* **279**, 49517–49522.

44. Synold, T. W., Dussault, I., and Forman, B. M. (2001) The orphan nuclear receptor SXR coordinately regulates drug metabolism and efflux. *Nat. Med.* **7**, 584–590.

45. Johnson, D. R., Li, C. W., Chen, L. Y., Ghosh, J. C., and Chen, J. D. (2006) Regulation and binding of pregnane X receptor by nuclear receptor corepressor silencing mediator of retinoid and thyroid hormone receptors (SMRT). *Mol. Pharmacol.* **69**, 99–108.

46. Takeshita, A., Taguchi, M., Koibuchi, N., and Ozawa, Y. (2002) Putative role of the orphan nuclear receptor SXR (steroid and xenobiotic receptor) in the mechanism of CYP3A4 inhibition by xenobiotics. *J. Biol. Chem.* **277**, 32453–32458.

47. Huang, H., Wang, H., Sinz, M., Zoeckler, M., Staudinger, J., Redinbo, M. R., Teotico, D. G., Locker, J., Kalpana, G. V., and Mani, S. (2007) Inhibition of drug metabolism by blocking the activation of nuclear receptors by ketoconazole. *Oncogene* **26**, 258–268.

48. Hayhurst, G. P., Lee, Y. H., Lambert, G., Ward, J. M., and Gonzalez, F. J. (2001) Hepatocyte nuclear factor 4alpha (nuclear receptor 2A1) is essential for maintenance of hepatic gene expression and lipid homeostasis. *Mol. Cell Biol.* **21**, 1393–1403.

49. Goodwin, B., Gauthier, K. C., Umetani, M., Watson, M. A., Lochansky, M. I., Collins, J. L., Leitersdorf, E., Mangelsdorf, D. J., Kliewer, S. A., and Repa, J. J. (2003) Identification of bile acid precursors as endogenous ligands for the nuclear xenobiotic pregnane X receptor. *Proc. Natl Acad. Sci. USA* **100**, 223–228.

50. Staudinger, J. L., Goodwin, B., Jones, S. A., Hawkins-Brown, D., MacKenzie, K. I., LaTour, A., Liu, Y., Klaassen, C. D., Brown, K. K., Reinhard, J., et al. (2001) The nuclear receptor PXR is a lithocholic acid sensor that protects against liver toxicity. *Proc. Natl Acad. Sci. USA* **98**, 3369–3374.

51. Goodwin, B., Jones, S. A., Price, R. R., Watson, M. A., McKee, D. D., Moore, L. B., Galardi, C., Wilson, J. G., Lewis, M. C., Roth, M. E., et al. (2000) A regulatory cascade of the nuclear receptors FXR, SHP-1, and LRH-1 represses bile acid biosynthesis. *Mol. Cell* **6**, 517–526.

52. Ourlin, J. C., Lasserre, F., Pineau, T., Fabre, J. M., Sa-Cunha, A., Maurel, P., Vilarem, M. J., and Pascussi, J. M. (2003) The small heterodimer partner interacts with the pregnane X receptor and represses its transcriptional activity. *Mol. Endocrinol.* **17**, 1693–1703.

53. Bae, Y., Kemper, J. K., and Kemper, B. (2004) Repression of CAR-mediated transactivation of CYP2B genes by the orphan nuclear receptor, short heterodimer partner (SHP). *DNA Cell Biol.* **23**, 81–91.

54. Li, T. and Chiang, J. Y. (2006) Rifampicin induction of CYP3A4 requires pregnane X receptor cross talk with hepatocyte nuclear factor 4alpha and coactivators, and suppression of small heterodimer partner gene expression. *Drug Metab. Dispos.* **34**, 756–764.

55. Ueda, A., Hamadeh, H. K., Webb, H. K., Yamamoto, Y., Sueyoshi, T., Afshari, C. A.,

Lehmann, J. M., and Negishi, M. (2002) Diverse roles of the nuclear orphan receptor CAR in regulating hepatic genes in response to phenobarbital. *Mol. Pharmacol.* **61**, 1–6.

56. Rosenfeld, J. M., Vargas, R., Jr., Xie, W., and Evans, R. M. (2003) Genetic profiling defines the xenobiotic gene network controlled by the nuclear receptor pregnane X receptor. *Mol. Endocrinol.* **17**, 1268–1282.

57. Miao, J., Fang, S., Bae, Y., and Kemper, J. K. (2006) Functional inhibitory cross-talk between constitutive androstane receptor and hepatic nuclear factor-4 in hepatic lipid/glucose metabolism is mediated by competition for binding to the DR1 motif and to the common coactivators, GRIP-1 and PGC-1alpha. *J. Biol. Chem.* **281**, 14537–14546.

58. Min, G., Kim, H., Bae, Y., Petz, L., and Kemper, J. K. (2002) Inhibitory cross-talk between estrogen receptor (ER) and constitutively activated androstane receptor (CAR). CAR inhibits ER-mediated signaling pathway by squelching p160 coactivators. *J. Biol. Chem.* **277**, 34626–34633.

59. Kodama, S., Koike, C., Negishi, M., and Yamamoto, Y. (2004) Nuclear receptors CAR and PXR cross talk with FOXO1 to regulate genes that encode drug-metabolizing and gluconeogenic enzymes. *Mol. Cell Biol.* **24**, 7931–7940.

60. Nakamura, K., Moore, R., Negishi, M., and Sueyoshi, T. (2007) Nuclear pregnane X receptor cross-talk with FoxA2 to mediate drug-induced regulation of lipid metabolism in fasting mouse liver. *J. Biol. Chem.* **282**, 9768–9776.

61. Gonzalez, M. M. and Carlberg, C. (2002) Cross-repression, a functional consequence of the physical interaction of non-liganded nuclear receptors and POU domain transcription factors. *J. Biol. Chem.* **277**, 18501–18509.

62. Zhou, J., Zhai, Y., Mu, Y., Gong, H., Uppal, H., Toma, D., Ren, S., Evans, R. M., and Xie, W. (2006) A novel pregnane X receptor-mediated and sterol regulatory element-binding protein-independent lipogenic pathway. *J. Biol. Chem.* **281**, 15013–15020.

63. Yamazaki, Y., Kakizaki, S., Horiguchi, N., Sohara, N., Sato, K., Takagi, H., Mori, M., and Negishi, M. (2007) The role of the nuclear receptor constitutive androstane receptor in the pathogenesis of non-alcoholic steatohepatitis. *Gut* **56**, 565–574.

64. Li, Y. C., Wang, D. P., and Chiang, J. Y. L. (1990) Regulation of cholesterol 7a-hydroxylase in the liver: cDNA cloning, sequencing and regulation of cholesterol 7a-hydroxylase mRNA. *J. Biol. Chem.* **265**, 12012–12019.

65. Sonoda, J., Xie, W., Rosenfeld, J. M., Barwick, J. L., Guzelian, P. S., and Evans, R. M. (2002) Regulation of a xenobiotic sulfonation cascade by nuclear pregnane X receptor (PXR). *Proc. Natl Acad. Sci. USA* **99**, 13801–13806.

66. Dussault, I., Yoo, H. D., Lin, M., Wang, E., Fan, M., Batta, A. K., Salen, G., Erickson, S. K., and Forman, B. M. (2003) Identification of an endogenous ligand that activates pregnane X receptor-mediated sterol clearance. *Proc. Natl Acad. Sci. USA* **100**, 833–838.

67. Trauner, M., Meier, P. J., and Boyer, J. L. (1999) Molecular regulation of hepatocellular transport systems in cholestasis. *J. Hepatol.* **31**, 165–178.

68. Zollner, G. and Trauner, M. (2006) Molecular mechanisms of cholestasis. *Wien. Med. Wochenschr.* **156**, 380–385.

69. Geier, A., Wagner, M., Dietrich, C. G., and Trauner, M. (2007) Principles of hepatic organic anion transporter regulation during cholestasis, inflammation and liver regeneration. *Biochim. Biophys. Acta* **1773**, 283–308.

70. Xie, W., Radominska-Pandya, A., Shi, Y., Simon, C. M., Nelson, M. C., Ong, E. S., Waxman, D. J., and Evans, R. M. (2001) An essential role for nuclear receptors

SXR/PXR in detoxification of cholestatic bile acids. *Proc. Natl Acad. Sci. USA* **98**, 3375–3380.

71. Schuetz, E. G., Strom, S., Yasuda, K., Lecureur, V., Assem, M., Brimer, C., Lamba, J., Kim, R. B., Ramachandran, V., Komoroski, B. J., et al. (2001) Disrupted bile acid homeostasis reveals an unexpected interaction among nuclear hormone receptors, transporters and cytochrome P450. *J. Biol. Chem.* **16**, 16.

72. Bhalla, S., Ozalp, C., Fang, S., Xiang, L., and Kemper, J. K. (2004) Ligand-activated pregnane X receptor interferes with HNF-4 signaling by targeting a common coactivator PGC-1alpha. Functional implications in hepatic cholesterol and glucose metabolism. *J. Biol. Chem.* **279**, 45139–45147.

73. Staudinger, J., Liu, Y., Madan, A., Habeebu, S., and Klaassen, C. D. (2001) Coordinate regulation of xenobiotic and bile acid homeostasis by pregnane X receptor. *Drug Metab. Dispos.* **29**, 1467–1472.

74. Shenoy, S. D., Spencer, T. A., Mercer-Haines, N. A., Abdolalipour, M., Wurster, W. L., Runge-Morris, M., and Kocarek, T. A. (2004) Induction of CYP3A by 2,3-oxidosqualene:lanosterol cyclase inhibitors is mediated by an endogenous squalene metabolite in primary cultured rat hepatocytes. *Mol. Pharmacol.* **65**, 1302–1312.

75. Shenoy, S. D., Spencer, T. A., Mercer-Haines, N. A., Alipour, M., Gargano, M. D., Runge-Morris, M., and Kocarek, T. A. (2004) CYP3A induction by liver x receptor ligands in primary cultured rat and mouse hepatocytes is mediated by the pregnane X receptor. *Drug Metab. Dispos.* **32**, 66–71.

76. Mitro, N., Vargas, L., Romeo, R., Koder, A., and Saez, E. (2007) T0901317 is a potent PXR ligand: implications for the biology ascribed to LXR. *FEBS Lett.* **581**, 1721–1726.

77. Sonoda, J., Chong, L. W., Downes, M., Barish, G. D., Coulter, S., Liddle, C., Lee, C. H., and Evans, R. M. (2005) Pregnane X receptor prevents hepatorenal toxicity from cholesterol metabolites. *Proc. Natl Acad. Sci. USA* **102**, 2198–2203.

78. Li, T., Chen, W., and Chiang, J. Y. (2007) PXR induces CYP27A1 and regulates cholesterol metabolism in the intestine. *J. Lipid Res.* **48**, 373–384.

79. Bachmann, K., Patel, H., Batayneh, Z., Slama, J., White, D., Posey, J., Ekins, S., Gold, D., and Sambucetti, L. (2004) PXR and the regulation of apoA1 and HDL-cholesterol in rodents. *Pharmacol. Res.* **50**, 237–246.

80. Claudel, T., Sturm, E., Duez, H., Torra, I. P., Sirvent, A., Kosykh, V., Fruchart, J. C., Dallongeville, J., Hum, D. W., Kuipers, F., et al. (2002) Bile acid-activated nuclear receptor FXR suppresses apolipoprotein A-I transcription via a negative FXR response element. *J. Clin. Invest.* **109**, 961–971.

81. Claudel, T., Staels, B., and Kuipers, F. (2005) The farnesoid X receptor: a molecular link between bile acid and lipid and glucose metabolism. *Arterioscler. Thromb. Vasc. Biol.* **25**, 2020–2030.

82. Masson, D., Lagrost, L., Athias, A., Gambert, P., Brimer-Cline, C., Lan, L., Schuetz, J. D., Schuetz, E. G., and Assem, M. (2005) Expression of the pregnane X receptor in mice antagonizes the cholic acid-mediated changes in plasma lipoprotein profile. *Arterioscler. Thromb. Vasc. Biol.* **25**, 2164–2169.

83. Nakae, J., Kitamura, T., Silver, D. L., and Accili, D. (2001) The forkhead transcription factor FoxO1 (Fkhr) confers insulin sensitivity onto glucose-6-phosphatase expression. *J. Clin. Invest.* **108**, 1359–1367.

84. Puigserver, P., Rhee, J., Donovan, J., Walkey, C. J., Yoon, J. C., Oriente, F., Kitamura, Y., Altomonte, J., Dong, H., Accili, D., et al. (2003) Insulin-regulated hepatic gluconeogenesis through FOXO1-PGC-1alpha interaction. *Nature* **423**, 550–555.

85. Maglich, J. M., Watson, J., McMillen, P. J., Goodwin, B., Willson, T. M., and Moore, J. T. (2004) The nuclear receptor CAR is a regulator of thyroid hormone metabolism during caloric restriction. *J. Biol. Chem.* **279**, 19832–19838.

86. Ding, X., Lichti, K., Kim, I., Gonzalez, F. J., and Staudinger, J. L. (2006) Regulation of constitutive androstane receptor and its target genes by fasting, cAMP, hepatocyte nuclear factor alpha, and the coactivator peroxisome proliferator-activated receptor gamma coactivator-1alpha. *J. Biol. Chem.* **281**, 26540–26551.

7

ANIMAL MODELS OF XENOBIOTIC NUCLEAR RECEPTORS AND THEIR UTILITY IN DRUG DEVELOPMENT

HAIBIAO GONG

Department of Biotechnology, LI-COR Biosciences, Lincoln, NE, USA

WEN XIE

Center for Pharmacogenetics, Department of Pharmaceutical Sciences, University of Pittsburgh, Pittsburgh, PA, USA

7.1 INTRODUCTION

The establishment of NRs pregnane X receptor (PXR) and constitutive androstane receptor (CAR) as "xenobiotic receptors" was published in 1998 [1, 2]. PXR and CAR were initially found to regulate the phase I CYP3A and CYP2B enzymes. Subsequent studies have shown that both receptors can also regulate the expression of phase II conjugating enzymes and "phase III" drug transporters; and for this reason, PXR and CAR have been proposed to function as "master" xenobiotic receptors.

Mouse models have played an important role in dissecting the function of xenobiotic receptors *in vivo*. Creation and characterization of these animal models have also suggested that many of them have potential utility in pharmaceutical development. These include the "loss-of-function" gene knock-out mice, "gain-of-function" transgenic mice, as well as "humanized" mice in which the mouse xenobiotic receptor genes are genetically replaced by their human counterparts. The humanized mice were created to address the species specificity of drug-induced regulation of drug-metabolizing enzymes and transporters. The humanized mice offer a unique model

Nuclear Receptors in Drug Metabolism   Edited by Wen Xie
Copyright © 2009 John Wiley & Sons, Inc.

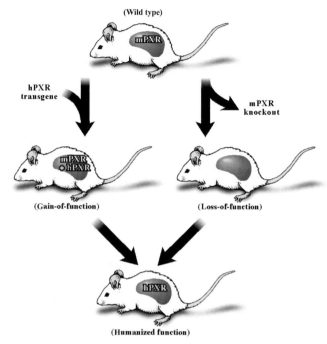

FIGURE 7.1   Strategies to create the loss-of-function knock-out, gain-of-function transgenic, and the combined "humanized" function models. (Adapted from Figure 1 in [3] with permission from the Academic Press Inc., Elsevier Science.)

system to dissect the drug-induced xenobiotic responses in a way similar to that happens in humans. The initial creation of xenobiotic receptor mouse models and subsequent use of these animals by many laboratories have greatly accelerated the research on the function of xenobiotic receptors in drug metabolism, drug–drug interaction, and xenobiotic receptors associated diseases [3–5]. Using PXR as an example, Figure 7.1 summarizes strategies to create the loss-of-function knock-out model, gain-of-function transgenic models, and the combined humanized function models [3, 6, 7].

In this chapter, we will review the creation of various mouse models of xenobiotic nuclear receptors, and the utilization of mouse models in drug metabolism studies and drug development. Special emphasis will put on the mouse models humanized for the nuclear receptors or the aryl hydrocarbon receptor (AhR), which is a nonnuclear receptor xenobiotic receptor.

7.2   PXR AND CAR LOSS-OF-FUNCTION (KNOCK-OUT) MOUSE MODELS

7.2.1   PXR Knock-Out Mice

The Evans lab at the Salk Institute first reported the creation of PXR knock-out mice [6]. In that case, to create PXR null mice, the mouse PXR genomic DNA was isolated by screening a 129/Sv genomic phage library (Stratagene, La Jolla, CA) using a PXR cDNA probe. A targeting vector was generated by replacing the second and third exons of PXR with a PGK-Neo selection marker, in conjunction with a negative selection marker (PGK-TK) (Figure 7.2a). The resulting mutant allele has a deletion of two exons spanning nucleotides 339–660 that include amino acids 63–170 of the PXR DNA binding domain (DBD) [2, 6].

The targeting vector was linearized by Not I digestion and transfected into J1 ES cells by electroporation. G418 (200 μg ml^{-1})- and ganciclovir (0.2 mM)-resistant ES clones were screened for designated homologous recombination by Southern blot analysis. PXR$^{+/-}$ ES cells were microinjected into C57BL6/J blastocysts, which were subsequently transplanted into the uteri of pseudopregnant ICR mice. Chimeric male progeny were crossed with C57B/6J females. Germline transmission of the mutant allele was detected in agouti progeny by Southern blot analysis (Figure 7.2b). Heterozygous (PXR$^{+/-}$) mice were crossbred to obtain the homozygous (PXR$^{-/-}$)

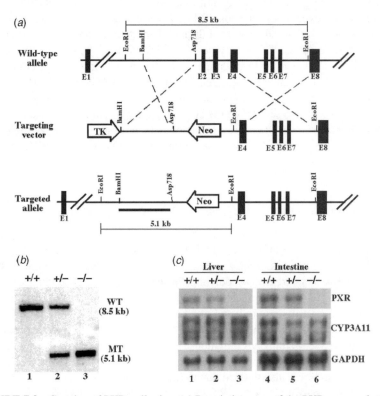

FIGURE 7.2 Creation of PXR null mice. (*a*) Restriction map of the PXR gene and strategy to generate PXR mutant allele. The PXR probes used for Southern blot and expected fragment sizes after EcoRI digestion are indicated. E, exon; Neo, neomycin resistance gene; TK, thymidine kinase promoter. (*b*) Southern blot of EcoRI-digested genomic DNA. WT, wild type; MT, Mutant. (*c*) Loss of PXR expression in the liver and small intestine of PXR$^{-/-}$ mice. (Adapted from Figure 1 in [6] with permission from the Nature Publishing Group.)

mice. The absence of PXR mRNA expression in the liver and small intestine of PXR null mice was confirmed by Northern blot analysis (Figure 7.2c).

7.2.2   Phenotype of the PXR Knock-Out Mice

The PXR null mice are both viable and fertile. The induction of *Cyp3a11* gene by pregnenolone-16α-carbonitrile (PCN) and dexamethasone (DEX), two potent mPXR agonists and *Cyp3a11* inducers, was lost in the PXR null mice, indicating that PXR is an essential mediator of *Cyp3a11* expression [6]. DEX has been long recognized as a rodent *Cyp3a* inducer and it has been controversial whether the DEX effect on *Cyp3a* gene expression is mediated by the glucocorticoid receptor (GR). Erin Schuetz and colleagues showed that the DEX-responsive activation of *Cyp3a11* was intact in the GR null mice [8]. Together, these results have conclusively demonstrated that PXR, rather than GR, is the mediator for the DEX effect on *Cyp3a* gene expression.

At the physiological level, PXR null mice exhibited a loss of protection against endogenous and exogenous toxicants. Bile acids, when accumulated, are toxic to the liver. Activation of PXR can promote bile acid detoxification and prevent toxicity by activating bile acid detoxifying enzymes. Four days of LCA treatment elicited more profound liver damage in PXR null mice as compared to their wild-type counterpart. Moreover, PCN pretreatment could not alleviate the LCA toxicity in PXR null mice, as it did in wild-type mice. A significant loss of PCN-mediated xenoprotection in PXR null mice was also seen when the animals were challenged with two prototypic xenotoxicants, the anesthetic tribromoethanol and the muscle relaxant zoxazolamine [6, 9]. It was reasoned that the loss of xenoprotection is due to lack of PXR-mediated xenobiotic responses.

7.2.3   Another Independently Developed PXR Null Mouse Line

Steven Kliewer's group, then at GlaxoSmithKline, reported the creation of another independent line of PXR null mice, which were also used to demonstrate the essential role of PXR in regulating *Cyp3a11* gene expression and protection of toxicants [10]. This mouse line was created by deleting the first exon of PXR coding sequence, which includes the translation start site and the first zinc finger of the PXR DBD (amino acids 1–63) [2]. As predicted, the knock-out mice express a shorter version of PXR transcript. It was found that neither PCN nor DEX induced the expression of *Cyp3a11* in the liver of this PXR null mouse line. The loss of PXR has no effect on the CAR pathway because in the same mice, phenobarbital (PB) still induced *Cyp3a11* gene expression. It was also noticed that the constitutive expression of *Cyp3a11* was increased about fourfold in this PXR knock-out line, suggesting a role of unliganded PXR in the suppression of CAR activity.

Also in this study, cotreatment with the PXR ligand PCN dramatically decreased the liver damage caused by LCA. However, treatment with PCN did not reverse the LCA toxicity in the PXR null mice. These results suggest a protective role of PXR in LCA toxicity [10].

7.2.4   CAR Knock-Out Mice

David Moore's group at the Baylor College of Medicine first reported the creation of the CAR knock-out mice. CAR knock-out mice were generated by replacing part of the DBD with the coding sequence for β-galactosidase next to the CAR promoter, and the lack of CAR mRNA expression in the liver and intestine of the CAR knock-out mice was confirmed by Northern blot analysis [11]. Although the deletion of CAR has no effect on any overt phenotype under normal physiological conditions, the induction of *Cyp2b10* mRNA by PB or TCPOBOP (1,4-bis[2-(3,5-dichloropyridyloxy)] benzene) is completely absent in the knock-out animals. The loss of CAR also alters the sensitivity of mice to some xenobiotics such as zoxazolamine and cocaine. Interestingly, in the same CAR knock-out mice, animal's sensitivity to zoxazolamine and cocaine was decreased and increased, respectively. The results from the CAR knock-out study clearly demonstrate that CAR is required for the responses to PB-like inducers of xenobiotic metabolism [11].

Subsequent study by the same group showed that activation of CAR prevented mice from acetaminophen (APAP) hepatotoxicity. The CAR knock-out mice were resistant to the combined treatment of CAR agonists and low doses of acetaminophen, which causes severe hepatic toxicity in wild-type mice. Furthermore, the CAR null mice were also resistant to the single treatment of toxic doses acetaminophen. The lack of response of CAR null mice to APAP was reasoned to be associated with the absence of the induction of APAP-metabolizing enzymes, such as CYP1A2, CYP3A11, and GST (glutathione *S*-transferases) Pi [12–15], by either CAR agonists or APAP itself. It was also shown in the same study that the inverse CAR agonist androstanol prevented APAP toxicity in the wild-type mice, but not in the CAR null mice, implying both negative and positive roles of CAR in the modulation of APAP toxicity [16].

Further studies using CAR knock-out mice revealed that CAR plays an important role in the metabolism of several endobiotics, such as bilirubin and bile acids. In wild-type mice, activation of CAR increases hepatic expression of enzymes involved in bilirubin clearance pathway, including Ugt1a1, Gsta1, Gsta2, Mrp2, and SLC21A6, leading to an increased clearance of bilirubin [16, 17]. The protective effect of CAR activator was not seen in CAR knock-out mice [17]. The loss of CAR activity also causes the mice hypersensitive to acute LCA treatment due to the loss of induction of LCA-metabolizing enzymes and transporters [18].

7.2.5   Another Independently Developed CAR Null Mouse Line

Another group, led by Jurgen Lehmann then at the Tularik, reported the creation of another independent line of CAR null mice. This CAR knock-out line has been used by the Masahiko Negishi's group at the National Institute of Environmental Health Sciences to show the diverse role of CAR in regulating hepatic genes in response to PB [19]. It was concluded in this study that CAR seems to have diverse roles, both as a positive and negative regulator, in the regulation of hepatic genes in response to PB beyond drug/steroid metabolism.

7.2.6   Combined Loss of PXR and CAR

Accumulating evidence has shown that PXR and CAR cross talk with each other in regulating drug-metabolizing enzymes and transporters. Both PXR and CAR regulate genes involved in xenobiotic metabolism, including phase I cytochrome P450 enzymes, phase II conjugating enzymes, and drug transporters. PXR and CAR cross-regulate each other's target genes by sharing DNA binding sites [20–22]. It is of great interest to investigate the overlapping and distinct sets of genes regulated by PXR and CAR. It is also important to examine the overlapping and differential regulatory role of these two receptors in physiological and pathological responses. This was achieved by comparative studies using PXR knock-out, CAR knock-out, and PXR/CAR double knock-out (DKO) mice.

As PXR and CAR share many of the target genes that are implicated in bile acid detoxification, the effect of individual or combined loss of these two receptors was examined. Following LCA i.p. ($250\,mg\,kg^{-1}$), the wild-type, PXR null, and CAR null mice showed no or little hepatotoxicity, suggesting the single knock-out mice had sufficient LCA detoxification to prevent hepatotoxicity at this dosage. In contrast, the livers of LCA-treated DKO males showed massive liver damage [23]. These results suggested that the combined, but not individual, loss of PXR and CAR resulted in a robust sensitivity to LCA hepatotoxicity. Further analysis revealed that the increased sensitivity was male specific and was associated with a profound decrease in the expression of bile acid transporters and defects in bile acid clearance [23]. The lack of female sensitivity has been reasoned to be due to the combined effect of a sustained sulfotransferase 2A9 (SULT2A9) expression and milder transporter suppression in this sex [23]. Interestingly, in an independent study using the same animals, it was reported that CAR knock-outs were more sensitive to LCA than the PXR knock-outs and this sensitivity was not further increased in the DKO mice [18]. The discrepancy between these two reports may be due to the combined effect of several factors, such as LCA dose, route of drug administration, and the sex of animals used. A third group showed that the hepatic damage from bile duct ligation (BDL), another model of cholestasis, was increased in both CAR knock-out and PXR knock-out mice [24].

The individual and combined effect of loss of PXR and CAR in bilirubin detoxification was also investigated using the knock-out models. The loss of PXR and/or CAR did not significantly alter the basal level of serum bilirubin. Upon bilirubin infusion, while the CAR null and the DKO mice showed similar sensitivity as the wild-type mice, the PXR null mice exhibited a surprisingly complete resistance to hyperbilirubinemia [25]. The increased bilirubin resistance in PXR null mice was associated with the induction of bilirubin-detoxifying genes, such as UGT1A1, OATP4/SLC21A6, GSTA2, and MRP2 (multidrug resistance associated protein 2). The expression of these genes remained largely unchanged in CAR null or DKO mice, consistent with the sensitivity to hyperbilirubinemia found in both genotypes [25]. To understand the increased bilirubin clearance in PXR null mice, we proposed that the ligand-free PXR functions as a suppressor to inhibit the constitutive activity of CAR. This hypothesis was supported by cell culture studies, in which ligand-free PXR specifically suppressed the ability of CAR to induce the MRP2, a bilirubin-detoxifying transporter.

This suppression was, at least in part, due to the disruption of ligand-independent recruitment of coactivator by CAR. We conclude that PXR plays both positive and negative roles in regulating bilirubin homeostasis, and this provides a novel mechanism that may govern receptor cross talk and the hierarchy of xeno- and endobiotic regulation. It is likely that deletion of PXR results in derepression that leads to the same bilirubin resistance phenotype as seen in the VP–CAR transgenic mice [25].

7.3   PXR AND CAR GAIN-OF-FUNCTION (TRANSGENIC) MOUSE MODELS

7.3.1   Albumin Promoter–Human PXR (Alb–hPXR) Transgenic Mice

The Alb–hPXR transgenic mice were created in order to express the hPXR in the mouse liver. To construct the Alb–hPXR transgene, the hPXR cDNA was cloned downstream of the mouse albumin (Alb) promoter/enhancer [26]. A SV40 intron/poly (A) sequence was placed downstream of hPXR cDNA (Figure 7.3a). The same promoter has been used by others and us to target the expression of various transgenes to the liver.

Transgene was excised and purified from the plasmid vector and microinjected into the pronuclei of fertilized mouse eggs. The injected eggs were then implanted into the oviduct of pseudopregnant female mice. Transgene positive mice were screened by PCR on tail DNA [6]. The transgene expression in the liver was confirmed by Northern blot analysis (Figure 7.3b). Because the transgenic mice harbor both mPXR and hPXR in the liver, they exhibit a chimeric Cyp3a11 response to both the rodent-specific inducer PCN and the human-specific inducer rifampicin [6]. The Alb–hPXR transgenic mice were subsequently used to create the hPXR-humanized mice (see Section 7.4.1).

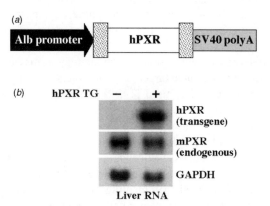

FIGURE 7.3   Creation of PXR transgenic mice. (*a*) Schematic representation of the Alb–hPXR transgene construct. (*b*) Northern blot analysis of mouse and human PXR gene expression in the liver of wild-type (−) and transgenic (+) mice. (Modified from Figure 3 in [6] with permission from the Nature Publishing Group.)

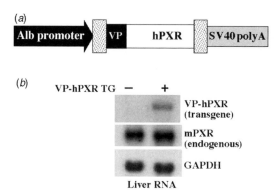

FIGURE 7.4 Creation of VP–PXR transgenic mice. Schematic representation of the Alb–VP–hPXR transgene construct, and transgene expression in the liver of transgenic mice.

7.3.2   Alb–VP–hPXR Transgenic Mice Expressing the Activated hPXR

Alb–VP–hPXR transgenic mice were created to express the constitutively activated hPXR (VP–hPXR) in the liver. VP–hPXR was generated by fusing the VP16 activation domain of the herpes simplex virus to the amino terminus of hPXR (Figure 7.4). It shared similar DNA binding specificity to hPXR [1]. A liver-specific transgenic mouse line was generated that expresses VP–hPXR [6] in the liver under the control of the Alb promoter. In these mice, hPXR is constitutively active even without the presence of PXR ligand (genetic activation). As expected, the *Cyp3a11* gene was constitutively induced in the liver of this mouse line. These mice are more resistant to the xenobiotic toxicants, such as tribromoethanol and zoxazolamine, presumably due to the activation of drug-metabolizing/detoxifying enzymes [6]. Microarray analysis showed that many xenobiotic genes, including phase I cytochrome P450 enzymes, phase II conjugating enzymes, and drug transporters, are regulated in the liver of these mice [27].

The VP-fusion receptors represent a unique strategy to dissect nuclear receptor function *in vivo*. Ligand-facilitated target gene identification using wild-type or gene knock-out mice has been widely used [19, 28]. We consider the use of the VP-fusion receptor transgenes to have unique advantages over drug treatment. This is particularly important since we now know that treatments with receptor pan-agonists, such as bile acids, may affect multiple receptors depending upon the tissue context [9, 10, 29]. Moreover, several lines of evidence suggest that ligand treatment may have additional transcriptional consequences independent of the presence of endogenous receptor. For example, Ueda et al. [19] identified 168 differentially expressed tags in response to PB treatment. However, nearly half of these tags were similarly affected in the CAR knock-out mice. Bypassing the requirement of ligand treatment, the VP fusion of receptors provides a unique strategy not only to study the biological consequences of receptor activation but also to identify target genes [27]. The utility and practicality of this strategy have been demonstrated in many published studies on

PXR [6, 9, 20, 27, 30, 31], CAR [25, 32], PPARδ/β (peroxisome proliferator-activated receptor) [33], PPARα [34], and LXR [35, 36].

The Alb–VP–hPXR transgenic mice have been used in several studies to determine the *in vivo* function of PXR. The effect of PXR activation by genetic means on bile acid toxicity was first investigated. VP–hPXR transgenic mice or their wild-type littermates were dosed with vehicle solvent or LCA (8 mg/day) before liver histological evaluation by H&E staining. The histological liver damage was scored by the appearance of areas of saponification/coagulative necrosis. After 4 days of treatment, 58% of wild-type mice exhibited areas of liver damage. In contrast, the VP–hPXR transgenic mice were completely resistant to LCA-induced liver damage [9], demonstrating that sustained activation of PXR is sufficient to prevent LCA-mediated histological liver damage. The induction of CYP3A11 expression was initially reasoned to be responsible for LCA resistance; however, the hydroxysteroid sulfotransferase SULT2A9 was later identified as a PXR target gene and was found to be induced in the VP–hPXR mice [37]. Since both CYP3A and SULT2A play a role in bile acid detoxification, it is likely that the LCA resistance resulted from a combined effect of CYP3A and SULT2A9 induction.

The genetic activation of PXR in transgenic mice was also shown to confer a resistance to hyperbilirubinemia. In this experiment, Alb–VP–hPXR transgenic mice and their wild-type controls were given a single dose of bilirubin ($10 \, \mathrm{mg \, kg^{-1}}$ body weight) via tail vein injection. One hour after injection, the serum levels of both total and conjugated bilirubin in VP–hPXR transgenic mice were less than half of the wild-type controls [38]. The enhanced bilirubin resistance in transgenic mice was thought to be due to the elevated expression of UGT1A1, the principal enzyme that facilitates bilirubin glucuronidation and subsequent clearance [9, 39].

In the third study, the Alb–VP–hPXR transgenic mice were used to show that activation of PXR disrupts glucocorticoid homeostasis. Steroid hormones, such as corticosteroids, are also substrates for UGTs and SULTs, the PXR target genes. It is conceivable that genetic activation of PXR and the resultant enzyme regulation may impact the hormonal homeostasis. To measure the effect of PXR activation on corticosterone levels, blood samples were collected from Alb–VP–hPXR transgenic mice and their wild-type controls. Corticosterone levels in plasma and urine were measured with a [^{125}I]-corticosterone RIA kit from ICN (Biomedical, Irvine, CA). The plasma and urinary corticosterone level were significantly higher in the VP–hPXR transgenic mice as compared to their wild-type littermates, indicating that the increased glucuronidation is associated with increased output of glucocorticoid [38].

A subsequent study showed that PXR may have a broader role in regulating the homeostasis of adrenal steroids that include glucocorticoid and mineralocorticoid [40]. Activation of PXR by genetic (using the Alb–VP–hXR transgene) or pharmacological (using a PXR agonist ligand) markedly increased plasma concentrations of corticosterone and aldosterone, the respective primary glucocorticoid and mineralocorticoid in rodents. The increased levels of corticosterone and aldosterone were associated with activation of adrenal steroidogenic enzymes, including CYP11a1, CYP11b1, CYP11b2, and 3β-Hsd. The PXR-activating transgenic mice also exhibited hypertrophy of the adrenal cortex, loss of glucocorticoid circadian rhythm, and lack of glucocorticoid responses to psychogenic stress. Interestingly, the transgenic mice had normal pituitary secretion of adrenocorticotropic hormone (ACTH) and the corticosterone suppressing effect of DEX was intact, suggesting a functional hypothalamus–pituitary–adrenal (HPA) axis despite a severe disruption of adrenal steroid homeostasis. The ACTH-independent hypercortisolism in the PXR-activating transgenic mice is reminiscent of the pseudo-Cushing's syndrome in patients. We propose that PXR is a potential endocrine disrupting factor that may have broad implications in steroid homeostasis and drug–hormone interactions.

The use of Alb–VP–hPXR transgenic mice also helped to show that PXR may have impact on hepatic steatosis [31]. This PXR-mediated triglyceride accumulation was independent of the activation of the lipogenic transcriptional factor sterol regulatory element binding protein 1c (SREBP-1c) and its primary lipogenic target enzymes, including fatty acid synthase (FAS) and acetyl CoA carboxylase 1 (ACC-1). Instead, the lipid accumulation in transgenic mice was associated with an increased expression of the free fatty acid transporter CD36 and several accessory lipogenic enzymes, such as stearoyl CoA desaturase-1 (SCD-1) and long chain free fatty acid elongase (FAE). Studies using transgenic and knock-out mice showed that PXR is both necessary and sufficient for CD36 activation. Promoter analyses revealed a DR-3 type of PXR response element in the mouse CD36 promoter, establishing CD36 as a direct transcriptional target of PXR. The activation of PXR was also associated with an inhibition of pro-β-oxidative genes, such as PPARα and thiolase, and an upregulation of PPARγ, a positive regulator of CD36. The cross-regulation of CD36 by PXR and PPARγ suggests that this fatty acid transporter may function as a common target of orphan nuclear receptors in their regulation of lipid homeostasis [31].

7.3.3   Fatty Acid Binding Protein Promoter–Human PXR (FABP–hPXR) Transgenic Mice

The FABP–hPXR transgenic mice is another hPXR transgenic line in which hPXR is targeted to both the liver and intestine, under the control of the rat fatty acid binding protein (FABP) promoter [31]. The same FABP promoter has been shown to target the expression of various transgenes to the liver and intestine [41–43]. As expected, the FABP–hPXR transgene targets the expression of hPXR to both the liver and all segments of the intestine [31]. The FABP–hPXR transgenic mice were first reported and used to show that a pharmacological activation of hPXR *in vivo* can also promote hepatic steatosis [31]. The FABP–hPXR transgenic mice were subsequently used to create the hPXR-humanized mice (see Section 7.4.1).

7.3.4   FABP–VP–hPXR Transgenic Mice

FABP–VP–hPXR transgenic mice were created to express the constitutively activated hPXR (VP–hPXR) in both the liver and intestine. Figure 7.5*a* shows the schematic representation of the transgene. Transgene expression was evaluated by Northern blot analysis using an hPXR-specific probe. As shown in Figure 7.5*b*, VP–hPXR was

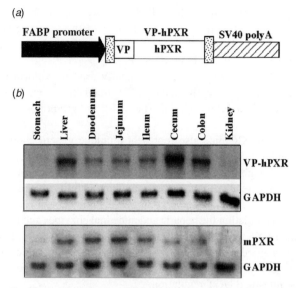

FIGURE 7.5   Creation of FABP–VP–hPXR transgenic mice. (*a*) Schematic representation of the transgene construct. (*b*) Northern blot analysis shows that the VP–hPXR transgene is expressed in the liver and throughout the intestinal tract. The membrane was probed with hPXR cDNA, mPXR, and GAPDH for the loading control.

expressed in the liver and throughout the intestinal tract, including the duodenum, jejunum, ileum, cecum, and colon. In contrast, the expression of hPXR was undetectable in the stomach and kidney. This pattern of transgene expression was similar to that of the endogenous mPXR (Figure 7.5*b*).

The FABP–VP–hPXR mice were first reported and used to show that activation of PXR sensitizes mice to oxidative stress [30]. It is found that expression of VP–hPXR in female transgenic mice resulted in a heightened sensitivity to paraquat, an oxidative xenobiotic toxicant [44]. The role of PXR in sensitizing mice to paraquat toxicity was confirmed by mPXR agonist PCN treatment. The PXR-induced paraquat sensitivity was associated with decreased activities of superoxide dismutase (SOD) and catalase (CAT), enzymes that scavenge superoxide and hydrogen peroxide, respectively [45, 46]. Paradoxically, the general expression and activity of glutathione *S*-transferases, a family of phase II enzymes that detoxify electrophilic and cytotoxic substrates [47], were also induced in these transgenic mice. Detailed analysis revealed that PXR regulates GST expression in an isozyme-, tissue-, and sex-specific manner, and this regulation is independent of the Nrf2/Keap1 pathway [30]. In the transgenic mice, the hepatic expression of GST Pi was modestly increased in females but was profoundly decreased in males, a clear example of gender-specific GST regulation. GST Pi regulation was also liver specific, as the transgene had little effect on intestinal expression of this isoform in either sex. GST Mu expression was increased in both the livers and intestines of transgenic mice of both genders, although the hepatic upregulation appeared to be more dramatic [30]. It was concluded in this study that PXR may have a novel function in mammalian oxidative stress response and this regulatory pathway may be implicated in carcinogenesis by sensitizing normal and cancerous tissues to oxidative cellular damage.

The FABP–VP–hPXR mice were subsequently used to confirm that, like their Alb–VP–hPXR counterparts, activation of hPXR in this transgenic line also resulted in the disruption of adrenal steroid homeostasis [40] and promotion of hepatic steatosis [31].

The FABP–VP–hPXR mice also represent a unique genetic model to study the comparative regulation of drug-metabolizing enzymes and transporters in the liver and intestine. Although liver and intestine share the expression of many enzymes and transporters, the relative expression and/or inducibility of these enzymes and transporters could be varied and we hypothesize that the nuclear receptors may have differential role in regulating the hepatic and intestinal enzymes and transporters. The above-mentioned PXR-mediated hepatic and intestinal regulation of GST isoforms represents such a good example [30]. Using the same transgenic model, we are currently investigating whether activation of PXR could have differential effect on the hepatic and intestinal UGT regulation.

7.3.5   Tetracycline-Inducible VP–CAR Transgenic Mice

The Alb and FABP promoters are used to create transgenic mice in which the transgenes are constitutively expressed in the target tissues as dictated by the promoters. In some cases, the creation of inducible transgenic mice may offer certain advantage and flexibility to facilitate the study of gene function, which has been exemplified in our creation of the tetracycline-inducible constitutively activated CAR (VP–CAR) transgenic mice. To create a transgenic mouse system that allowed conditional expression of VP–CAR in the liver, two lineages of transgenic mice were used, as diagramed in Figure 7.6a. First, we created the TetRE (TRE)–VP–CAR transgene that encodes VP–CAR under the control of a minimal cytomegalovirus promoter and the TRE [48]. The TRE–VP–CAR mice were then bred with the Lap (CEBP, CCAAT/enhancer-binding protein-β)-tTA activator line (Jackson Laboratory) to generate bitransgenic animals. Driven by the liver-specific Lap (CEBP) promoter, the Lap-tTA transgene directs the expression of the tetracycline-responsive transcriptional activator (tTA) exclusively in the hepatocytes [49]. We anticipated that tTA bound to TRE and consequently induced the expression of VP–CAR only in the absence of doxycycline (Dox). Addition of Dox (supplied in drinking at a concentration of $2\,mg\,ml^{-1}$) will result in the displacement of tTA from TRE and will silence VP–CAR expression (Tet-Off). Shown in Figure 7.6b is Northern blot analysis demonstrating the expected transgene expression and Dox regulation. As expected, the expression of *Cyp2b10* is dramatically induced in the liver of this mouse model [32].

Using this mouse model, we were able to show a novel CAR-mediated and CYP3A-independent pathway of bile acid detoxification [32]. When the TRE–VP–CAR/Lap-tTA bitransgenic mice or control littermates were dosed with vehicle solvent or LCA for 4 days before liver histological evaluation, the VP–CAR

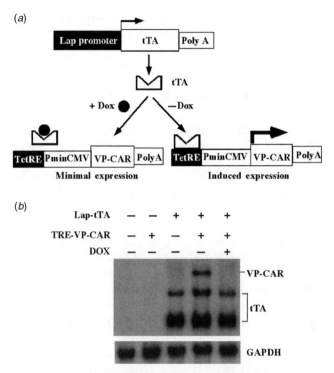

FIGURE 7.6 Creation of transgenic mice that harbor conditional expression of the activated CAR in the liver. (*a*) Schematic outline of the Lap-tTA/TetRE (TRE)–VP–CAR two-component Tet-Off transgenic system. The Lap-tTA transgene directs the expression of the tTA activator to the liver. The binding of tTA to the TRE and the induction of the transgene VP–CAR should only occur in the absence of Dox. (*b*) Liver-specific conditional expression of VP–CAR. Liver RNAs of mice with indicated genotypes were subjected to Northern blot analysis. The mouse in the rightmost lane was subjected to 5 days of Dox treatment. (Modified from Figure 1 in [32] with permission from the Amer. Soc. Pharmacology Experimental Therapeutics.)

mice were completely resistant to LCA-induced liver damage, demonstrating that sustained activation of CAR is sufficient to prevent LCA-mediated cholestasis. This effect is CAR activation dependent, as treatment of Dox blocked protection, presumably due to the silence of VP–CAR transgene expression [32]. The LCA resistance in VP–CAR mice was associated with an induction of SULT2A9 but not CYP3A11 [32]. Results from the VP–CAR transgenic mice have also led to the identification of SULT2A9 as a direct CAR target gene [32].

The VP–CAR mice were also used to demonstrate the effect of CAR activation on bilirubin clearance. No differences in serum bilirubin levels were seen in untreated VP–CAR or wild-type mice. However, upon bilirubin injection, the levels of total bilirubin in the transgenic mice were less than half of those in wild-type mice [25]. The protection was VP–CAR dependent, as no protection was seen in mice treated with Dox for 7 days prior to bilirubin injection [25]. Consistent with their resistance to hyperbilirubinemia, the VP–CAR mice had increased expression of genes encoding bilirubin-detoxifying enzymes and transporters such as UGT1A1, OATP4, GSTA2, OATP2, and MRP2 [25].

7.4   HUMANIZED MOUSE MODELS

The responses to xenobiotics exhibit significant species specificities presumably due to the divergence of the LBD sequences and the resulting ligand binding properties. As a result, it is not uncommon that the information obtained from the rodent models cannot be applied to humans. It has been perceived that rodent models have limited utility in predicting drug-related human effects due to significant species differences in drug-metabolizing enzymes, transporters, and nuclear hormone receptors. For example, PCN is an effective CYP3A inducer in rat but not humans, and rifampicin induces CYP3A in humans but not in rats [50]. These findings have been attributed to the species differences in the effect of several drugs on CYP3A expression mediated by the nuclear hormone receptor PXR [50, 51]. Human hepatocytes are considered the most relevant system to evaluate or predict human metabolism or effects of a new drug. A significant disadvantage of this system is the lack of routine availability of good quality human liver tissue or cells, as well as the known interindividual variation in the expression of hepatic drug-metabolizing enzymes [6, 7, 52, 53]. To address these issues, humanized animal models were developed. These mouse models can be used to evaluate the potential effect of xenobiotics, especially drug–drug interactions, in humans using animal models.

There are several strategies to generate humanized mice. The most common method is to introduce a transgene that contains a human cDNA under the control of tissue-specific promoter, such as liver-specific Alb promoter or the liver- and intestine-specific FABP promoter. This is a straightforward and relatively efficient strategy to create humanized mice. However, the expression level of the transgenes may not truly reflect the endogenous human gene expression. The second approach is to introduce the human genomic sequences into the mouse genome. In the latter case, usually sequences from bacterial artificial chromosome (BAC) clones that contain both coding sequences and regulatory elements are used. The transgenic mice that contain human gene can then be bred with mouse receptor gene knock-out mice, in which the corresponding endogenous mouse gene is deleted, leading to the production of humanized mice expressing only the human gene. Another potential strategy is to knock-in the human gene in the mouse gene locus, in which the endogenous gene is disrupted and replaced with the human gene. Unlike the promoter-based transgenic mice, the expression of human gene in the knock-in mice will be under the control of the endogenous gene regulatory elements [3, 5].

7.4.1   Humanized PXR Mice Carrying Human PXR cDNA

7.4.1.1 Generation of hPXR Humanized Mice. As diagramed in Figure 7.1, the Alb–hPXR or FABP–hPXR transgene (see Sections 7.3.1 and 7.3.3, respectively)

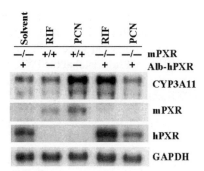

FIGURE 7.7   Drug response profile in the humanized mice. Mice with indicated genotypes were treated with a single dose of RIF (5 mg kg^{-1}) or PCN (40 mg kg^{-1}) 24 h prior to liver harvesting. Liver RNA was isolated and subjected to Northern blot analysis with indicated probes. (Adapted from Figure 3 in [6] with permission from the Nature Publishing Group.)

was bred into a PXR null background, and the resulting PXR null/hPXR transgenic mice lack mPXR but have hPXR transgene expressed in the liver (when the Alb–hPXR transgene was used) or both the liver and intestine (when the FABP–hPXR transgene was used).

7.4.1.2 Humanized Drug Response Profile in the Humanized Mice.   The drug induction of *Cyp3a11* was compared between the wild-type and humanized mice by Northern blot analysis (Figure 7.7). In the Alb–hPXR-humanized mice, *Cyp3a11* was no longer induced by PCN but was efficiently induced by the human-specific PXR inducer rifampicin (RIF). In contrast, no induction occurred in wild-type mice by RIF. This result firmly established that the species difference in the inducibility of *Cyp3a11* is attributed to the xenobiotic receptor, instead of the promoter structure of the *Cyp3a11* gene [6]. Expressing hPXR and responding to human-specific PXR inducer exclusively, the humanized mice represent a major step toward generating humanized toxicological models to predict xenobiotic enzyme inducibility and drug–drug interactions.

7.4.2   Humanized PXR Mice Carrying Human PXR Genomic DNA

Frank Gonzalez's laboratory at the National Cancer Institute reported the creation of humanized PXR mouse model by BAC transgenesis in PXR null mice [54]. To create the transgenic mice, a BAC clone containing the complete hPXR sequence, as well as 5′- and 3′-flanking sequences, was linearized and microinjected into fertilized FVB/N mouse eggs. The resulting transgenic mice were bred with PXR null mice [10]. Quantitative PCR revealed that hPXR was expressed in liver, duodenum, jejunum, and ileum of the humanized mice. Similar to the previously described humanized mice [6], these mice also mimicked the human response to PXR ligands treatment [54]. In the same study, it was shown that in rifampicin-pretreated PXR-humanized mice, an approximately 60% decrease was observed for both the maximal midazolam serum concentration (C(max)) and the area under the concentration–time curve, as a result of a threefold increase in midazolam $1'$-hydroxylation. These results illustrate the potential utility of the PXR-humanized mice in the investigation of drug–drug interactions mediated by CYP3A and suggest that the PXR-humanized mouse model would be an appropriate *in vivo* tool for evaluation of the overall pharmacokinetic consequences of human PXR activation by drugs [54].

7.4.3 Humanized CAR Mice

David Moore's group reported the creation of the humanized CAR mice. CAR also shows species difference in ligand specificity. TCPOBOP activates mouse CAR potently but does not activate the human CAR [55]. On the contrary, CITCO (6-(4-chlorophenyl)imidazo[2,1-*b*][1,3]thiazole-5-carbaldehyde O-(3,4-dichlorobenzyl)oxime) was identified as a specific human CAR agonist [56].

The humanized CAR mice were created using a strategy similar to that used for the creation of humanized PXR mice. A transgenic mouse line expressing hCAR cDNA driven by the liver-specific Alb promoter was created. This line was used to breed with CAR knock-out mice to produce the humanized CAR mice only expressing hCAR in the liver [16]. PB treatment of the humanized CAR mice induces the expression of CAR target genes *Cyp1a2* and *Cyp3a11*, leading to an increased sensitivity of these mice to APAP. High dose of APAP also increases expression of *Cyp1a2*, *Cyp3a11*, and *Gstpi*, indicating that hCAR is activated by APAP [16].

The humanized CAR mice have also been used to study species-specific CAR ligands. By comparing the responses in wild-type mice and humanized CAR mice, it was found that the widely used antiemetic meclizine acts on mCAR and hCAR oppositely [57]. Meclizine increases mCAR transactivation and stimulates binding of mCAR to the steroid receptor coactivator 1. It also activates the expression of CAR target genes in wild-type mice. In contrast, meclizine suppresses hCAR transactivation and inhibits the PB-induced expression of CAR target genes in primary hepatocytes derived from humanized CAR mice. Moreover, meclizine prevents APAP-induced liver toxicity in humanized CAR mice, which is also opposite to its role on APAP toxicity in wild-type mice [57].

7.4.4 Humanized PPARα Mice

The species difference between mouse and human PPARα is exemplified by their different ability to mediate the hepatocarcinogenesis in response to sustained agonists exposure. Chronic treatment of rodents with PPARα agonists results in the formation of hepatocellular carcinomas. However, epidemiological evidence shows that humans appear to be resistant to these peroxisome proliferators [58]. To investigate the mechanism of the difference why rodents and humans respond differently to these chemicals, a humanized PPARα mouse line was generated. The humanized PPARα mice, designated hPPAR-Tet-Off, were obtained by breeding PPARα null mice with the TRE–hPPARα/LAP-tTA double transgenic mouse. The resulting mice lack the expression of mouse PPARα but express the human PPARα in a liver-specific manner

[59]. The humanized mice express the hPPARα protein at a level comparable to the mPPARα level in wild-type mouse. Similar to the wild-type mice, the humanized mice are responsive to PPARα ligand Wy-14643 treatment, which induces the expression of gene encoding peroxisomal and mitochondrial fatty acid metabolizing enzymes. As a result, the serum triglycerides in the humanized PPARα mice are decreased, also resembling the wild mice [59]. However, the humanized mice respond differently to PPARα activators when the ability to form hepatocarcinoma was evaluated. While the wild-type mice exhibit hepatocellular proliferation as evidenced by the elevated expression of cell cycle control genes, increased BrdU incorporation into hepatocyte nuclei, and hepatomegaly, the humanized mice do not exhibit these responses [59, 60]. These findings suggest that the species specificity of human and rodent responses to peroxisome proliferators is likely determined by the differences in PPARα sequence and structure [61], which results in the functional differences and regulation of different panels of genes.

7.4.5 Humanized AhR Mice

The PAS domain transcriptional factor AhR is a nonnuclear receptor xenobiotic receptor. In the early 1990s, the AhR and its partner Ah receptor nuclear translocator (Arnt) protein were identified as a transcriptional sensor mediating the induction of CYP1A and 1B1 genes by dioxin and related polycyclic aromatic hydrocarbons [62, 63].

It has been noticed that the susceptibility to environmental toxicants polycyclic aromatic hydrocarbons (PAHs) and halogenated aromatic hydrocarbons (HAHs), such as 2,3,7,8-tetrachlorodibenzo-p-dioxin (TCDD), is largely influenced by functional polymorphisms of the AhR [64, 65]. To gain insight into the hazards to human health posed by these compounds interacting with the hAhR, a humanized mouse model that harbors the hAhR cDNA was created [66]. This was made possible by replacing part of the mAhR sequence, including exons 1 and 2, by hAhR cDNA through homologous recombination. In the resulting targeted genome locus, the hAhR was under the control of the endogenous mouse AhR promoter (Figure 7.8). As a result, in the hAhR knock-in mice, the expression level of hAhR is comparable to that of endogenous murine AhR expression level [66]. In response to TCDD treatment, the hAhR knock-in mice exhibited weaker induction of AhR target genes such as *Cyp1a1* and *Cyp1a2* than did both C57BL/6J mice and DBA/2 mice. Consistent with the gene regulation, the maternal exposure to TCDD caused less severe abnormalities in hAhR knock-in mice than in C57BL/6J mice and DBA/2 mice [66]. These results suggest that the hAhR expressed in mice retain a functional human specificity in response to chemical stimulation.

7.4.6 Cytochrome P450 Humanized Mice

It has been recognized that creation of mice humanized for the human cytochrome P450 (CYP) enzymes is also important to study the function as well as regulation of the human CYP enzymes. CYP enzymes are the most important contributors to

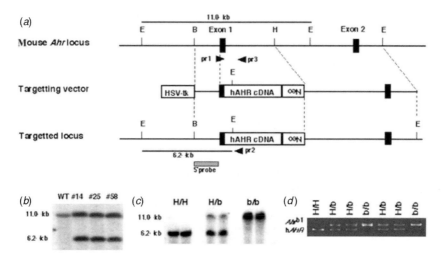

FIGURE 7.8   Generation of the hAhR knock-in mouse. (*a*) Strategy for hAHR cDNA knock-in by homologous recombination. E, H, and B are restriction sites for EcoRI, HindIII, and BamHI, respectively. The 5′-genomic probe used for Southern blot analysis is indicated by the hatched box. The positions of wild-type (pr 3) and mutant allele-specific (pr 2) primers and the common primer (pr 1) used in the genotyping PCR are indicated by arrowheads. (*b*) Southern blot analyses of three recombinant ES clones. Genomic DNA was digested by EcoRI generating 11.0- and 6.2-kb bands for the wild-type and targeted alleles, respectively, by using the 5′-genomic probe. (*c*) Genotyping of the Ahr gene by Southern blot analysis. (*d*) Genotyping of littermates from the intercrosses of heterozygotes. PCR fragments of wild type amplified with pr 1 and pr 3 (Ahrb-1; 280 bp) and mutant allele with pr 1 and pr 2 (hAHR; 240 bp) were shown. (Adapted from Figure 1 in [66] with permission from the Natl. Acad. Sciences.)

the metabolism of xenobiotics, including clinically administrated drugs and environmental toxicants. A number of CYP enzymes are also involved in the metabolism of endobiotics, such as bile acids and steroid hormones. Many CYP enzymes have been identified to be targets of xenobiotic nuclear receptors [3, 5, 67–69]. Since early 2000s, various humanized mice that express human CYP enzymes have been generated. Examples include CYP1A1 [70], CYP1A2 [70, 71], CYP1B1 [72], CYP2D6 [73], CYP2E1 [74, 75], CYP3A4 [76, 77], CYP3A7 [78], CYP4B1 [79], CYP7A1 [80], CYP19 [81], and CYP27 [82]. These mouse models are very useful in the understanding of human responses to the chemicals metabolized by these enzymes.

7.5   UTILITY OF XENOBIOTIC MOUSE MODELS IN PHARMACEUTICAL DEVELOPMENT

In addition to their use to dissect the transcriptional control of drug-metabolizing enzymes and transporters, the use of mouse models, especially the humanized mice, offers a unique opportunity to study the *in vivo* relationship between drug concentrations and the corresponding physiological effects. The effects of a drug candidate could be evaluated in such a model by administering several doses of the drug candidate and measuring the effects on the RNA expression or activity of various enzymes. Using this approach we have demonstrated that increasing concentrations of rifampicin $(1–10\,\mathrm{mg\,kg^{-1}})$ lead to corresponding increases in *CYP3A11* mRNA levels in hPXR mice [6]. During the drug development process, this type of study could be employed with new drug candidates to assess the upregulation of drug-metabolizing enzymes or transporters. By administering drug candidates to animals for several days at doses yielding plasma exposures similar to those predicted to be efficacious in humans, the effect of the candidate drugs on hPXR-mediated enzyme and transporter expression can be assessed in this *in vivo* setting.

The study of drug dose (or plasma concentration) and their corresponding pharmacological effects by mathematical modeling are often referred to as pharmacokinetic–pharmacodynamic analysis. This is very useful to predict effective plasma drug concentrations and determine the drug dosage [83]. However, the prediction of the human xenobiotic responses on the basis of rodent experiment is unreliable given the species specificity of xenobiotic response. In view of this, the creation of mouse models with humanized xenobiotic responses is of practical use in pharmaceutical research and development. The humanized mouse models enable the study of drug action and metabolism in the exclusive presence of human receptors. For example, the humanized PXR or CAR mice can be used to study the correlation between the dosage/concentration of many drugs and their regulatory effects on CYP3A and 2B genes in a manner similar to that happens in humans.

Compared to the *in vitro* cell cultures, these *in vivo* models reflect the drug responses under the physiological conditions [6, 11, 16, 84]. These models offer a dynamic system incorporating drug absorption, distribution, metabolism, and elimination, albeit in mouse, in contrast to the static system of cell culture where cells are continually exposed to drug and perhaps their metabolites. In addition, the concentration–effect relationship can be expanded to compare the *in vivo* liver/plasma concentrations with concentrations used in the *in vitro* human hepatocyte system and their corresponding *in vivo* and *in vitro* effects on gene transcription, enzyme induction, or toxicity. This type of comparison may actually improve the extrapolation of *in vitro* human hepatocyte concentration–induction effects to the prediction of induction and toxicological responses in patients.

7.6 CLOSING REMARKS

The generation and utilization of xenobiotic receptor mouse models have greatly advanced the research on the transcriptional regulation of drug-metabolizing enzymes and transporters. These mice have offered a unique and valuable vehicle to dissect the complexity of xenobiotic receptor-mediated gene regulation and the implication of such regulation in drug metabolism and many pathophysiological events. Moreover, these genetic models, combined with the pharmacological tools, provide valuable tools to investigate the function of these receptors in the metabolism of xenobiotics and endobiotics.

There are still outstanding challenges ahead. These may include, but are not limited to, the following:

(1) To create xenobiotic receptor knock-in mice that express the wild-type or activated xenobiotic nuclear receptors under the control of the endogenous/natural receptor promoters. The knock-in mice will allow normalization of the expression of the transgene to that of the endogenous/wild-type receptors. The knock-in may also help to reveal the function of receptors outside of the liver and intestine.

(2) To create xenobiotic receptor reporter transgenic mice that also express a luciferase reporter gene under the control of the xenobiotic receptor responsive elements. Such nuclear receptor reporter transgenic mice have been reported for a number of receptors, such as the estrogen receptor (ER) [85], farnesoid X receptor (FXR) [86], and PPAR [87]. These mice can be used to monitor the *in vivo* responses of these receptors to both xenobiotics and endobiotics.

(3) To create humanized mice for both the xenobiotic receptors and their target drug-metabolizing enzymes.

(4) To create transgenic mice expressing the polymorphic variants of xenobiotic receptors. Since the cloning and characterization of PXR and CAR as xenobiotic receptors, considerable progresses have been made to identify the polymorphic variants of PXR and CAR, which may account for the interindividual variation in drug metabolism phenotype. Transgenic mice expressing these polymorphic variants would be important to examine the functional relevance of these polymorphisms *in vivo*.

REFERENCES

1. Blumberg, B., Sabbagh, W., Jr., Juguilon, H., Bolado, J., Jr., van Meter, C. M., Ong, E. S., and Evans, R. M. (1998) SXR, a novel steroid and xenobiotic-sensing nuclear receptor. *Genes Dev.* **12**, 3195–3205.

2. Kliewer, S. A., Moore, J. T., Wade, L., Staudinger, J. L., Watson, M. A., Jones, S. A., McKee, D. D., Oliver, B. B., Willson, T. M., Zetterstrom, R. H., et al. (1998) An orphan nuclear receptor activated by pregnanes defines a novel steroid signaling pathway. *Cell* **92**, 73–82.

3. Gong, H., Sinz, M. W., Feng, Y., Chen, T., Venkataramanan, R., and Xie, W. (2005) Animal models of xenobiotic receptors in drug metabolism and diseases. *Methods Enzymol.* **400**, 598–618.

4. Xie, W. and Evans, R. M. (2002) Pharmaceutical use of mouse models humanized for the xenobiotic receptor. *Drug Discov. Today* **7**, 509–515.

5. Gonzalez, F. J. and Yu, A. M. (2006) Cytochrome P450 and xenobiotic receptor humanized mice. *Annu. Rev. Pharmacol. Toxicol.* **46**, 41–64.

6. Xie, W., Barwick, J. L., Downes, M., Blumberg, B., Simon, C. M., Nelson, M. C., Neuschwander-Tetri, B. A., Brunt, E. M., Guzelian, P. S., and Evans, R. M. (2000) Humanized xenobiotic response in mice expressing nuclear receptor SXR. *Nature* **406**, 435–439.

7. Xie, W. and Evans, R. M. (2001) Orphan nuclear receptors: the exotics of xenobiotics. *J. Biol. Chem.* **276**, 37739–37742.

8. Schuetz, E. G., Schmid, W., Schutz, G., Brimer, C., Yasuda, K., Kamataki, T., Bornheim, L., Myles, K., and Cole, T. J. (2000) The glucocorticoid receptor is essential for induction of cytochrome P-4502B by steroids but not for drug or steroid induction of CYP3A or P-450 reductase in mouse liver. *Drug Metab. Dispos.* **28**, 268–278.

9. Xie, W., Radominska-Pandya, A., Shi, Y., Simon, C. M., Nelson, M. C., Ong, E. S., Waxman, D. J., and Evans, R. M. (2001) An essential role for nuclear receptors SXR/PXR in detoxification of cholestatic bile acids. *Proc. Natl Acad. Sci. USA* **98**, 3375–3380.

10. Staudinger, J. L., Goodwin, B., Jones, S. A., Hawkins-Brown, D., MacKenzie, K. I., LaTour, A., Liu, Y., Klaassen, C. D., Brown, K. K., Reinhard, J., et al. (2001) The nuclear receptor PXR is a lithocholic acid sensor that protects against liver toxicity. *Proc. Natl Acad. Sci. USA* **98**, 3369–3374.

11. Wei, P., Zhang, J., Egan-Hafley, M., Liang, S., and Moore, D. D. (2000) The nuclear receptor CAR mediates specific xenobiotic induction of drug metabolism. *Nature* **407**, 920–923.

12. Snawder, J. E., Roe, A. L., Benson, R. W., and Roberts, D. W. (1994) Loss of CYP2E1 and CYP1A2 activity as a function of acetaminophen dose: relation to toxicity. *Biochem. Biophys. Res. Commun.* **203**, 532–539.

13. Patten, C. J., Thomas, P. E., Guy, R. L., Lee, M., Gonzalez, F. J., Guengerich, F. P., and Yang, C. S. (1993) Cytochrome P450 enzymes involved in acetaminophen activation by rat and human liver microsomes and their kinetics. *Chem. Res. Toxicol.* **6**, 511–518.

14. Henderson, C. J., Wolf, C. R., Kitteringham, N., Powell, H., Otto, D., and Park, B. K. (2000) Increased resistance to acetaminophen hepatotoxicity in mice lacking glutathione S-transferase Pi. *Proc. Natl Acad. Sci. USA* **97**, 12741–12745.

15. Coles, B., Wilson, I., Wardman, P., Hinson, J. A., Nelson, S. D., and Ketterer, B. (1988) The spontaneous and enzymatic reaction of N-acetyl-p-benzoquinoneimine with glutathione: a stopped-flow kinetic study. *Arch. Biochem. Biophys.* **264**, 253–260.

16. Zhang, J., Huang, W., Chua, S. S., Wei, P., and Moore, D. D. (2002) Modulation of acetaminophen-induced hepatotoxicity by the xenobiotic receptor CAR. *Science* **298**, 422–424.

17. Huang, W., Zhang, J., Chua, S. S., Qatanani, M., Han, Y., Granata, R., and Moore, D. D. (2003) Induction of bilirubin clearance by the constitutive androstane receptor (CAR). *Proc. Natl Acad. Sci. USA* **100**, 4156–4161.

18. Zhang, J., Huang, W., Qatanani, M., Evans, R. M., and Moore, D. D. (2004) The constitutive androstane receptor and pregnane X receptor function coordinately to prevent bile acid-induced hepatotoxicity. *J. Biol. Chem.* **279**, 49517–49522.

19. Ueda, A., Hamadeh, H. K., Webb, H. K., Yamamoto, Y., Sueyoshi, T., Afshari, C. A., Lehmann, J. M., and Negishi, M. (2002) Diverse roles of the nuclear orphan receptor CAR in regulating hepatic genes in response to phenobarbital. *Mol. Pharmacol.* **61**, 1–6.

20. Xie, W., Barwick, J. L., Simon, C. M., Pierce, A. M., Safe, S., Blumberg, B., Guzelian, P. S., and Evans, R. M. (2000) Reciprocal activation of xenobiotic response genes by nuclear receptors SXR/PXR and CAR. *Genes Dev.* **14**, 3014–3023.

21. Goodwin, B., Moore, L. B., Stoltz, C. M., McKee, D. D., and Kliewer, S. A. (2001) Regulation of the human CYP2B6 gene by the nuclear pregnane X receptor. *Mol. Pharmacol.* **60**, 427–431.

22. Sueyoshi, T., Kawamoto, T., Zelko, I., Honkakoski, P., and Negishi, M. (1999) The repressed nuclear receptor CAR responds to phenobarbital in activating the human CYP2B6 gene. *J. Biol. Chem.* **274**, 6043–6046.

23. Uppal, H., Toma, D., Saini, S. P., Ren, S., Jones, T. J., and Xie, W. (2005) Combined loss of orphan receptors PXR and CAR heightens sensitivity to toxic bile acids in mice. *Hepatology* **41**, 168–176.

24. Stedman, C. A., Liddle, C., Coulter, S. A., Sonoda, J., Alvarez, J. G., Moore, D. D., Evans, R. M., and Downes, M. (2005) Nuclear receptors constitutive androstane receptor and pregnane X receptor ameliorate cholestatic liver injury. *Proc. Natl Acad. Sci. USA* **102**, 2063–2068.

25. Saini, S. P., Mu, Y., Gong, H., Toma, D., Uppal, H., Ren, S., Li, S., Poloyac, S. M., and Xie, W. (2005) Dual role of orphan nuclear receptor pregnane X receptor in bilirubin detoxification in mice. *Hepatology* **41**, 497–505.

26. Pinkert, C. A., Ornitz, D. M., Brinster, R. L., and Palmiter, R. D. (1987) An albumin enhancer located 10 kb upstream functions along with its promoter to direct efficient, liver-specific expression in transgenic mice. *Genes Dev.* **1**, 268–276.

27. Rosenfeld, J. M., Vargas, R., Jr., Xie, W., and Evans, R. M. (2003) Genetic profiling defines the xenobiotic gene network controlled by the nuclear receptor pregnane X receptor. *Mol. Endocrinol.* **17**, 1268–1282.

28. Maglich, J. M., Stoltz, C. M., Goodwin, B., Hawkins-Brown, D., Moore, J. T., and Kliewer, S. A. (2002) Nuclear pregnane X receptor and constitutive androstane receptor regulate overlapping but distinct sets of genes involved in xenobiotic detoxification. *Mol. Pharmacol.* **62**, 638–646.

29. Makishima, M., Lu, T. T., Xie, W., Whitfield, G. K., Domoto, H., Evans, R. M., Haussler, M. R., and Mangelsdorf, D. J. (2002) Vitamin D receptor as an intestinal bile acid sensor. *Science* **296**, 1313–1316.

30. Gong, H., Singh, S. V., Singh, S. P., Mu, Y., Lee, J. H., Saini, S. P., Toma, D., Ren, S., Kagan, V. E., Day, B. W., et al. (2006) Orphan nuclear receptor pregnane X receptor sensitizes oxidative stress responses in transgenic mice and cancerous cells. *Mol. Endocrinol.* **20**, 279–290.

31. Zhou, J., Zhai, Y., Mu, Y., Gong, H., Uppal, H., Toma, D., Ren, S., Evans, R. M., and Xie, W. (2006) A novel pregnane X receptor-mediated and sterol regulatory element-binding protein-independent lipogenic pathway. *J. Biol. Chem.* **281**, 15013–15020.

32. Saini, S. P., Sonoda, J., Xu, L., Toma, D., Uppal, H., Mu, Y., Ren, S., Moore, D. D., Evans, R. M., and Xie, W. (2004) A novel constitutive androstane receptor-mediated and CYP3A-independent pathway of bile acid detoxification. *Mol. Pharmacol.* **65**, 292–300.

33. Wang, Y. X., Lee, C. H., Tiep, S., Yu, R. T., Ham, J., Kang, H., and Evans, R. M. (2003) Peroxisome-proliferator-activated receptor delta activates fat metabolism to prevent obesity. *Cell* **113**, 159–170.

34. Yang, Q., Yamada, A., Kimura, S., Peters, J. M., and Gonzalez, F. J. (2006) Alterations in skin and stratified epithelia by constitutively activated PPARalpha. *J. Invest. Dermatol.* **126**, 374–385.

35. Uppal, H., Saini, S. P., Moschetta, A., Mu, Y., Zhou, J., Gong, H., Zhai, Y., Ren, S., Michalopoulos, G. K., Mangelsdorf, D. J., et al. (2007) Activation of LXRs prevents bile acid toxicity and cholestasis in female mice. *Hepatology* **45**, 422–432.

36. Zhou, J., Febbraio, M., Wada, T., Zhai, Y., Kuruba, R., He, J., Lee, J. H., Khadem, S., Ren, S., Li, S., et al. (2008) Hepatic fatty acid transporter Cd36 is a common target of LXR, PXR, and PPARgamma in promoting steatosis. *Gastroenterology* **134**, 556–567.

37. Sonoda, J., Xie, W., Rosenfeld, J. M., Barwick, J. L., Guzelian, P. S., and Evans, R. M. (2002) Regulation of a xenobiotic sulfonation cascade by nuclear pregnane X receptor (PXR). *Proc. Natl Acad. Sci. USA* **99**, 13801–13806.

38. Xie, W., Yeuh, M. F., Radominska-Pandya, A., Saini, S. P., Negishi, Y., Bottroff, B. S., Cabrera, G. Y., Tukey, R. H., and Evans, R. M. (2003) Control of steroid, heme, and carcinogen metabolism by nuclear pregnane X receptor and constitutive androstane receptor. *Proc. Natl Acad. Sci. USA* **100**, 4150–4155.

39. Tukey, R. H. and Strassburg, C. P. (2000) Human UDP-glucuronosyltransferases: metabolism, expression, and disease. *Annu. Rev. Pharmacol. Toxicol.* **40**, 581–616.

40. Zhai, Y., Pai, H. V., Zhou, J., Amico, J. A., Vollmer, R. R., and Xie, W. (2007) Activation of pregnane X receptor disrupts glucocorticoid and mineralocorticoid homeostasis. *Mol. Endocrinol.* **21**, 138–147.

41. Simon, T. C., Cho, A., Tso, P., and Gordon, J. I. (1997) Suppressor and activator functions mediated by a repeated heptad sequence in the liver fatty acid-binding protein gene (Fabpl). Effects on renal, small intestinal, and colonic epithelial cell gene expression in transgenic mice. *J. Biol. Chem.* **272**, 10652–10663.

42. Wong, M. H., Rubinfeld, B., and Gordon, J. I. (1998) Effects of forced expression of an NH2-terminal truncated beta-catenin on mouse intestinal epithelial homeostasis. *J. Cell. Biol.* **141**, 765–777.

43. Sweetser, D. A., Birkenmeier, E. H., Hoppe, P. C., McKeel, D. W., and Gordon, J. I. (1988) Mechanisms underlying generation of gradients in gene expression within the intestine: an analysis using transgenic mice containing fatty acid binding protein-human growth hormone fusion genes. *Genes Dev.* **2**, 1318–1332.

44. Suntres, Z. E. (2002) Role of antioxidants in paraquat toxicity. *Toxicology* **180**, 65–77.

45. Rojkind, M., Dominguez-Rosales, J. A., Nieto, N., and Greenwel, P. (2002) Role of hydrogen peroxide and oxidative stress in healing responses. *Cell Mol. Life. Sci.* **59**, 1872–1891.

46. Forsberg, L., de Faire, U., and Morgenstern, R. (2001) Oxidative stress, human genetic variation, and disease. *Arch. Biochem. Biophys.* **389**, 84–93.

47. Hayes, J. D. and Pulford, D. J. (1995) The glutathione S-transferase supergene family: regulation of GST and the contribution of the isoenzymes to cancer chemoprotection and drug resistance. *Crit. Rev. Biochem. Mol. Biol.* **30**, 445–600.

48. Xie, W., Chow, L. T., Paterson, A. J., Chin, E., and Kudlow, J. E. (1999) Conditional expression of the ErbB2 oncogene elicits reversible hyperplasia in stratified epithelia and up-regulation of TGFalpha expression in transgenic mice. *Oncogene* **18**, 3593–3607.

49. Kistner, A., Gossen, M., Zimmermann, F., Jerecic, J., Ullmer, C., Lubbert, H., and Bujard, H. (1996) Doxycycline-mediated quantitative and tissue-specific control of gene expression in transgenic mice. *Proc. Natl Acad. Sci. USA* **93**, 10933–10938.

50. Kocarek, T. A., Schuetz, E. G., Strom, S. C., Fisher, R. A., and Guzelian, P. S. (1995) Comparative analysis of cytochrome P4503A induction in primary cultures of rat, rabbit, and human hepatocytes. *Drug Metab. Dispos.* **23**, 415–421.

51. Jones, S. A., Moore, L. B., Shenk, J. L., Wisely, G. B., Hamilton, G. A., McKee, D. D., Tomkinson, N. C., LeCluyse, E. L., Lambert, M. H., Willson, T. M., et al. (2000) The pregnane X receptor: a promiscuous xenobiotic receptor that has diverged during evolution. *Mol. Endocrinol.* **14**, 27–39.

52. Xie, W., Uppal, H., Saini, S. P., Mu, Y., Little, J. M., Radominska-Pandya, A., and Zemaitis, M. A. (2004) Orphan nuclear receptor-mediated xenobiotic regulation in drug metabolism. *Drug Discov. Today* **9**, 442–449.

53. Mills, J. B., Rose, K. A., Sadagopan, N., Sahi, J., and de Morais, S. M. (2004) Induction of drug metabolism enzymes and MDR1 using a novel human hepatocyte cell line. *J. Pharmacol. Exp. Ther.* **309**, 303–309.

54. Ma, X., Shah, Y., Cheung, C., Guo, G. L., Feigenbaum, L., Krausz, K. W., Idle, J. R., and Gonzalez, F. J. (2007) The PREgnane X receptor gene-humanized mouse: a model for investigating drug–drug interactions mediated by cytochromes P450 3A. *Drug Metab. Dispos.* **35**, 194–200.

55. Moore, L. B., Parks, D. J., Jones, S. A., Bledsoe, R. K., Consler, T. G., Stimmel, J. B., Goodwin, B., Liddle, C., Blanchard, S. G., Willson, T. M., et al. (2000) Orphan nuclear receptors constitutive androstane receptor and pregnane X receptor share xenobiotic and steroid ligands. *J. Biol. Chem.* **275**, 15122–15127.

56. Maglich, J. M., Parks, D. J., Moore, L. B., Collins, J. L., Goodwin, B., Billin, A. N., Stoltz, C. A., Kliewer, S. A., Lambert, M. H., Willson, T. M., et al. (2003) Identification of a novel human constitutive androstane receptor (CAR) agonist and its use in the identification of CAR target genes. *J. Biol. Chem.* **278**, 17277–17283.

57. Huang, W., Zhang, J., Wei, P., Schrader, W. T., and Moore, D. D. (2004) Meclizine is an agonist ligand for mouse constitutive androstane receptor (CAR) and an inverse agonist for human CAR. *Mol. Endocrinol.* **18**, 2402–2408.

58. Ashby, J., Brady, A., Elcombe, C. R., Elliott, B. M., Ishmael, J., Odum, J., Tugwood, J. D., Kettle, S., and Purchase, I. F. (1994) Mechanistically-based human hazard assessment of peroxisome proliferator-induced hepatocarcinogenesis. *Hum. Exp. Toxicol.* **13** (Suppl. 2), S1–S117.

59. Cheung, C., Akiyama, T. E., Ward, J. M., Nicol, C. J., Feigenbaum, L., Vinson, C., and Gonzalez, F. J. (2004) Diminished hepatocellular proliferation in mice humanized for the nuclear receptor peroxisome proliferator-activated receptor alpha. *Cancer Res.* **64**, 3849–3854.

60. Gonzalez, F. J. (2007) Animal models for human risk assessment: the peroxisome proliferator-activated receptor alpha-humanized mouse. *Nutr. Rev.* **65**, S2–S6.

61. Sher, T., Yi, H. F., McBride, O. W., and Gonzalez, F. J. (1993) cDNA cloning, chromosomal mapping, and functional characterization of the human peroxisome proliferator activated receptor. *Biochemistry* **32**, 5598–5604.

62. Hoffman, E. C., Reyes, H., Chu, F. F., Sander, F., Conley, L. H., Brooks, B. A., and Hankinson, O. (1991) Cloning of a factor required for activity of the Ah (dioxin) receptor. *Science* **252**, 954–958.

63. Reyes, H., Reisz-Porszasz, S., and Hankinson, O. (1992) Identification of the Ah receptor nuclear translocator protein (Arnt) as a component of the DNA binding form of the Ah receptor. *Science* **256**, 1193–1195.

64. Ema, M., Ohe, N., Suzuki, M., Mimura, J., Sogawa, K., Ikawa, S., and Fujii-Kuriyama, Y. (1994) Dioxin binding activities of polymorphic forms of mouse and human arylhydro-carbon receptors. *J. Biol. Chem.* **269**, 27337–27343.

65. Fujii-Kuriyama, Y., Ema, M., Mimura, J., Matsushita, N., and Sogawa, K. (1995) Poly-morphic forms of the Ah receptor and induction of the CYP1A1 gene. *Pharmacogenetics* **5**, S149–S153.

66. Moriguchi, T., Motohashi, H., Hosoya, T., Nakajima, O., Takahashi, S., Ohsako, S., Aoki, Y., Nishimura, N., Tohyama, C., Fujii-Kuriyama, Y., et al. (2003) Distinct response to dioxin in an arylhydrocarbon receptor (AHR)-humanized mouse. *Proc. Natl Acad. Sci. USA* **100**, 5652–5657.

67. Gonzalez, F. J. and Gelboin, H. V. (1994) Role of human cytochromes P450 in the metabolic activation of chemical carcinogens and toxins. *Drug Metab. Rev.* **26**, 165–183.

68. Gong, H. and Xie, W. (2004) Orphan nuclear receptors, PXR and LXR: new ligands and therapeutic potential. *Expert Opin. Ther. Targets* **8**, 49–54.

69. Sonoda, J., Rosenfeld, J. M., Xu, L., Evans, R. M., and Xie, W. (2003) A nuclear receptor-mediated xenobiotic response and its implication in drug metabolism and host protection. *Curr. Drug Metab.* **4**, 59–72.

70. Jiang, Z., Dalton, T. P., Jin, L., Wang, B., Tsuneoka, Y., Shertzer, H. G., Deka, R., and Nebert, D. W. (2005) Toward the evaluation of function in genetic variability: character-izing human SNP frequencies and establishing BAC-transgenic mice carrying the human CYP1A1_CYP1A2 locus. *Hum. Mutat.* **25**, 196–206.

71. Ueno, T., Tamura, S., Frels, W. I., Shou, M., Gonzalez, F. J., and Kimura, S. (2000) A transgenic mouse expressing human CYP1A2 in the pancreas. *Biochem. Pharmacol.* **60**, 857–863.

72. Hwang, D. Y., Chae, K. R., Shin, D. H., Hwang, J. H., Lim, C. H., Kim, Y. J., Kim, B. J., Goo, J. S., Shin, Y. Y., Jang, I. S., et al. (2001) Xenobiotic response in humanized double transgenic mice expressing tetracycline-controlled transactivator and human CYP1B1. *Arch. Biochem. Biophys.* **395**, 32–40.

73. Corchero, J., Granvil, C. P., Akiyama, T. E., Hayhurst, G. P., Pimprale, S., Feigenbaum, L., Idle, J. R., and Gonzalez, F. J. (2001) The CYP2D6 humanized mouse: effect of the human CYP2D6 transgene and HNF4alpha on the disposition of debrisoquine in the mouse. *Mol. Pharmacol.* **60**, 1260–1267.

74. Cheung, C., Yu, A. M., Ward, J. M., Krausz, K. W., Akiyama, T. E., Feigenbaum, L., and Gonzalez, F. J. (2005) The cyp2e1-humanized transgenic mouse: role of cyp2e1 in acetaminophen hepatotoxicity. *Drug Metab. Dispos.* **33**, 449–457.

75. Morgan, K., French, S. W., and Morgan, T. R. (2002) Production of a cytochrome P450 2E1 transgenic mouse and initial evaluation of alcoholic liver damage. *Hepatology* **36**, 122–134.

76. Granvil, C. P., Yu, A. M., Elizondo, G., Akiyama, T. E., Cheung, C., Feigenbaum, L., Krausz, K. W., and Gonzalez, F. J. (2003) Expression of the human CYP3A4 gene in the small intestine of transgenic mice: in vitro metabolism and pharmacokinetics of midazo-lam. *Drug Metab. Dispos.* **31**, 548–558.

77. Yu, A. M., Fukamachi, K., Krausz, K. W., Cheung, C., and Gonzalez, F. J. (2005) Poten-tial role for human cytochrome P450 3A4 in estradiol homeostasis. *Endocrinology* **146**, 2911–2919.

78. Li, Y., Yokoi, T., Kitamura, R., Sasaki, M., Gunji, M., Katsuki, M., and Kamataki, T. (1996) Establishment of transgenic mice carrying human fetus-specific CYP3A7. *Arch. Biochem. Biophys.* **329**, 235–240.

79. Imaoka, S., Hayashi, K., Hiroi, T., Yabusaki, Y., Kamataki, T., and Funae, Y. (2001) A transgenic mouse expressing human CYP4B1 in the liver. *Biochem. Biophys. Res. Commun.* **284**, 757–762.

80. Chen, J. Y., Levy-Wilson, B., Goodart, S., and Cooper, A. D. (2002) Mice expressing the human CYP7A1 gene in the mouse CYP7A1 knock-out background lack induction of CYP7A1 expression by cholesterol feeding and have increased hypercholesterolemia when fed a high fat diet. *J. Biol. Chem.* **277**, 42588–42595.

81. Li, X., Nokkala, E., Yan, W., Streng, T., Saarinen, N., Warri, A., Huhtaniemi, I., Santti, R., Makela, S., and Poutanen, M. (2001) Altered structure and function of reproductive organs in transgenic male mice overexpressing human aromatase. *Endocrinology* **142**, 2435–2442.

82. Meir, K., Kitsberg, D., Alkalay, I., Szafer, F., Rosen, H., Shpitzen, S., Avi, L. B., Staels, B., Fievet, C., Meiner, V., et al. (2002) Human sterol 27-hydroxylase (CYP27) overexpressor transgenic mouse model. Evidence against 27-hydroxycholesterol as a critical regulator of cholesterol homeostasis. *J. Biol. Chem.* **277**, 34036–34041.

83. Meibohm, B. and Derendorf, H. (2002) Pharmacokinetic/pharmacodynamic studies in drug product development. *J. Pharm. Sci.* **91**, 18–31.

84. Qatanani, M., Wei, P., and Moore, D. D. (2004) Alterations in the distribution and orexigenic effects of dexamethasone in CAR-null mice. *Pharmacol. Biochem. Behav.* **78**, 285–291.

85. Ciana, P., Di Luccio, G., Belcredito, S., Pollio, G., Vegeto, E., Tatangelo, L., Tiveron, C., and Maggi, A. (2001) Engineering of a mouse for the in vivo profiling of estrogen receptor activity. *Mol. Endocrinol.* **15**, 1104–1113.

86. Houten, S. M., Volle, D. H., Cummins, C. L., Mangelsdorf, D. J., and Auwerx, J. (2007) In vivo imaging of farnesoid X receptor activity reveals the ileum as the primary bile acid signaling tissue. *Mol. Endocrinol.* **21**, 1312–1323.

87. Ciana, P., Biserni, A., Tatangelo, L., Tiveron, C., Sciarroni, A. F., Ottobrini, L., and Maggi, A. (2007) A novel peroxisome proliferator-activated receptor responsive element-luciferase reporter mouse reveals gender specificity of peroxisome proliferator-activated receptor activity in liver. *Mol. Endocrinol.* **21**, 388–400.

8

NUCLEAR RECEPTORS AND DRUG–DRUG INTERACTIONS WITH PRESCRIPTION AND HERBAL MEDICINES

ROMMEL G. TIRONA AND RICHARD B. KIM

Division of Clinical Pharmacology, Department of Medicine and Department of Physiology and Pharmacology, Schulich School of Medicine and Dentistry, The University of Western Ontario, Ontario, Canada

8.1 INTRODUCTION

Despite the best intentions of health care providers, the Institute of Medicine estimated that over 2 million adverse drug reactions occur yearly in the United States alone (www.iom.org). It is often stated that adverse drug reactions are the fourth leading cause of death among hospitalized American patients [1]. These statistics are even more alarming when you consider that less than 10% of all adverse drug reactions are reported. Drug–drug interactions are thought to account for 20–30% of all adverse drug reactions [2]. Many of clinical effects of these drug–drug interactions result from alterations in the pharmacokinetics or disposition profiles of drugs. A significant proportion of these drug–drug interactions involve inhibition of drug-metabolizing enzymes leading to unexpected increase in systemic drug levels of sufficient magnitude, which then lead to overt toxicities. However, less considered and hence least likely to be reported are drug–drug interactions that occur through induction of drug elimination pathways. Generally, this type of interaction results in decreased plasma drug levels and often manifest as loss of therapeutic efficacy or a withdrawal syndrome. There is little doubt, however, that a better understanding

Nuclear Receptors in Drug Metabolism   Edited by Wen Xie
Copyright © 2009 John Wiley & Sons, Inc.

of the mechanistic basis for such induction-related interactions has the potential to provide a rationale for prescribing drug combinations likely to be devoid of unwanted loss in drug efficacy.

The adaptive response to drug exposure that triggers an increase in enzymatic capacity for drug removal was first described in 1960 [3]. Conney and colleagues noted that pretreatment of rats with the barbiturate drug, phenobarbital, increased hepatic drug metabolic activity and shortened the duration of hypnotic effects [3]. These findings led to the observations in 1963 that in humans, phenobarbital pretreatment lowers plasma levels of coumarin and phenytoin [4]. In 1973, Remmer and colleagues were able to demonstrate directly that the hepatic activities of the drug-metabolizing enzymes, the cytochromes P450 (CYP) that were obtained by needle biopsy, were increased in patients who were treated with phenobarbital, phenytoin, and rifampin [5]. A transcriptional mechanism of the inductive response by phenobarbital was described in 1981 by Adesnik and colleagues [6]. However, it was not until studies in the late 1990s that identified the ligand-activated transcription factors pregnane X receptor (PXR) [7–11] and constitutive androstane receptor (CAR) [12–15] did the molecular basis of induction-type drug–drug interactions become clarified. A number of human drug-metabolizing enzymes and transporters have convincingly been shown to be directly regulated by ligand-activated PXR and CAR (Table 8.1). These include members of the CYP, UDP-glucuronosyltransferase (UGT), and sulfotransferase (SULT) families of enzymes as well as the drug transporters, P-glycoprotein (P-gp) encoded by the multidrug resistance 1 (*MDR1, ABCB1*) gene and multidrug resistance associated protein 2 (*MRP2, ABCC2*). While induction of a number of these PXR/CAR target genes can be linked to specific drug interactions, for a few such as SULT2A1 and MRP2, little evidence is available to implicate them in clinically relevant adverse drug effects (see Section 8.4).

TABLE 8.1   Human Drug Disposition Genes
Regulated by PXR and CAR

Oxidative Metabolism (Phase I)
- CYP2B6 [16, 17]
- CYP2C9 [18–20]
- CYP2C19 [21, 22]
- CYP3A4 [9]
- CYP3A7 [23, 24]
Conjugation (Phase II)
- UGT1A1 [25, 26]
- UGT1A3 [27]
- UGT1A4 [27]
- SULT2A1 [28, 29]
Transport (Phase III)
- MDR1 [30]
- MRP2 [31]
- MRP4 [32]
- OATP1A2 [33]

In this chapter, we will focus on elements of prescription drug–drug and herbal–drug interactions. The emphasis will be on compiling essential lists of prescription and herbal medicines that are significant inducers of drug disposition pathways and understanding the pharmacology that determines why these drugs are prone to causing induction-type drug–drug interactions. We will also outline aspects of the clinical drug interactions including features of drugs that are particularly susceptible to drug interactions with inducing agents and give several examples of clinical consequences. The time course of induction and deinduction will also be discussed in the context of therapeutic management of patients taking PXR-/CAR-activating agents.

8.2 PRESCRIPTION DRUGS/DRUG CLASSES INVOLVED IN INDUCTIVE DRUG INTERACTIONS

To the benefit of patients and relief of clinicians, the number of prescription drugs that are involved in clinically relevant induction-type drug–drug interactions is relatively small (Table 8.2) and amount to less than 20 medicines. To simplify this further, these drugs fall into only five major categories of therapeutic agents: anticonvulsants, rifamycin antimycobacterials, human immunodeficiency virus (HIV)

TABLE 8.2   Essential Inducers List—Prescription Medicines

Anticonvulsants
- Phenobarbital
- Carbamazepine
- Oxcarbazepine
- Phenytoin
- Valproic acid
- Lamotrigine
- Topiramate
- Felbamate

Rifamycin Antibiotics
- Rifampin
- Rifabutin

HIV Protease Inhibitors
- Ritonavir
- Nelfinavir
- Lopinavir
- Tipranavir
- Amprenavir
- Atazanavir

Non-Nucleoside Reverse Transcriptase Inhibitors
- Efavirenz
- Nevirapine

Other
- Bosentan

protease inhibitors, non-nucleoside reverse transcriptase inhibitors (NNRTIs), and miscellaneous. For some drug categories such as the anticonvulsants, HIV protease inhibitors, and NNRTIs, the induction effects are not considered "a class effect" and alternative medicines that lack such inductive effects are available. Although this suggests that interaction of inducing drugs with nuclear receptors is not generally related to the intended therapeutic effects, there is at least one exception to this case in the rifamycin antibiotics.

8.2.1   Anticonvulsants

The anticonvulsant medications are the most notorious of drugs that cause inductive drug–drug interactions. In 1965, phenobarbital was first shown to affect the metabolism of coadministered drugs (dicumarol) in humans [34]. Demonstration that the chronic dosing causes autoinduction of the metabolizing enzymes involved in a drug's elimination in humans was first reported in 1975 with carbamazepine [35]. Many anticonvulsants such as phenobarbital [8, 13, 14, 36], carbamazepine [37], phenytoin [38], topiramate [39], and valproic acid [40] induce drug disposition genes by activating either or both CAR and PXR. For some inducer drugs like lamotrigine, felbamate, and oxcarbazepine, the nuclear receptors involved have not been described in the literature. The relative contributions of CAR and PXR to the inductive response appear to be drug and gene dependent. For example, phenobarbital largely acts through the CAR pathway and CYP2B6 is more sensitive to activation by CAR than PXR.

8.2.2   Rifamycin Antibiotics

The first indications that rifamycin antibiotics caused induction of drug metabolism were the descriptions in the early 1970s that rifampin interferes with the anticoagulation response to acenocoumarol [41] and the hormonal effects of oral contraceptives [42]. Subsequently, the inductive response by rifampin on microsomal enzymes was shown in biopsied human livers [5]. At therapeutic doses, rifampin appears to cause more profound drug interactions than the newer agent rifabutin [43] although clinically significant effects are observed with both drugs. The differing effects between rifampin and rifabutin on the magnitude of inductive responses likely result from differences in the attained plasma levels of these drugs during therapy (Table 8.3). Both drugs interact with equal potency toward PXR but the concentrations of rifampin *in vivo* are high enough to activate the receptor, whereas rifabutin levels are somewhat lower and thus tend to result in a more attenuated inductive response. Interestingly, the rifamycin drug rifaximin, used in the treatment of inflammatory bowel disease, is not known to cause drug interactions *in vivo* [57] despite that rifaximin can activate PXR *in vitro* [58]. In human PXR transgenic mice, rifaximin was shown to induce intestinal but not hepatic drug metabolic enzymes and transporters [58]. A lack of significant drug interactions in humans to date may be explained by the lack of oral absorption and exposure to liver. These studies also reveal that the therapeutic

TABLE 8.3 Drug Used in HIV Infection and Reports of Inductive Drug Interactions

| | Clinical Reports: Induction Drug–Drug Interactions | | | | | |
| | PXR/CAR (CYP3A4) | | | CAR/PXR (CYP2B6) | | |
Drug	Induction Interaction	Drug(s) Affected	Reference	Induction Interaction	Drug(s) Affected	Reference
HIV-PIs						
Amprenavir (APV)	Yes	Delavirdine	Product Monograph	Yes	Methadone	[45]
Atazanavir (AZV)	No	–	Product Monograph	No	–	Product Monograph
Indinavir (IDV)	No	–	Product Monograph	No	Methadone	[46]
Lopinavir (LPV)	No	–	Product Monograph	Yes	Methadone Bupropion	[47]
Nelfinavir (NFV)	Yes	Simvastatin, Atorvastatin, Delavirdine	Product Monograph [48]	Yes	Methadone	[49]
Ritonavir (RTV)	Yes	Alprazolam, Ethinyl Estradiol	Product Monograph	Yes	Methadone Bupropion	[50]
Saquinavir (SQV)	No	–	Product Monograph	No/Yes	Efavirenz	Product Monograph [51]
NNRTIs						
Delavirdine (DLV)	No	–	Product Monograph	No	–	Product Monograph
Efavirenz (EFV)	Yes	SQV, AZV, IDV	Product Monograph	Yes	Methadone	[52]
Nevirapine (NVP)	Yes	IDV	[53]	Yes	Methadone	[54]
Anti-Infectives						
Rifabutin	Yes	Many	[43]	No	–	[55]
Rifampin	Yes	Many	[43]	Yes	Methadone	[56]

response of rifaximin in intestinal inflammation is likely not due to antimicrobial activity but due to PXR-mediated interference with NF-κβ signaling [59, 60].

8.2.3   HIV Protease Inhibitors (HIV-PI)

A drug–drug interaction with ritonavir that caused enhanced clearance of oral contraceptives was the first indication that the classes of HIV-PI drugs were inducers of drug metabolism [61]. Today, a number of peptidomimetic HIV-PI such as amprenavir, atazanavir, lopinavir, nelfinavir, and ritonavir are known inducers of drug-metabolizing enzymes and transporters [62]. These drugs activate PXR [62] *in vitro* and induce the expression of PXR target genes in cultured human hepatocytes [63]. The new nonpeptidic HIV-PI, tipranavir, is also an inducer [64]. Induction of drug metabolism is not typically observed during treatment with HIV-PIs due to the fact that "boosted" therapy with the combination of ritonavir and another HIV-PI is standard of practice. Ritonavir is a potent mechanism-based inhibitor of CYP3A4 [65] and therefore the net effect seen in patients is that of CYP3A4 inhibition and not induction (Table 8.3). Interestingly, HIV-PIs do not significantly inhibit CYP2B6-mediated drug biotransformation and hence induction-type drug interactions involving CYP2B6-metabolized drugs have been documented (Table 8.3).

8.2.4   Non-Nucleoside Reverse Transcriptase Inhibitors

Treatment with the NNRTIs efavirenz and nevirapine was noted to cause methadone withdrawal symptoms in patients with HIV [66, 52]. Subsequently, both NNRTIs were shown to induce drug-metabolizing enzymes in cultured hepatocytes due to activation of CAR and weak activation of PXR [37].

8.2.5   Bosentan

The plasma levels of the dual endothelin receptor antagonist bosentan decrease with multiple dosing [67] as a result of PXR-mediated autoinduction of CYP3A4 [68].

8.3   HERBAL MEDICINES INVOLVED IN INDUCTIVE DRUG INTERACTIONS

It is estimated that 40 million adult Americans use herbal medicines [69] and that 43% of individuals who take herbal medicines are also taking prescription drugs [70]. The use of herbal medicines is particularly high in cancer (63%) [71] and HIV-infected (75%) (www.who.int/mediacentre/factsheets/fs134/en/) patients; two populations in which prescription drugs with narrow therapeutic margins are often used. Herbal–drug interactions of clinical significance are estimated to occur in 45% of individuals using prescription and complementary medicines [70]. Among the most well documented of herbal medicines that cause drug interactions is St. John's wort, a natural product that induces drug-metabolizing enzymes and transporters. A

TABLE 8.4 Induction-Type Herb–Drug Interactions

Herb	Drugs Affected	Likely Induced Genes
Hypericum perforatum (St. John's wort)	Cyclosporine [72], indinavir [73], omeprazole [74], digoxin [75, 76], imatinib [77], oral contraceptives [78], tacrolimus [79], warfarin [80], theophylline [81], midazolam [82], fexofenadine [82], alprazolam [83], quazepam [84], talinolol [85], ivabradine [86], irinotecan [87]	CYP3A4, P-glycoprotein, CYP2C9, CYP2C19, CYP1A2, CYP2E1
Echinacea purpurea or *angustifolia*	Midazolam [88]	CYP3A4
Allium sativum (garlic)	Saquinavir [89]	MDR1
Ginkgo biloba	Omeprazole [90]	CYP2C19
Panax quinquefolius (American Ginseng)	Warfarin [91]	CYP2C9

few other herbal medicines such as Echinacea, garlic, Ginkgo, and Ginseng have also been shown to cause induction-type drug interactions but information from the literature are less compelling than with St. John's wort regarding their clinical significance (Table 8.4).

8.3.1 St. John's Wort

Preparations of St. John's wort (*Hypericum perforatum*) are commonly used in the treatment of mild depression [92] and are the second leading herbal remedies sold in the United States behind Echinacea. Drug interactions with St. John's wort began to surface in 1999 when case reports and studies demonstrated that plasma levels of theophylline [81], digoxin [75], cyclosporine [72], and indinavir [73] were reduced in patients taking this herbal medication. It became apparent that St. John's wort interacted with medicines by inducing of intestinal P-gp and hepatic/intestinal CYP3A4 [76]. Moore and colleagues isolated the inducing component in St. John's wort, hyperforin, and demonstrated that it was a potent activator of PXR [93]. Since that time, a significant number of published reports detailing St. John's wort–drug interactions (Table 8.4) have cemented the notion that there are liabilities to self-administration of this herbal medicine. Hyperforin content in various St. John's wort preparations varies widely [94]. The hyperforin content in different products directly relates to the magnitude of drug interaction with cyclosporine [79], digoxin [95], and midazolam [96]. In light of its considerable use in the population, St. John's wort likely causes the most drug interactions that involve induction mechanisms of all medicines—prescription or herbal. While the product monographs of most affected prescription medicines have warnings regarding risks of coadministration with

St. John's wort, few, if any, cautionary statements are found on bottles of St. John's wort preparations. This has prompted some discussion regarding the need for additional regulation in sales of these herbal products [97].

8.3.2   Other Herbal Medicines

Garlic is thought to have beneficial effects by reducing plasma lipids, decreasing blood pressure, and acting as an antithrombotic. There are a few indications that garlic causes induction-type herb–drug interactions. For example, coadministration of garlic supplements significantly reduced the plasma levels of saquinavir, a CYP3A4-metabolized drug [89]. In addition, sulfur compounds in garlic oil such as diallyl sulfide activate CAR to induce CYP2B enzymes [98, 99]. Echinacea is among the most used herbal medicines and used in the treatment of common colds. One study demonstrated that hepatic CYP3A4 activity, as assessed by clearance of intravenous midazolam, was increased after treatment with Echinacea [88]. The effect on the gut appeared to oppose that of the liver in that intestinal CYP3A4 activity decreased resulting overall in a reduction of oral midazolam bioavailability with Echinacea treatment. Ginko is used to treat dementia and intermittent claudication. One study demonstrated that Ginko induces CYP2C19 activity by showing that there was a significant reduction in omeprazole (a CYP2C19 substrate) plasma levels with concomitant treatment [90]. However, an inductive response of Ginko toward CYP3A4 was not observed in another study that examined the effects of this herbal medicine on alprazolam pharmacokinetics [100]. Ginseng, often used as a tonic and mood enhancer, has been shown to interact with the CYP2C9 substrate drug warfarin [91]. In that study, Ginseng treatment reduced the anticoagulant effects of warfarin by decreasing the plasma drug levels [91] suggesting that it induces CYP2C9. By contrast, other investigators found no effect of Ginseng on warfarin pharmacokinetics or pharmacodynamics [101]. These conflicting results may be due to variability component species in products that are said to contain Ginseng [102] and underscore the difficulties in assessing the drug interaction potential of herbal medicines.

8.4   PHARMACOLOGY OF INDUCTION

It is the interest of the pharmaceutical industry to develop new medicines that are efficacious and safe. Hence, there is much effort to predict whether drugs in development may have the potential for significant drug–drug interactions including those that involve induction. With the knowledge that just a few key ligand-activated nuclear receptors such as PXR and CAR participate in most induction drug–drug interactions and that species differences in inductive response exist, the types of predictive *in vitro* models that have utility are relatively small. The Food and Drug Administration draft guidance for industry on drug interactions (http://www.fda.gov/CDER/GUIDANCE/6695dft.pdf) suggests the use of primary human hepatocyte culture model to assess induction potential of drugs under development. Furthermore, the guidance suggests utility of cell-based nuclear receptor (PXR) activation assays with some qualifications. The ability of such models to predict the inductive response by drugs on enzyme and transporter expression is based on classic pharmacological principles.

In theory, the magnitude of pharmacological effect elicited by drug–receptor interactions is determined by potency (EC_{50}) and efficacy (E_{max}). Moreover, the concentration of inducing agent at the receptor site affects the degree of inductive response. The pharmacological response is often described using a hyperbolic function such as the Hill equation, which takes into account the aforementioned variables. For induction *in vivo*, we require an understanding of nuclear receptor ligand concentrations within the major absorptive and eliminating organs such as the intestine, liver, and kidney. While the concentration of chemicals in these organs is typically not measured, often plasma concentration can be used as a surrogate. Surprisingly, there is a lack of human data in the literature to directly show that the inductive response is related to the plasma concentrations of a given agent. However, it is generally considered that for the most part, drug concentrations in plasma are proportional to drug dose. With this in mind, studies with St. John's wort have convincingly demonstrated that dose of this herbal is related to the magnitude of changes in digoxin plasma exposures [95].

Given that there is clear evidence to demonstrate that a dose–response effect exists, it follows that inducer concentrations in plasma and the potency of interaction with nuclear receptors would predict the magnitude of induction. A compilation of *in vivo* drug concentrations and *in vitro* determined interaction affinity with PXR is shown in Table 8.5 using data from the literature as well as our own unpublished results. Here, the maximal concentration in plasma after oral administration (C_{max}) is considered since it resembles the concentrations of drug that could be achieved within hepatocytes. The affinity of drugs toward PXR is estimated by EC_{50} determinations from cell-based reporter assays. Hence, on the basis of pharmacological principles, the magnitude of the inductive response *in vivo* should be predicted by the ratio of C_{max} to EC_{50}. A measure of inductive response *in vivo* is the change in area-under-the-plasma- concentration-time profile (AUC) of a probe compound before and during treatment with an inducing agent. Probe compounds are typically those whose pharmacokinetics is determined largely by a single elimination process. For instance, midazolam is cleared mainly by CYP3A4 metabolism and is considered as a sensitive probe for CYP3A4 enzyme activity *in vivo*. CYP3A4 expression is generally thought to be among the most highly sensitive responses to PXR activation [115]. Currently, there are no studies that show that C_{max}/EC_{50} values for a specific inducer given at different doses are directly related to the change in enzymatic/transporter activity as assessed using probe substrate pharmacokinetics. However, if you consider different drugs and their C_{max}/EC_{50} values in relation to *in vivo* CYP3A4 activity as determined using midazolam pharmacokinetics or another phenotypic measure such as ^{14}C-erythromycin breath test, the validity of the pharmacological response paradigm for nuclear receptor mediated induction can be evaluated (Table 8.5).

For drugs examined in this exercise, the phenotypic measure of induction is CYP3A4 activity and since these responses are estimated by various techniques (midazolam pharmacokinetics, erythromycin breath test), it is difficult to relate the

TABLE 8.5 PXR Induction Metric and CYP3A4 Induction *In Vivo*

Drug	C_{max} *In Vivo*[a] (μM)	PXR EC_{50} (μM)	C_{max}/EC_{50}	Clinical Report of CYP3A4 Induction		
				Interacting Drug	$\Delta AUC^{b}/\Delta ERBT^{c}$ ΔConcentration	Reference
Marked CYP3A4 Induction						
Amprenavir	10.6	5.2[d]	2.03	Delavirdine	↓ 61% AUC	[44]
Avasimibe	11.9	1	11.9	Midazolam	↓ 85% AUC	[104]
Carbamazepine	39	0.9	43	Midazolam	↓ 94% AUC	[105, 106]
Hyperforin	0.38	0.023	16.5	Midazolam	↓ 52% AUC	[93, 107]
Phenobarbital	56	10	5.6	Ethinyl Estradiol	↓ 70% Concentration	[105, 106, 108]
Phenytoin	30	25	1.2	Midazolam	↓ 94% AUC	[105, 106]
Rifampin	16.5	1.9[d]	8.7	Midazolam	↓ 96% AUC	[109]
Ritonavir	16	0.5[d]	32	Meperidine	↓ 62% AUC	Product Monograph
Moderate CYP3A4 Induction						
Bosentan	2.9	19.9	1.5×10^{-1}	Simvastatin	↓ 34% AUC	[68, 110]
Efavirenz	12.6	13[d]	1.0	Erythromycin	↑ 55% ERBT	[111]
Nelfinavir	6.0	6.1[d]	1.0	Ethinyl Estradiol	↓ 42% AUC	Product Monograph
Rifabutin	0.47	1.6[d]	8.8×10^{-2}	HIV-PIs	↓ 15–40% AUC	[43]
Topiramate	16.2	500	3.2×10^{-2}	Ethinyl Estradiol	↓ 18–32% AUC	[39, 112]
Troglitazone	6.8	3	2.2	Simvastatin	↓ 40% AUC	[113, 114]
No CYP3A4 Induction						
Lovastatin	1.2×10^{-4}	1	1.2×10^{-4}	No reports	—	[8]
Nifedipine	1.2×10^{-4}	4.3	2.7×10^{-5}	No reports	—	[10]
Simvastatin	1.1×10^{-4}	0.8	1.4×10^{-4}	No reports	—	[105]

[a]From [103] or Product Monograph.
[b]AUC, area-under-the-plasma-concentration-time curve.
[c]ERBT, erythromycin breath test.
[d]Unpublished data.

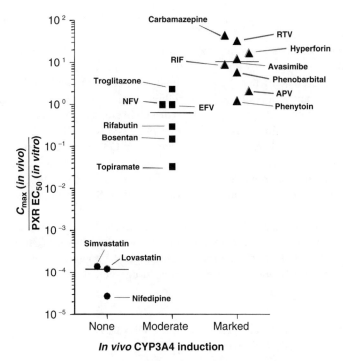

FIGURE 8.1 *In vivo* induction follows classic pharmacological principles. C_{max}/EC_{50} predicts whether a PXR activator causes significant drug–drug interactions.

in vivo effects of a series of inducers using a single continuous variable. Therefore, an arbitrary categorization is presented whereby drugs that all activate PXR *in vitro* are segregated into those that cause marked, moderate, and no induction of CYP3A4 activity *in vivo* (Table 8.5). When shown graphically, the C_{max}/EC_{50} values for drugs in each induction category cluster together (Figure 8.1). For example, drugs that cause marked induction of CYP3A4 have C_{max}/EC_{50} values of greater than 1 and those that induce moderately have ratios between 0.1 and 1 (Figure 8.1). PXR-activating drugs that do not cause CYP3A4 induction *in vivo* have C_{max}/EC_{50} values much less than 0.1. This analysis reveals that nuclear receptor mediated induction *in vivo* is like all pharmacology, based on drug concentrations and affinity toward the target receptors.

What also emerges from the exercise is that while many drugs can activate PXR, not all will cause clinically relevant drug interactions. This is particularly evident for certain drug classes such as calcium channel blocking drugs [116], HMG-CoA reductase inhibitors (statins) [117], and thiazolidinediones (glitazones) [118]. Another limitation of the C_{max}/EC_{50} value in predicting changes in CYP3A4 activity is the potential that the PXR agonist is also an inhibitor of CYP3A4. This is the case for ritonavir, which is a potent CYP3A4 inhibitor [119] and a PXR activator. Numerous drug interaction studies have clearly shown that ritonavir-mediated CYP3A4 inhibition overcomes enzyme induction to result in higher plasma levels of most coadministered drugs.

For a number of reasons, plasma drug concentrations may not entirely predict intracellular concentrations and hence that which nuclear receptors would be exposed. One reason being that drug accumulation in cells is modulated by the actions of membrane transporters that enhance cellular drug uptake or efflux. Indeed, when rifampin uptake into cells is enhanced *in vitro* by overexpression of a hepatic uptake transporter, organic anion transporting polypeptide 1B1 (OATP1B1), greater PXR activation was observed [120]. However, an impact of OATP1B1 transport of rifampin on CYP3A4 induction *in vivo* remains uncertain [121]. Another interesting example is that of rifaximin, a drug that is capable of activating PXR *in vitro* [58] but is not absorbed into the circulation after oral administration. Studies in hPXR-humanized mice show that rifaximin induces intestinal CYP3A expression, but in humans rifaximin does not induce CYP3A4 activity *in vivo* as assessed by the pharmacokinetics of oral contraceptives [57]. These findings question whether rifaximin causes intestine-specific CYP3A4 induction in humans and highlight the importance of understanding inducer tissue distribution to predict inductive drug interactions.

8.5 CLINICAL ASPECTS OF INDUCTION-TYPE DRUG INTERACTIONS

8.5.1 Enzymes, Transporters, and Drugs Highly Sensitive to Induction

Drugs that interact with nuclear receptors to cause induction of target metabolic enzymes and membrane transporters have differing effects on the pharmacokinetics of coadministered medications. Whether induction has clinical relevance to the coadministered drug will be determined by the pharmacological dose–response relationship as well as the therapeutic index. Modulation of drug disposition pathways, be it metabolism or transport, by induction and inhibition mechanisms will manifest as a marked change in plasma levels when a given drug is eliminated by a relatively few mechanisms. For example, drugs that are metabolized by multiple CYP enzymes, some of which may not be inducible such as CYP2D6, would be less prone to significant changes in plasma levels when coadministered with an inducing agent. Those drugs that are cleared solely by an induction-sensitive pathway would be particularly susceptible to significant changes in plasma drug concentrations. While many drug-metabolizing enzymes and transporters have been shown to be regulated by nuclear receptor mediated pathways *in vitro* [122], for a number of those identified, there is no evidence that induction occurs *in vivo* from a gene expression or functional perspective. For instance, the sulfation enzyme SULT2A1 is known to be regulated by PXR [28], CAR [28], and VDR [123], but the functional effect of nuclear receptor agonists has not been directly demonstrated in humans *in vivo*. In addition, the efflux transporter MRP4 is regulated by CAR [32] but demonstration of nuclear receptor mediated induction in humans is lacking. However, for most inducible metabolizing enzymes and transporters, there are several examples of substrate drugs sensitive to induction and many of these have become *in vivo* probes to evaluate the drug interaction potential of medicines. However, for a number of enzymes and transporters, there is only sparse evidence for induction of gene expression *in vivo*.

8.5.1.1 CYP2B6. Bupropion is a drug indicated for depression and smoking cessation that is metabolized primarily by hydroxylation via CYP2B6 [124, 125]. Recent studies revealed that bupropion plasma concentrations decline significantly when coadministered with the nuclear receptor agonists rifampin [126] and lopinavir/ritonavir [127]. The latter example is interesting in that it is one of a few drug interactions with the HIV-PI that results in decreased blood levels of coadministered drugs. This interaction indicates that CYP2B6 induction occurs with lopinavir/ritonavir therapy and that these inducers do not significantly inhibit CYP2B6 enzyme activity as they do for CYP3A4.

8.5.1.2 CYP2C9. The anticoagulant warfarin, a narrow therapeutic index drug, is a good example of a CYP2C9-metabolized drug that is sensitive to inducers [18–20]. It is administered as a racemic mixture with the S-enantiomer being the more pharmacologically active species. The plasma levels of both enantiomers are reduced by coadministration of rifampin [128] and St. John's wort [101] to result in decreased anticoagulant response.

8.5.1.3 CYP2C19. Studies have shown that PXR and CAR regulate hepatic CYP2C19 expression *in vitro* [21, 22]. The metabolism and plasma levels of the CYP2C19 substrates omeprazole [74] and voriconazole [129] are affected by the PXR agonists St. John's wort and rifampin, demonstrating that this enzyme is inducible *in vivo*.

8.5.1.4 CYP3A4. CYP3A4 is responsible for the majority of human drug metabolism and is also among the most sensitive enzymes to the inducing effects of PXR and CAR agonists. Therefore, inductive drug interactions are most commonly observed with CYP3A4 substrate drugs. Despite the large number of drugs metabolized by this pathway, resources are available to guide clinicians in determining those drug interactions that are most clinically relevant (e.g., http://medicine.iupui.edu/flockhart/table.htm). The most commonly used probe drug to assess CYP3A4 activity is the benzodiazepine, midazolam. The magnitude of induction responses by different PXR activators is evidenced by the 16-fold and 2.7-fold decrease in oral midazolam AUC with rifampin [130] and St. John's wort [82] pretreatment.

8.5.1.5 UGTs. There are 17 known UDP-glucuronosyltransferase proteins that belong to two subfamilies (UGT1 and UTGT2). The expression of several UGT1A enzymes is known to be up-regulated by PXR/CAR agonists *in vitro* [25, 26]. An example that gives evidence that UGT1A induction occurs *in vivo* comes from a study examining the pharmacokinetics of the anticonvulsant drug lamotrigine before and after treatment with the known PXR agonists lopinavir/ritonavir. Consistent with the drug being primarily eliminated by UGT1A4-mediated glucuronidation [131], lamotrigine AUC was reduced 50% with lopinavir/ritonavir treatment [132]. Although the molecular mechanisms have not been elucidated, it appears likely that other UGTs including UGT2B7 are regulated by nuclear receptors. For instance, the plasma levels of the UGT2B7-metabolized drug zidovudine [133] are significantly reduced in patients taking concomitant rifampin [134, 135].

8.5.1.6 MDR1. The efflux transporter, P-gp, is expressed in a number of tissues including the intestine, liver, kidney, and brain capillaries. Molecular studies have shown that the *MDR1* gene is regulated by PXR [30] and CAR [136]. In a classic study by Greiner and colleagues [137], the inducibility of P-gp *in vivo* was demonstrated on both gene expression and functional levels. They observed that the AUC of the P-gp substrate drug, digoxin, was decreased by 30% when subjects were pretreated with rifampin. Although the magnitude of change in digoxin plasma levels may appear modest, the effect is clinically significant since this drug has a narrow therapeutic range. Importantly, they show that intestinal P-gp expression is induced by rifampin treatment to cause decreased absorption of digoxin [137].

8.5.1.7 MRP2. The efflux transporter MRP2, encoded by the *ABCC2* gene, is highly expressed in the liver, small intestine, and kidney. Studies in humans have shown that intestinal MRP2 expression is induced by treatment with rifampin [138]. Furthermore, hepatic MRP2 was up-regulated in nonhuman primates after rifampin administration [139]. Subsequent studies determined that *ABCC2* is regulated by the nuclear receptors PXR, CAR, and FXR [31]. At present, examples of the influence of MRP2 induction on drug pharmacokinetics are lacking in the literature. This is likely due to the fact that MRP2 substrates are often drug metabolites such as glucuronides and the enzymatic formation of such phase II metabolites is inducible. This emphasizes the difficulty arising in understanding the effects of concomitant induction of metabolism and transport on pharmacokinetics. Notwithstanding, it is well known that PXR/CAR agonists such as phenobarbital and rifampin are often used to treat jaundice. The efficacy of these treatments is thought to be brought about by induction of MRP2, which then enhances the elimination of glucuronidated bilirubin metabolites into bile.

8.5.2 Therapeutic Implications of Induction-Type Drug Interactions

The most common impact of induction-mediated drug interactions is attenuation or loss of therapeutic effects. This occurs because induction of drug-metabolizing enzymes and transporters typically results in decreased plasma concentrations of the concomitantly administered drug. Some of the first indications that induction causes clinically relevant effects come from case reports of oral contraceptive failure in women taking rifampin [42]. Later studies demonstrated clearly that ethinyl estradiol plasma levels were significantly reduced by rifampin treatment [140]. In other early reports, rifampin coadministration was found to cause withdrawal symptoms in patients taking methadone [56]. Not surprisingly, St. John's wort, use was later found to cause methadone withdrawal [141]. The withdrawal symptoms occur due to rapid decline in methadone levels during induction. Another related example of the therapeutic impact of induction-type drug interactions is that of loss of opioid analgesia. Fromm and colleagues demonstrated that the analgesic effect of morphine is completely lost in subjects treated with rifampin [142]. Rifampin coadministration reduced morphine plasma levels, although there were no changes in morphine glucuronide recovery in urine indicating the glucuronidation pathway was not induced [142]. A potential explanation for these findings comes from a study using the PXR-humanized mouse model. In this report, rifampin-treated mice exhibited upregulation of P-gp expression and activity at the level of the blood–brain barrier [143]. Furthermore, this increased brain capillary P-gp activity resulted in reduced antinociceptive effects of methadone in rifampin-treated animals [143]. Hence, induction of brain MDR1 expression by PXR/CAR activators may also affect the central nervous system effects of P-gp substrates. Lastly, the profound effects of induction-mediated drug interactions have resurfaced in recent years with case reports of St. John's wort treatment causing transplant rejection in patients taking cyclosporine [72, 144, 145]. These reports helped bring public attention to the notion that herbal medicines are not innocuous, and that their use can be the cause of significant drug interactions.

In the treatment of particular conditions such as epilepsy and HIV infection, induction of drug-metabolizing enzymes and transporters is more often the rule than the exception. This is because our drug armamentarium for these conditions largely contains PXR and/or CAR activators (Table 8.2). For HIV infection, the HIV-PIs are the mainstay of highly active antiretroviral therapy (HAART) and antimycobacterial drugs such as the rifamycins are commonly prescribed in the treatment and prophylaxis of opportunistic infections. In the treatment of epilepsy, the majority of anticonvulsant medications are PXR/CAR activators (Table 8.2). A common theme in both conditions is that of the treatment with combined inducing agents such as dual HIV-PIs and multiple antiepileptic drugs, which may act mutually to alter drug levels. Furthermore, the polypharmacy encountered in the treatment of these conditions results in a complex mixture of enzyme inducers and inhibitors that make prediction of net drug effects (loss of efficacy or toxicity) extremely difficult [146, 147].

8.5.3 Time Course of Induction and Deinduction

Understanding the time course of the induction and deinduction response has direct implications in pharmacotherapy. First, the clinician can determine whether an observed clinical effect could be caused by an induction drug interaction by knowing the time line of expected inductive responses. Furthermore, if patients are knowingly administered inducing agents, the clinician can establish the time required before dose adjustments are necessary for concomitantly prescribe medicines. For a drug with a known autoinduction response (e.g., carbamazepine), the time course of induction will determine the time to reach steady-state plasma levels with initial dosing in inducer naive patients and when dose changes are best done. Lastly, understanding the deinduction time course will allow the clinician to estimate when to expect rebound changes in concomitant drug levels after inducers are discontinued.

At the molecular level, the rapid effect of exposure to PXR and CAR activators on target gene mRNA expression occurs within a few hours followed by changes in protein content. How these changes in gene expression upon inducer treatment relate to the time course of effects on the pharmacokinetics of the coadministered drug has been well studied. Early studies in dogs showed that the half-life of amobarbital was shorter during the second dose of the drug given 24 h after the initial dose, indicating that functional autoinduction is rapidly apparent within a day [148]. Later, carbamazepine levels in children were shown to decrease rapidly after the first few doses and steady-state levels were achieved within a month [149]. Using urinary recovery of the 6β-hydroxylated metabolite of cortisol as a marker of CYP3A4 activity, the inductive effects of rifampin administration could be observed within a day, reaching a peak and plateau within 5 days [150]. After rifampin discontinuation, urinary 6β-hydroxycortisol levels began to decline within 2 days achieving preinduction levels in 7 days. These data suggest that functional deinduction *in vivo* is rather rapid, probably owing to the relatively short half-life of rifampin (~3 h), rapid decrease in CYP3A4 gene transcription, and increased protein turnover. The available pharmacokinetic data from the literature regarding the time course of induction/deinduction are largely reflective of CYP3A4 activities and relatively little is known about response times for other drug-metabolizing enzymes and transporters.

8.6   INHIBITION OF NUCLEAR RECEPTORS IN CLINICAL DRUG INTERACTIONS

Several drugs are known to antagonize the actions of PXR including the antifungal ketoconazole [151, 152] and the antineoplastic agent Trabectedin (ET-742) [153]. Furthermore, dietary constituents such as sulforaphane [154], which is found in broccoli, and the soy-based lipid stigmasterol [155] are functional PXR antagonists. Inhibition of PXR activity has been considered a novel strategy to overcome chemotherapeutic drug resistance by attenuating the expression of drug-metabolizing enzymes and transporters in tumors [151, 152]. However, studies so far have not examined the effects of PXR antagonism *in vivo* although it is likely that many other drug and dietary compounds inhibit PXR to alter the expression of drug disposition genes and pharmacokinetics. Nevertheless, there have been no reports of attenuated inductive response of prototypical nuclear receptor agonists such as rifampin or St. John's wort by coadministered drugs or dietary constituents.

8.7   NUCLEAR RECEPTOR MEDIATED DRUG SIDE EFFECTS

There is a growing appreciation that activation of PXR or CAR is the cause of drug side effects. The clinical indication for many PXR/CAR agonist drugs requires either prolonged or lifelong treatment, as is the case for rifampin to treat tuberculosis, or HIV-PIs or antiepileptic drugs. Such chronic PXR/CAR activation appears to have deleterious effects. For example, PXR activation by rifampin is thought to cause osteomalacia [156] due to induction of vitamin D metabolizing enzymes [157, 158]. Furthermore, rifampin treatment has been associated with pseudo-Cushing's syndrome [159] potentially due to the endocrine disrupting effects of PXR activation leading to hypercortisolism [160]. Another, potential side effect of PXR/CAR activation is hypothyroidism. Indeed chronic anticonvulsant [161, 162] and rifampin [163, 164] use has been associated with hypothyroidism. The underlying mechanism for thyroid hormone dysregulation appears to be induction of hepatic thyroid hormone metabolism by glucuronidation and sulfation [165] or biliary clearance of free hormone [166].

8.8   PERSPECTIVES

The examples of drug–drug and drug–herbal interactions frequently encountered in the clinic have well-established mechanisms involving nuclear receptor mediated up-regulation of gene expression. Among all the currently prescribed drugs, only a small and manageable group has been implicated in these untoward drug effects. With herbal medicines, we probably understand the potential for induction of drug-metabolizing enzymes and transporters for the most popular remedies but know less about others. Due to the lack of product standardization, we must be careful not to generalize the drug interaction potential of specific herbal preparations. We should have comfort in knowing that many of the inductive responses of drugs and herbal medicines are dependent on achieved tissue and plasma levels of the inducer as well as the attendant nuclear receptor pharmacology. The predictability in clinical responses that are afforded by *in vitro* models of receptor action should now allow for the development of novel drugs that are devoid of inductive drug interactions. However, this has not prevented the introduction of new drugs to the market capable of this type of drug interaction. A recent example being the market release of the nonpeptidic HIV-PI tipranavir [167] in 2005. Therefore, clinicians must remain aware of potential inductive drug interactions for new medicines. This may be all the more important since PXR and CAR are potential novel drug targets [168]. Overall, while there are aspects of nuclear receptor biology that remain to be determined, the science has significantly matured since the first description of inductive response in 1960 to the current characterization of PXR/CAR biology. It is expected that further progress in this field will ensure the development of safer new medicines and better management of drugs currently in clinical use.

REFERENCES

1. Lazarou, J., Pomeranz, B. H., and Corey, P. N. (1998) Incidence of adverse drug reactions in hospitalized patients: a meta-analysis of prospective studies. *JAMA* **279**, 1200–1205.

2. Kohler, G. I., Bode-Boger, S. M., Busse, R., Hoopmann, M., Welte, T., and Boger, R. H. (2000) Drug–drug interactions in medical patients: effects of in-hospital treatment and relation to multiple drug use. *Int. J. Clin. Pharmacol. Ther.* **38**, 504–513.

3. Conney, A. H., Davison, C., Gastel, R., and Burns, J. J. (1960) Adaptive increases in drug-metabolizing enzymes induced by phenobarbital and other drugs. *J. Pharmacol. Exp. Ther.* **130**, 1–8.

4. Cucinell, S. A., Koster, R., Conney, A. H., and Burns, J. J. (1963) Stimulatory effect of phenobarbital on the metabolism of diphenylhydantoin. *J. Pharmacol. Exp. Ther.* **141**, 157–160.

5. Remmer, H., Schoene, B., and Fleischmann, R. A. (1973) Induction of the unspecific microsomal hydroxylase in the human liver. *Drug Metab. Dispos.* **1**, 224–230.

6. Adesnik, M., Bar-Nun, S., Maschio, F., Zunich, M., Lippman, A., and Bard, E. (1981) Mechanism of induction of cytochrome P-450 by phenobarbital. *J. Biol. Chem.* **256**, 10340–10345.

7. Kliewer, S. A., Moore, J. T., Wade, L., Staudinger, J. L., Watson, M. A., Jones, S. A., McKee, D. D., Oliver, B. B., Willson, T. M., Zetterstrom, R. H., et al. (1998) An orphan nuclear receptor activated by pregnanes defines a novel steroid signaling pathway. *Cell* **92**, 73–82.

8. Lehmann, J. M., McKee, D. D., Watson, M. A., Willson, T. M., Moore, J. T., and Kliewer, S. A. (1998) The human orphan nuclear receptor PXR is activated by compounds that regulate CYP3A4 gene expression and cause drug interactions. *J. Clin. Invest.* **102**, 1016–1023.

9. Goodwin, B., Hodgson, E., and Liddle, C. (1999) The orphan human pregnane X receptor mediates the transcriptional activation of CYP3A4 by rifampicin through a distal enhancer module. *Mol. Pharmacol.* **56**, 1329–1339.

10. Bertilsson, G., Heidrich, J., Svensson, K., Asman, M., Jendeberg, L., Sydow-Backman, M., Ohlsson, R., Postlind, H., Blomquist, P., and Berkenstam, A. (1998) Identification of a human nuclear receptor defines a new signaling pathway for CYP3A induction. *Proc. Natl Acad. Sci. USA* **95**, 12208–12213.

11. Blumberg, B., Sabbagh, W., Jr., Juguilon, H., Bolado, J., Jr., van Meter, C. M., Ong, E. S., and Evans, R. M. (1998) SXR, a novel steroid and xenobiotic-sensing nuclear receptor. *Genes Dev.* **12**, 3195–3205.

12. Forman, B. M., Tzameli, I., Choi, H. S., Chen, J., Simha, D., Seol, W., Evans, R. M., and Moore, D. D. (1998) Androstane metabolites bind to and deactivate the nuclear receptor CAR-beta. *Nature* **395**, 612–615.

13. Sueyoshi, T., Kawamoto, T., Zelko, I., Honkakoski, P., and Negishi, M. (1999) The repressed nuclear receptor CAR responds to phenobarbital in activating the human CYP2B6 gene. *J. Biol. Chem.* **274**, 6043–6046.

14. Honkakoski, P., Zelko, I., Sueyoshi, T., and Negishi, M. (1998) The nuclear orphan receptor CAR-retinoid X receptor heterodimer activates the phenobarbital-responsive enhancer module of the CYP2B gene. *Mol. Cell. Biol.* **18**, 5652–5658.

15. Baes, M., Gulick, T., Choi, H. S., Martinoli, M. G., Simha, D., and Moore, D. D. (1994) A new orphan member of the nuclear hormone receptor superfamily that interacts with a subset of retinoic acid response elements. *Mol. Cell. Biol.* **14**, 1544–1551.

16. Goodwin, B., Moore, L. B., Stoltz, C. M., McKee, D. D., and Kliewer, S. A. (2001) Regulation of the human CYP2B6 gene by the nuclear pregnane X receptor. *Mol. Pharmacol.* **60**, 427–431.

17. Wang, H., Faucette, S., Sueyoshi, T., Moore, R., Ferguson, S., Negishi, M., and LeCluyse, E. L. (2003) A novel distal enhancer module regulated by pregnane X receptor/constitutive androstane receptor is essential for the maximal induction of CYP2B6 gene expression. *J. Biol. Chem.* **278**, 14146–14152.

18. Chen, Y., Ferguson, S. S., Negishi, M., and Goldstein, J. A. (2004) Induction of human CYP2C9 by rifampicin, hyperforin, and phenobarbital is mediated by the pregnane X receptor. *J. Pharmacol. Exp. Ther.* **308**, 495–501.

19. Gerbal-Chaloin, S., Daujat, M., Pascussi, J. M., Pichard-Garcia, L., Vilarem, M. J., and Maurel, P. (2002) Transcriptional regulation of CYP2C9 gene. Role of glucocorticoid receptor and constitutive androstane receptor. *J. Biol. Chem.* **277**, 209–217.

20. Ferguson, S. S., LeCluyse, E. L., Negishi, M., and Goldstein, J. A. (2002) Regulation of human CYP2C9 by the constitutive androstane receptor: discovery of a new distal binding site. *Mol. Pharmacol.* **62**, 737–746.

21. Chen, Y., Ferguson, S. S., Negishi, M., and Goldstein, J. A. (2003) Identification of constitutive androstane receptor and glucocorticoid receptor binding sites in the CYP2C19 promoter. *Mol. Pharmacol.* **64**, 316–324.

22. Gerbal-Chaloin, S., Pascussi, J. M., Pichard-Garcia, L., Daujat, M., Waechter, F., Fabre, J. M., Carrere, N., and Maurel, P. (2001) Induction of CYP2C genes in human hepatocytes in primary culture. *Drug Metab. Dispos.* **29**, 242–251.

23. Pascussi, J. M., Jounaidi, Y., Drocourt, L., Domergue, J., Balabaud, C., Maurel, P., and Vilarem, M. J. (1999) Evidence for the presence of a functional pregnane X receptor response element in the CYP3A7 promoter gene. *Biochem. Biophys. Res. Commun.* **260**, 377–381.

24. Bertilsson, G., Berkenstam, A., and Blomquist, P. (2001) Functionally conserved xenobiotic responsive enhancer in cytochrome P450 3A7. *Biochem. Biophys. Res. Commun.* **280**, 139–144.

25. Sugatani, J., Kojima, H., Ueda, A., Kakizaki, S., Yoshinari, K., Gong, Q. H., Owens, I. S., Negishi, M., and Sueyoshi, T. (2001) The phenobarbital response enhancer module in the human bilirubin UDP-glucuronosyltransferase UGT1A1 gene and regulation by the nuclear receptor CAR. *Hepatology* **33**, 1232–1238.

26. Sugatani, J., Yamakawa, K., Tonda, E., Nishitani, S., Yoshinari, K., Degawa, M., Abe, I., Noguchi, H., and Miwa, M. (2004) The induction of human UDP-glucuronosyltransferase 1A1 mediated through a distal enhancer module by flavonoids and xenobiotics. *Biochem. Pharmacol.* **67**, 989–1000.

27. Gardner-Stephen, D., Heydel, J. M., Goyal, A., Lu, Y., Xie, W., Lindblom, T., Mackenzie, P., and Radominska-Pandya, A. (2004) Human PXR variants and their differential effects on the regulation of human UDP-glucuronosyltransferase gene expression. *Drug Metab. Dispos.* **32**, 340–347.

28. Echchgadda, I., Song, C. S., Oh, T., Ahmed, M., De La Cruz, I. J., and Chatterjee, B. (2007) The xenobiotic-sensing nuclear receptors pregnane X receptor, constitutive androstane receptor, and orphan nuclear receptor hepatocyte nuclear factor 4{alpha} in the regulation of human steroid-/bile acid-sulfotransferase. *Mol. Endocrinol.* **21**, 2099–2111.

29. Fang, H. L., Strom, S. C., Ellis, E., Duanmu, Z., Fu, J., Duniec-Dmuchowski, Z., Falany, C. N., Falany, J. L., Kocarek, T. A., and Runge-Morris, M. (2007) Positive and negative regulation of human hepatic hydroxysteroid sulfotransferase (SULT2A1) gene transcription by rifampicin: roles of hepatocyte nuclear factor 4{alpha} and pregnane X receptor. *J. Pharmacol. Exp. Ther.* **323**, 586–598.

30. Geick, A., Eichelbaum, M., and Burk, O. (2001) Nuclear receptor response elements mediate induction of intestinal MDR1 by rifampin. *J. Biol. Chem.* **276**, 14581–14587.

31. Kast, H. R., Goodwin, B., Tarr, P. T., Jones, S. A., Anisfeld, A. M., Stoltz, C. M., Tontonoz, P., Kliewer, S., Willson, T. M., and Edwards, P. A. (2002) Regulation of multidrug resistance-associated protein 2 (ABCC2) by the nuclear receptors pregnane X receptor, farnesoid X-activated receptor, and constitutive androstane receptor. *J. Biol. Chem.* **277**, 2908–2915.

32. Assem, M., Schuetz, E. G., Leggas, M., Sun, D., Yasuda, K., Reid, G., Zelcer, N., Adachi, M., Strom, S., Evans, R. M., et al. (2004) Interactions between hepatic Mrp4 and Sult2a as revealed by the constitutive androstane receptor and Mrp4 knockout mice. *J. Biol. Chem.* **279**, 22250–22257.

33. Miki, Y., Suzuki, T., Kitada, K., Yabuki, N., Shibuya, R., Moriya, T., Ishida, T., Ohuchi, N., Blumberg, B., and Sasano, H. (2006) Expression of the steroid and xenobiotic receptor and its possible target gene, organic anion transporting polypeptide-A, in human breast carcinoma. *Cancer Res.* **66**, 535–542.

34. Cucinell, S. A., Conney, A. H., Sansur, M., and Burns, J. J. (1965) Drug interactions in man. I. Lowering effect of phenobarbital on plasma levels of bishydroxycoumarin (dicumarol) and diphenylhydantoin (dilantin). *Clin. Pharmacol. Ther.* **6**, 420–429.

35. Rawlins, M. D., Collste, P., Bertilsson, L., and Palmer, L. (1975) Distribution and elimination kinetics of carbamazepine in man. *Eur. J. Clin. Pharmacol.* **8**, 91–96.

36. Moore, L. B., Parks, D. J., Jones, S. A., Bledsoe, R. K., Consler, T. G., Stimmel, J. B., Goodwin, B., Liddle, C., Blanchard, S. G., Willson, T. M., et al. (2000) Orphan nuclear receptors constitutive androstane receptor and pregnane X receptor share xenobiotic and steroid ligands. *J. Biol. Chem.* **275**, 15122–15127.

37. Faucette, S. R., Zhang, T. C., Moore, R., Sueyoshi, T., Omiecinski, C. J., LeCluyse, E. L., Negishi, M., and Wang, H. (2007) Relative activation of human pregnane X receptor versus constitutive androstane receptor defines distinct classes of CYP2B6 and CYP3A4 inducers. *J. Pharmacol. Exp. Ther.* **320**, 72–80.

38. Wang, H., Faucette, S., Moore, R., Sueyoshi, T., Negishi, M., and LeCluyse, E. (2004) Human constitutive androstane receptor mediates induction of CYP2B6 gene expression by phenytoin. *J. Biol. Chem.* **279**, 29295–29301.

39. Nallani, S. C., Glauser, T. A., Hariparsad, N., Setchell, K., Buckley, D. J., Buckley, A. R., and Desai, P. B. (2003) Dose-dependent induction of cytochrome P450 (CYP) 3A4 and activation of pregnane X receptor by topiramate. *Epilepsia* **44**, 1521–1528.

40. Cerveny, L., Svecova, L., Anzenbacherova, E., Vrzal, R., Staud, F., Dvorak, Z., Ulrichova, J., Anzenbacher, P., and Pavek, P. (2007) Valproic acid induces CYP3A4 and MDR1 gene expression by activation of constitutive androstane receptor and pregnane X receptor pathways. *Drug Metab. Dispos.* **35**, 1032–1041.

41. Michot, F., Burgi, M., and Buttner, J. (1970) Rimactan (rifampicin) and anticoagulant therapy. *Schweiz. Med. Wochenschr.* **100**, 583–584.

42. Reimers, D. and Jezek, A. (1971) The simultaneous use of rifampicin and other antitubercular agents with oral contraceptives. *Prax. Pneumol.* **25**, 255–262.

43. Finch, C. K., Chrisman, C. R., Baciewicz, A. M., and Self, T. H. (2002) Rifampin and rifabutin drug interactions: an update. *Arch. Intern. Med.* **162**, 985–992.

44. Justesen, U. S., Klitgaard, N. A., Brosen, K., and Pedersen, C. (2003) Pharmacokinetic interaction between amprenavir and delavirdine after multiple-dose administration in healthy volunteers. *Br. J. Clin. Pharmacol.* **55**, 100–106.

45. Hendrix, C. W., Wakeford, J., Wire, M. B., Lou, Y., Bigelow, G. E., Martinez, E., Christopher, J., Fuchs, E. J., and Snidow, J. W. (2004) Pharmacokinetics and pharmacodynamics of methadone enantiomers after coadministration with amprenavir in opioid-dependent subjects. *Pharmacotherapy* **24**, 1110–1121.

46. Cantilena, L., McCrae, J., Blazes, D., Winchell, G., Carides, A., Royce, C., and Deutsch, P. (1999) Lack of pharmacokinetic interaction between indinavir and methadone. *Clin. Pharmacol. Ther.* **65**, 135.

47. McCance-Katz, E. F., Rainey, P. M., Friedland, G., and Jatlow, P. (2003) The protease inhibitor lopinavir–ritonavir may produce opiate withdrawal in methadone-maintained patients. *Clin. Infect. Dis.* **37**, 476–482.

48. Hsyu, P. H., Schultz-Smith, M. D., Lillibridge, J. H., Lewis, R. H., and Kerr, B. M. (2001) Pharmacokinetic interactions between nelfinavir and 3-hydroxy-3-methylglutaryl coenzyme A reductase inhibitors atorvastatin and simvastatin. *Antimicrob. Agents Chemother.* **45**, 3445–3450.

49. Hsyu, P. H., Lillibridge, J. H., Maroldo, L., Weiss, W. R., and Kerr, B. M. (2000) Pharmacokinetic (PK) and pharmacodynamic (PD) interactions between nelfinavir and methadone. *7th CROI*, Abstract 87.

50. Hsu, A., Granneman, G. R., Carothers, L., Dennis, S., Chiu, L., Valdes, J., and Sun, E. (1998) Ritonavir does not increase methadone exposure in healthy volunteers. *5th CROI*, Abstract 324.

51. Shelton, M. J., Cloen, D., DiFrancesco, R., Berenson, C. S., Esch, A., de Caprariis, P. J., Palic, B., Schur, J. L., Bugge, C. J., Ljungqvist, A., et al. (2004) The effects of once-daily saquinavir/minidose ritonavir on the pharmacokinetics of methadone. *J. Clin. Pharmacol.* **44**, 293–304.

52. Clarke, S. M., Mulcahy, F. M., Tjia, J., Reynolds, H. E., Gibbons, S. E., Barry, M. G., and Back, D. J. (2001) The pharmacokinetics of methadone in HIV-positive patients receiving the non-nucleoside reverse transcriptase inhibitor efavirenz. *Br. J. Clin. Pharmacol.* **51**, 213–217.

53. Murphy, R. L., Sommadossi, J. P., Lamson, M., Hall, D. B., Myers, M., and Dusek, A. (1999) Antiviral effect and pharmacokinetic interaction between nevirapine and indinavir in persons infected with human immunodeficiency virus type 1. *J. Infect. Dis.* **179**, 1116–1123.

54. Clarke, S. M., Mulcahy, F. M., Tjia, J., Reynolds, H. E., Gibbons, S. E., Barry, M. G., and Back, D. J. (2001) Pharmacokinetic interactions of nevirapine and methadone and guidelines for use of nevirapine to treat injection drug users. *Clin. Infect. Dis.* **33**, 1595–1597.

55. Brown, L. S., Sawyer, R. C., Li, R., Cobb, M. N., Colborn, D. C., and Narang, P. K. (1996) Lack of a pharmacologic interaction between rifabutin and methadone in HIV-infected former injecting drug users. *Drug Alcohol Depend.* **43**, 71–77.

56. Kreek, M. J., Garfield, J. W., Gutjahr, C. L., and Giusti, L. M. (1976) Rifampin-induced methadone withdrawal. *N. Engl. J. Med.* **294**, 1104–1106.

57. Trapnell, C. B., Connolly, M., Pentikis, H., Forbes, W. P., and Bettenhausen, D. K. (2007) Absence of effect of oral rifaximin on the pharmacokinetics of ethinyl estradiol/norgestimate in healthy females. *Ann. Pharmacother.* **41**, 222–228.

58. Ma, X., Shah, Y. M., Guo, G. L., Wang, T., Krausz, K. W., Idle, J. R., and Gonzalez, F. J. (2007) Rifaximin is a gut-specific human pregnane X receptor activator. *J. Pharmacol. Exp. Ther.* **322**, 391–398.

59. Shah, Y. M., Ma, X., Morimura, K., Kim, I., and Gonzalez, F. J. (2007) Pregnane X receptor activation ameliorates DSS-induced inflammatory bowel disease via inhibition of NF-kappaB target gene expression. *Am. J. Physiol. Gastrointest. Liver Physiol.* **292**, G1114–G1122.

60. Zhou, C., Tabb, M. M., Nelson, E. L., Grun, F., Verma, S., Sadatrafiei, A., Lin, M., Mallick, S., Forman, B. M., Thummel, K. E., et al. (2006) Mutual repression between steroid and xenobiotic receptor and NF-kappaB signaling pathways links xenobiotic metabolism and inflammation. *J. Clin. Invest.* **116**, 2280–2289.

61. Ouellet, D., Hsu, A., Qian, J., Locke, C. S., Eason, C. J., Cavanaugh, J. H., Leonard, J. M., and Granneman, G. R. (1998) Effect of ritonavir on the pharmacokinetics of ethinyl oestradiol in healthy female volunteers. *Br. J. Clin. Pharmacol.* **46**, 111–116.

62. Dussault, I., Lin, M., Hollister, K., Wang, E. H., Synold, T. W., and Forman, B. M. (2001) Peptide mimetic HIV protease inhibitors are ligands for the orphan receptor SXR. *J. Biol. Chem.* **276**, 33309–33312.

63. Dixit, V., Hariparsad, N., Li, F., Desai, P., Thummel, K. E., and Unadkat, J. D. (2007) Cytochrome P450 enzymes and transporters induced by anti-HIV protease inhibitors in human hepatocytes: implications for predicting clinical drug interactions. *Drug Metab. Dispos.* **35**, 1853–1859.

64. Mukwaya, G., MacGregor, T., Hoelscher, D., Heming, T., Legg, D., Kavanaugh, K., Johnson, P., Sabo, J. P., and McCallister, S. (2005) Interaction of ritonavir-boosted tipranavir with loperamide does not result in loperamide-associated neurologic side effects in healthy volunteers. *Antimicrob. Agents Chemother.* **49**, 4903–4910.

65. Koudriakova, T., Iatsimirskaia, E., Utkin, I., Gangl, E., Vouros, P., Storozhuk, E., Orza, D., Marinina, J., and Gerber, N. (1998) Metabolism of the human immunodeficiency virus protease inhibitors indinavir and ritonavir by human intestinal microsomes and expressed cytochrome P4503A4/3A5: mechanism-based inactivation of cytochrome P4503A by ritonavir. *Drug Metab. Dispos.* **26**, 552–561.

66. Heelon, M. W. and Meade, L. B. (1999) Methadone withdrawal when starting an antiretroviral regimen including nevirapine. *Pharmacotherapy* **19**, 471–472.

67. van Giersbergen, P. L., Halabi, A., and Dingemanse, J. (2002) Single- and multiple-dose pharmacokinetics of bosentan and its interaction with ketoconazole. *Br. J. Clin. Pharmacol.* **53**, 589–595.

68. van Giersbergen, P. L., Gnerre, C., Treiber, A., Dingemanse, J., and Meyer, U. A. (2002) Bosentan, a dual endothelin receptor antagonist, activates the pregnane X nuclear receptor. *Eur. J. Pharmacol.* **450**, 115–121.

69. Kennedy, J. (2005) Herb and supplement use in the US adult population. *Clin. Ther.* **27**, 1847–1858.

70. Peng, C. C., Glassman, P. A., Trilli, L. E., Hayes-Hunter, J., and Good, C. B. (2004) Incidence and severity of potential drug-dietary supplement interactions in primary care patients: an exploratory study of 2 outpatient practices. *Arch. Intern. Med.* **164**, 630–636.

71. Sparreboom, A., Cox, M. C., Acharya, M. R., and Figg, W. D. (2004) Herbal remedies in the United States: potential adverse interactions with anticancer agents. *J. Clin. Oncol.* **22**, 2489–2503.

72. Ruschitzka, F., Meier, P. J., Turina, M., Luscher, T. F., and Noll, G. (2000) Acute heart transplant rejection due to Saint John's wort. *Lancet* **355**, 548–549.

73. Piscitelli, S. C., Burstein, A. H., Chaitt, D., Alfaro, R. M., and Falloon, J. (2000) Indinavir concentrations and St. John's wort. *Lancet* **355**, 547–548.

74. Wang, L. S., Zhou, G., Zhu, B., Wu, J., Wang, J. G., Abd El-Aty, A. M., Li, T., Liu, J., Yang, T. L., Wang, D., et al. (2004) St. John's wort induces both cytochrome P450 3A4-catalyzed sulfoxidation and 2C19-dependent hydroxylation of omeprazole. *Clin. Pharmacol. Ther.* **75**, 191–197.

75. Johne, A., Brockmoller, J., Bauer, S., Maurer, A., Langheinrich, M., and Roots, I. (1999) Pharmacokinetic interaction of digoxin with an herbal extract from St. John's wort (Hypericum perforatum). *Clin. Pharmacol. Ther.* **66**, 338–345.

76. Durr, D., Stieger, B., Kullak-Ublick, G. A., Rentsch, K. M., Steinert, H. C., Meier, P. J., and Fattinger, K. (2000) St. John's wort induces intestinal P-glycoprotein/MDR1 and intestinal and hepatic CYP3A4. *Clin. Pharmacol. Ther.* **68**, 598–604.

77. Frye, R. F., Fitzgerald, S. M., Lagattuta, T. F., Hruska, M. W., and Egorin, M. J. (2004) Effect of St. John's wort on imatinib mesylate pharmacokinetics. *Clin. Pharmacol. Ther.* **76**, 323–329.

78. Schwarz, U. I., Buschel, B., and Kirch, W. (2003) Unwanted pregnancy on self-medication with St. John's wort despite hormonal contraception. *Br. J. Clin. Pharmacol.* **55**, 112–113.

79. Mai, I., Stormer, E., Bauer, S., Kruger, H., Budde, K., and Roots, I. (2003) Impact of St. John's wort treatment on the pharmacokinetics of tacrolimus and mycophenolic acid in renal transplant patients. *Nephrol. Dial. Transplant.* **18**, 819–822.

80. Yue, Q. Y., Bergquist, C., and Gerden, B. (2000) Safety of St. John's wort (Hypericum perforatum). *Lancet* **355**, 576–577.

81. Nebel, A., Schneider, B. J., Baker, R. K., and Kroll, D. J. (1999) Potential metabolic interaction between St. John's wort and theophylline. *Ann. Pharmacother.* **33**, 502.

82. Dresser, G. K., Schwarz, U. I., Wilkinson, G. R., and Kim, R. B. (2003) Coordinate induction of both cytochrome P4503A and MDR1 by St. John's wort in healthy subjects. *Clin. Pharmacol. Ther.* **73**, 41–50.

83. Markowitz, J. S., Donovan, J. L., DeVane, C. L., Taylor, R. M., Ruan, Y., Wang, J. S., and Chavin, K. D. (2003) Effect of St. John's wort on drug metabolism by induction of cytochrome P450 3A4 enzyme. *JAMA* **290** 1500–1504.

84. Kawaguchi, A., Ohmori, M., Tsuruoka, S., Nishiki, K., Harada, K., Miyamori, I., Yano, R., Nakamura, T., Masada, M., and Fujimura, A. (2004) Drug interaction between St. John's wort and quazepam. *Br. J. Clin. Pharmacol.* **58**, 403–410.

85. Schwarz, U. I., Hanso, H., Oertel, R., Miehlke, S., Kuhlisch, E., Glaeser, H., Hitzl, M., Dresser, G. K., Kim, R. B., and Kirch, W. (2007) Induction of intestinal P-glycoprotein by St. John's wort reduces the oral bioavailability of talinolol. *Clin. Pharmacol. Ther.* **81**, 669–678.

86. Portoles, A., Terleira, A., Calvo, A., Martinez, I., and Resplandy, G. (2006) Effects of Hypericum perforatum on ivabradine pharmacokinetics in healthy volunteers: an open-label, pharmacokinetic interaction clinical trial. *J. Clin. Pharmacol.* **46**, 1188–1194.

87. Mathijssen, R. H., Verweij, J., de Bruijn, P., Loos, W. J., and Sparreboom, A. (2002) Effects of St. John's wort on irinotecan metabolism. *J. Natl Cancer. Inst.* **94**, 1247–1249.

88. Gorski, J. C., Huang, S. M., Pinto, A., Hamman, M. A., Hilligoss, J. K., Zaheer, N. A., Desai, M., Miller, M., and Hall, S. D. (2004) The effect of echinacea (Echinacea purpurea root) on cytochrome P450 activity in vivo. *Clin. Pharmacol. Ther.* **75**, 89–100.

89. Piscitelli, S. C., Burstein, A. H., Welden, N., Gallicano, K. D., and Falloon, J. (2002) The effect of garlic supplements on the pharmacokinetics of saquinavir. *Clin. Infect. Dis.* **34**, 234–238.

90. Yin, O. Q., Tomlinson, B., Waye, M. M., Chow, A. H., and Chow, M. S. (2004) Pharmacogenetics and herb–drug interactions: experience with Ginkgo biloba and omeprazole. *Pharmacogenetics* **14**, 841–850.

91. Yuan, C. S., Wei, G., Dey, L., Karrison, T., Nahlik, L., Maleckar, S., Kasza, K., Ang-Lee, M., and Moss, J. (2004) Brief communication: American ginseng reduces warfarin's effect in healthy patients: a randomized, controlled trial. *Ann. Intern. Med.* **141**, 23–27.

92. Butterweck, V. (2003) Mechanism of action of St. John's wort in depression: what is known? *CNS Drugs* **17**, 539–562.

93. Moore, L. B., Goodwin, B., Jones, S. A., Wisely, G. B., Serabjit-Singh, C. J., Willson, T. M., Collins, J. L., and Kliewer, S. A. (2000) St. John's wort induces hepatic drug metabolism through activation of the pregnane X receptor. *Proc. Natl Acad. Sci. USA* **97**, 7500–7502.

94. Godtel-Armbrust, U., Metzger, A., Kroll, U., Kelber, O., and Wojnowski, L. (2007) Variability in PXR-mediated induction of CYP3A4 by commercial preparations and dry extracts of St. John's wort. *Naunyn Schmiedebergs Arch. Pharmacol.* **375**, 377–382.

95. Mueller, S. C., Uehleke, B., Woehling, H., Petzsch, M., Majcher-Peszynska, J., Hehl, E. M., Sievers, H., Frank, B., Riethling, A. K., and Drewelow, B. (2004) Effect of St. John's wort dose and preparations on the pharmacokinetics of digoxin. *Clin. Pharmacol. Ther.* **75**, 546–557.

96. Mueller, S. C., Majcher-Peszynska, J., Uehleke, B., Klammt, S., Mundkowski, R. G., Miekisch, W., Sievers, H., Bauer, S., Frank, B., Kundt, G., et al. (2006) The extent of induction of CYP3A by St. John's wort varies among products and is linked to hyperforin dose. *Eur. J. Clin. Pharmacol.* **62**, 29–36.

97. Tirona, R. G. and Bailey, D. G. (2006) Herbal product-drug interactions mediated by induction. *Br. J. Clin. Pharmacol.* **61**, 677–681.

98. Zhang, P., Noordine, M. L., Cherbuy, C., Vaugelade, P., Pascussi, J. M., Duee, P. H., and Thomas, M. (2006) Different activation patterns of rat xenobiotic metabolism genes by two constituents of garlic. *Carcinogenesis* **27**, 2090–2095.

99. Fisher, C. D., Augustine, L. M., Maher, J. M., Nelson, D. M., Slitt, A. L., Klaassen, C. D., Lehman-McKeeman, L. D., and Cherrington, N. J. (2007) Induction of drug-metabolizing enzymes by garlic and allyl sulfide compounds via activation of constitutive androstane receptor and nuclear factor E2-related factor 2. *Drug Metab. Dispos.* **35**, 995–1000.

100. Markowitz, J. S., Donovan, J. L., Lindsay DeVane, C., Sipkes, L., and Chavin, K. D. (2003) Multiple-dose administration of Ginkgo biloba did not affect cytochrome P-450 2D6 or 3A4 activity in normal volunteers. *J. Clin. Psychopharmacol.* **23**, 576–581.

101. Jiang, X., Williams, K. M., Liauw, W. S., Ammit, A. J., Roufogalis, B. D., Duke, C. C., Day, R. O., and McLachlan, A. J. (2004) Effect of St. John's wort and ginseng on the pharmacokinetics and pharmacodynamics of warfarin in healthy subjects. *Br. J. Clin. Pharmacol.* **57**, 592–599.

102. Plotnikoff, G. A., McKenna, D., Watanabe, K., and Blumenthal, M. (2004) Ginseng and warfarin interactions. *Ann. Intern. Med.* **141**, 1917–2024.

103. Hardman, J. G., Limbird, L. E., and Gilman, A. G. (2001) *Goodman and Gilman's The Pharmacological Basis of Therapeutics*, 10th edn., McGraw-Hill, New York.

104. Sahi, J., Milad, M. A., Zheng, X., Rose, K. A., Wang, H., Stilgenbauer, L., Gilbert, D., Jolley, S., Stern, R. H., and LeCluyse, E. L. (2003) Avasimibe induces CYP3A4 and

multiple drug resistance protein 1 gene expression through activation of the pregnane X receptor. *J. Pharmacol. Exp. Ther.* **306**, 1027–1034.

105. El-Sankary, W., Gibson, G. G., Ayrton, A., and Plant, N. (2001) Use of a reporter gene assay to predict and rank the potency and efficacy of CYP3A4 inducers. *Drug Metab. Dispos.* **29**, 1499–1504.

106. Backman, J. T., Olkkola, K. T., Ojala, M., Laaksovirta, H., and Neuvonen, P. J. (1996) Concentrations and effects of oral midazolam are greatly reduced in patients treated with carbamazepine or phenytoin. *Epilepsia* **37**, 253–257.

107. Wang, Z., Gorski, J. C., Hamman, M. A., Huang, S. M., Lesko, L. J., and Hall, S. D. (2001) The effects of St. John's wort (Hypericum perforatum) on human cytochrome P450 activity. *Clin. Pharmacol. Ther.* **70**, 317–326.

108. Back, D. J., Bates, M., Bowden, A., Breckenridge, A. M., Hall, M. J., Jones, H., MacIver, M., Orme, M., Perucca, E., Richens, A., et al. (1980) The interaction of phenobarbital and other anticonvulsants with oral contraceptive steroid therapy. *Contraception* **22**, 495–503.

109. Backman, J. T., Olkkola, K. T., and Neuvonen, P. J. (1996) Rifampin drastically reduces plasma concentrations and effects of oral midazolam. *Clin. Pharmacol. Ther.* **59**, 7–13.

110. Dingemanse, J., Schaarschmidt, D., and van Giersbergen, P. L. (2003) Investigation of the mutual pharmacokinetic interactions between bosentan, a dual endothelin receptor antagonist, and simvastatin. *Clin. Pharmacokinet.* **42**, 293–301.

111. Mouly, S., Lown, K. S., Kornhauser, D., Joseph, J. L., Fiske, W. D., Benedek, I. H., and Watkins, P. B. (2002) Hepatic but not intestinal CYP3A4 displays dose-dependent induction by efavirenz in humans. *Clin. Pharmacol. Ther.* **72**, 1–9.

112. Rosenfeld, W. E., Doose, D. R., Walker, S. A., and Nayak, R. K. (1997) Effect of topiramate on the pharmacokinetics of an oral contraceptive containing norethindrone and ethinyl estradiol in patients with epilepsy. *Epilepsia* **38**, 317–323.

113. Jones, S. A., Moore, L. B., Shenk, J. L., Wisely, G. B., Hamilton, G. A., McKee, D. D., Tomkinson, N. C., LeCluyse, E. L., Lambert, M. H., Willson, T. M., et al. (2000) The pregnane X receptor: a promiscuous xenobiotic receptor that has diverged during evolution. *Mol. Endocrinol.* **14**, 27–39.

114. Prueksaritanont, T., Vega, J. M., Zhao, J., Gagliano, K., Kuznetsova, O., Musser, B., Amin, R. D., Liu, L., Roadcap, B. A., Dilzer, S., et al. (2001) Interactions between simvastatin and troglitazone or pioglitazone in healthy subjects. *J. Clin. Pharmacol.* **41**, 573–581.

115. Goodwin, B., Redinbo, M. R., and Kliewer, S. A. (2002) Regulation of cyp3a gene transcription by the pregnane x receptor. *Annu. Rev. Pharmacol. Toxicol.* **42**, 1–23.

116. Drocourt, L., Pascussi, J. M., Assenat, E., Fabre, J. M., Maurel, P., and Vilarem, M. J. (2001) Calcium channel modulators of the dihydropyridine family are human pregnane X receptor activators and inducers of CYP3A, CYP2B, and CYP2C in human hepatocytes. *Drug Metab. Dispos.* **29**, 1325–1331.

117. Kocarek, T. A., Dahn, M. S., Cai, H., Strom, S. C., and Mercer-Haines, N. A. (2002) Regulation of CYP2B6 and CYP3A expression by hydroxymethylglutaryl coenzyme A inhibitors in primary cultured human hepatocytes. *Drug Metab. Dispos.* **30**, 1400–1405.

118. Sahi, J., Black, C. B., Hamilton, G. A., Zheng, X., Jolley, S., Rose, K. A., Gilbert, D., LeCluyse, E. L., and Sinz, M. W. (2003) Comparative effects of thiazolidinediones on in vitro P450 enzyme induction and inhibition. *Drug Metab. Dispos.* **31**, 439–446.

119. Kumar, G. N., Rodrigues, A. D., Buko, A. M., and Denissen, J. F. (1996) Cytochrome P450-mediated metabolism of the HIV-1 protease inhibitor ritonavir (ABT-538) in human liver microsomes. *J. Pharmacol. Exp. Ther.* **277**, 423–431.

120. Tirona, R. G., Leake, B. F., Wolkoff, A. W., and Kim, R. B. (2003) Human organic anion transporting polypeptide-C (SLC21A6) is a major determinant of rifampin-mediated pregnane X receptor activation. *J. Pharmacol. Exp. Ther.* **304**, 223–228.

121. Niemi, M., Kivisto, K. T., Diczfalusy, U., Bodin, K., Bertilsson, L., Fromm, M. F., and Eichelbaum, M. (2006) Effect of SLCO1B1 polymorphism on induction of CYP3A4 by rifampicin. *Pharmacogenet. Genomics* **16**, 565–568.

122. Tirona, R. G. and Kim, R. B. (2005) Nuclear receptors and drug disposition gene regulation. *J. Pharm. Sci.* **94**, 1169–1186.

123. Echchgadda, I., Song, C. S., Roy, A. K., and Chatterjee, B. (2004) Dehydroepiandrosterone sulfotransferase is a target for transcriptional induction by the vitamin D receptor. *Mol. Pharmacol.* **65**, 720–729.

124. Hesse, L. M., Venkatakrishnan, K., Court, M. H., von Moltke, L. L., Duan, S. X., Shader, R. I., and Greenblatt, D. J. (2000) CYP2B6 mediates the in vitro hydroxylation of bupropion: potential drug interactions with other antidepressants. *Drug Metab. Dispos.* **28**, 1176–1183.

125. Faucette, S. R., Hawke, R. L., Lecluyse, E. L., Shord, S. S., Yan, B., Laethem, R. M., and Lindley, C. M. (2000) Validation of bupropion hydroxylation as a selective marker of human cytochrome P450 2B6 catalytic activity. *Drug Metab. Dispos.* **28**, 1222–1230.

126. Loboz, K. K., Gross, A. S., Williams, K. M., Liauw, W. S., Day, R. O., Blievernicht, J. K., Zanger, U. M., and McLachlan, A. J. (2006) Cytochrome P450 2B6 activity as measured by bupropion hydroxylation: effect of induction by rifampin and ethnicity. *Clin. Pharmacol. Ther.* **80**, 75–84.

127. Hogeland, G. W., Swindells, S., McNabb, J. C., Kashuba, A. D., Yee, G. C., and Lindley, C. M. (2007) Lopinavir/ritonavir reduces bupropion plasma concentrations in healthy subjects. *Clin. Pharmacol. Ther.* **81**, 69–75.

128. Heimark, L. D., Gibaldi, M., Trager, W. F., O'Reilly, R. A., and Goulart, D. A. (1987) The mechanism of the warfarin-rifampin drug interaction in humans. *Clin. Pharmacol. Ther.* **42**, 388–394.

129. Rengelshausen, J., Banfield, M., Riedel, K. D., Burhenne, J., Weiss, J., Thomsen, T., Walter-Sack, I., Haefeli, W. E., and Mikus, G. (2005) Opposite effects of short-term and long-term St. John's wort intake on voriconazole pharmacokinetics. *Clin. Pharmacol. Ther.* **78**, 25–33.

130. Floyd, M. D., Gervasini, G., Masica, A. L., Mayo, G., George, A. L., Jr., Bhat, K., Kim, R. B., and Wilkinson, G. R. (2003) Genotype-phenotype associations for common CYP3A4 and CYP3A5 variants in the basal and induced metabolism of midazolam in European- and African-American men and women. *Pharmacogenetics* **13**, 595–606.

131. Green, M. D. and Tephly, T. R. (1996) Glucuronidation of amines and hydroxylated xenobiotics and endobiotics catalyzed by expressed human UGT1.4 protein. *Drug Metab. Dispos.* **24**, 356–363.

132. van der Lee, M. J., Dawood, L., ter Hofstede, H. J., de Graaff-Teulen, M. J., van Ewijk-Beneken Kolmer, E. W., Caliskan-Yassen, N., Koopmans, P. P., and Burger, D. M. (2006) Lopinavir/ritonavir reduces lamotrigine plasma concentrations in healthy subjects. *Clin. Pharmacol. Ther.* **80**, 159–168.

133. Barbier, O., Turgeon, D., Girard, C., Green, M. D., Tephly, T. R., Hum, D. W., and Belanger, A. (2000) 3'-Azido-3'-deoxythymidine (AZT) is glucuronidated by human UDP-glucuronosyltransferase 2B7 (UGT2B7). *Drug Metab. Dispos.* **28**, 497–502.

134. Burger, D. M., Meenhorst, P. L., Koks, C. H., and Beijnen, J. H. (1993) Pharmacokinetic interaction between rifampin and zidovudine. *Antimicrob. Agents Chemother.* **37**, 1426–1431.

135. Gallicano, K. D., Sahai, J., Shukla, V. K., Seguin, I., Pakuts, A., Kwok, D., Foster, B. C., and Cameron, D. W. (1999) Induction of zidovudine glucuronidation and amination pathways by rifampicin in HIV-infected patients. *Br. J. Clin. Pharmacol.* **48**, 168–179.

136. Burk, O., Arnold, K. A., Geick, A., Tegude, H., and Eichelbaum, M. (2005) A role for constitutive androstane receptor in the regulation of human intestinal MDR1 expression. *Biol. Chem.* **386**, 503–513.

137. Greiner, B., Eichelbaum, M., Fritz, P., Kreichgauer, H. P., von Richter, O., Zundler, J., and Kroemer, H. K. (1999) The role of intestinal P-glycoprotein in the interaction of digoxin and rifampin. *J. Clin. Invest.* **104**, 147–153.

138. Fromm, M. F., Kauffmann, H. M., Fritz, P., Burk, O., Kroemer, H. K., Warzok, R. W., Eichelbaum, M., Siegmund, W., and Schrenk, D. (2000) The effect of rifampin treatment on intestinal expression of human MRP transporters. *Am. J. Pathol.*, **157**, 1575–1580.

139. Kauffmann, H. M., Keppler, D., Gant, T. W., and Schrenk, D. (1998) Induction of hepatic mrp2 (cmrp/cmoat) gene expression in nonhuman primates treated with rifampicin or tamoxifen. *Arch. Toxicol.* **72**, 763–768.

140. Bolt, H. M., Bolt, M., and Kappus, H. (1977) Interaction of rifampicin treatment with pharmacokinetics and metabolism of ethinyloestradiol in man. *Acta Endocrinol. (Copenh.)* **85**, 189–197.

141. Eich-Hochli, D., Oppliger, R., Golay, K. P., Baumann, P., and Eap, C. B. (2003) Methadone maintenance treatment and St. John's wort—a case report. *Pharmacopsychiatry* **36**, 35–37.

142. Fromm, M. F., Eckhardt, K., Li, S., Schanzle, G., Hofmann, U., Mikus, G., and Eichelbaum, M. (1997) Loss of analgesic effect of morphine due to coadministration of rifampin. *Pain* **72**, 261–267.

143. Bauer, B., Yang, X., Hartz, A. M., Olson, E. R., Zhao, R., Kalvass, J. C., Pollack, G. M., and Miller, D. S. (2006) In vivo activation of human pregnane X receptor tightens the blood–brain barrier to methadone through P-glycoprotein up-regulation. *Mol. Pharmacol.* **70**, 1212–1219.

144. Barone, G. W., Gurley, B. J., Ketel, B. L., Lightfoot, M. L., and Abul-Ezz, S. R. (2000) Drug interaction between St. John's wort and cyclosporine. *Ann. Pharmacother.* **34**, 1013–1016.

145. Breidenbach, T., Hoffmann, M. W., Becker, T., Schlitt, H., and Klempnauer, J. (2000) Drug interaction of St. John's wort with cyclosporine. *Lancet* **355**, 1912.

146. Patsalos, P. N. and Perucca, E. (2003) Clinically important drug interactions in epilepsy: general features and interactions between antiepileptic drugs. *Lancet Neurol.* **2**, 347–356.

147. Back, D. J. (2006) Drug–drug interactions that matter. *Top. HIV Med.* **14**, 88–92.

148. Reilly, P. A., Inaba, T., Kadar, D., and Endrenyi, L. (1978) Enzyme induction following a single dose of amobarbital in dogs. *J. Pharmacokinet. Biopharm.* **6**, 305–313.

149. Bertilsson, L., Hojer, B., Tybring, G., Osterloh, J., and Rane, A. (1980) Autoinduction of carbamazepine metabolism in children examined by a stable isotope technique. *Clin. Pharmacol. Ther.* **27**, 83–88.

150. Tran, J. Q., Kovacs, S. J., McIntosh, T. S., Davis, H. M., and Martin, D. E. (1999) Morning spot and 24-hour urinary 6 beta-hydroxycortisol to cortisol ratios: intraindividual variability and correlation under basal conditions and conditions of CYP 3A4 induction. *J. Clin. Pharmacol.* **39**, 487–494.

151. Wang, H., Huang, H., Li, H., Teotico, D. G., Sinz, M., Baker, S. D., Staudinger, J., Kalpana, G., Redinbo, M. R., and Mani, S. (2007) Activated pregnenolone X-receptor is a target for ketoconazole and its analogs. *Clin. Cancer. Res.* **13**, 2488–2495.

152. Huang, H., Wang, H., Sinz, M., Zoeckler, M., Staudinger, J., Redinbo, M. R., Teotico, D. G., Locker, J., Kalpana, G. V., and Mani, S. (2007) Inhibition of drug metabolism by blocking the activation of nuclear receptors by ketoconazole. *Oncogene* **26**, 258–268.

153. Synold, T. W., Dussault, I., and Forman, B. M. (2001) The orphan nuclear receptor SXR coordinately regulates drug metabolism and efflux. *Nat. Med.* **7**, 584–590.

154. Zhou, C., Poulton, E. J., Grun, F., Bammler, T. K., Blumberg, B., Thummel, K. E., and Eaton, D. L. (2007) The dietary isothiocyanate sulforaphane is an antagonist of the human steroid and xenobiotic nuclear receptor. *Mol. Pharmacol.* **71**, 220–229.

155. Carter, B. A., Prendergast, D. R., Taylor, O. A., Zimmerman, T. L., Furstenberg, R. V., Moore, D. D., and Karpen, S. J. (2007) Stigmasterol, a soy lipid-derived phytosterol, is an antagonist of the bile acid nuclear receptor FXR. *Pediatr. Res.* **62**, 301–306.

156. Shah, S. C., Sharma, R. K., Hemangini, and Chitle, A. R. (1981) Rifampicin induced osteomalacia. *Tubercle* **62**, 207–209.

157. Pascussi, J. M., Robert, A., Nguyen, M., Walrant-Debray, O., Garabedian, M., Martin, P., Pineau, T., Saric, J., Navarro, F., Maurel, P., et al. (2005) Possible involvement of pregnane X receptor-enhanced CYP24 expression in drug-induced osteomalacia. *J. Clin. Invest.* **115**, 177–186.

158. Zhou, C., Assem, M., Tay, J. C., Watkins, P. B., Blumberg, B., Schuetz, E. G., and Thummel, K. E. (2006) Steroid and xenobiotic receptor and vitamin D receptor crosstalk mediates CYP24 expression and drug-induced osteomalacia. *J. Clin. Invest.* **116**, 1703–1712.

159. Terzolo, M., Borretta, G., Ali, A., Cesario, F., Magro, G., Boccuzzi, A., Reimondo, G., and Angeli, A. (1995) Misdiagnosis of Cushing's syndrome in a patient receiving rifampicin therapy for tuberculosis. *Horm. Metab. Res.* **27**, 148–150.

160. Zhai, Y., Pai, H. V., Zhou, J., Amico, J. A., Vollmer, R. R., and Xie, W. (2007) Activation of pregnane X receptor disrupts glucocorticoid and mineralocorticoid homeostasis. *Mol. Endocrinol.* **21**, 138–147.

161. Eiris-Punal, J., Del Rio-Garma, M., Del Rio-Garma, M. C., Lojo-Rocamonde, S., Novo-Rodriguez, I., and Castro-Gago, M. (1999) Long-term treatment of children with epilepsy with valproate or carbamazepine may cause subclinical hypothyroidism. *Epilepsia* **40**, 1761–1766.

162. Kodama, S., Tanaka, K., Konishi, H., Momota, K., Nakasako, H., Nakayama, S., Yagi, J., and Koderazawa, K. (1989) Supplementary thyroxine therapy in patients with hypothyroidism induced by long-term anticonvulsant therapy. *Acta Paediatr. Jpn.* **31**, 555–562.

163. Takasu, N., Kinjou, Y., Kouki, T., Takara, M., Ohshiro, Y., and Komiya, I. (2006) Rifampin-induced hypothyroidism. *J. Endocrinol. Invest.* **29**, 645–649.

164. Kim, D. L., Song, K. H., Lee, J. H., Lee, K. Y., and Kim, S. K. (2007) Rifampin-induced hypothyroidism without underlying thyroid disease. *Thyroid* **17**, 793–795.

165. Qatanani, M., Zhang, J., and Moore, D. D. (2005) Role of the constitutive androstane receptor in xenobiotic-induced thyroid hormone metabolism. *Endocrinology* **146**, 995–1002.

166. Wong, H., Lehman-McKeeman, L. D., Grubb, M. F., Grossman, S. J., Bhaskaran, V. M., Solon, E. G., Shen, H. S., Gerson, R. J., Car, B. D., Zhao, B., et al. (2005) Increased hepatobiliary clearance of unconjugated thyroxine determines DMP 904-induced alterations in thyroid hormone homeostasis in rats. *Toxicol. Sci.* **84**, 232–242.

167. Vourvahis, M. and Kashuba, A. D. (2007) Mechanisms of pharmacokinetic and pharmacodynamic drug interactions associated with ritonavir-enhanced tipranavir. *Pharmacotherapy* **27**, 888–909.

168. Wipf, P., Gong, H., Janjic, J. M., Li, S., Day, B. W., and Xie, W. (2007) New opportunities for pregnane X receptor (PXR) targeting in drug development. Lessons from enantio- and species-specific PXR ligands identified from a discovery library of amino acid analogues. *Mini. Rev. Med. Chem.* **7**, 617–625.

9

GENETIC VARIANTS OF XENOBIOTIC RECEPTORS AND THEIR IMPLICATIONS IN DRUG METABOLISM AND PHARMACOGENETICS

JATINDER LAMBA AND ERIN G. SCHUETZ

Department of Pharmaceutical Sciences, St. Jude Children's Research Hospital, Memphis, TN, USA

9.1 PXR (PREGNANE X RECEPTOR) BACKGROUND

Human *PXR* was originally cloned in 1998 by Lehmann et al. [1] and simultaneously by Blumberg et al. [2] as SXR (steroid and xenobiotic receptor) and by Bertilsson et al. [3] and termed as PAR. Ligand-activated PXR mediates steroid and drug induction of many genes in liver and intestine. The initial target gene identified was *CYP3A4*, although many other drug detoxification genes including *MDR1/ABCB1* (encoding P-glycoprotein), *CYP2B6*, *CYP2C9*, and others have now been identified as targets of ligand-activated PXR. CYP3A4 catalyzes the oxidative metabolism of half of clinically administered drugs and P-gp effluxes many drugs from human hepatocytes and enterocytes. Hence, administration of PXR ligands and subsequent induction of CYP3A4 and P-gp lead to accelerated drug clearance of their substrates by either autoinduction or transinduction, one of the major mechanisms for drug interactions. There are numerous reports in the literature of PXR ligands, in particular rifampin, causing accelerated CYP3A4 substrate clearance, and requiring dose adjustments to maintain therapeutic efficacy.

Nuclear Receptors in Drug Metabolism   Edited by Wen Xie
Copyright © 2009 John Wiley & Sons, Inc.

There is significant variation among humans in the magnitude of rifampin-mediated induction of CYP3A4 and P-gp in humans [4–7]. Thus, it was reasoned that *PXR* genetic variation could be a contributor to individual variation in the extent of CYP3A4 and P-gp induction. It was also possible that sequence variations in *PXR* could affect basal expression of CYP3A or P-gp or a growing number of other ADME (absorption, distribution, metabolism, and excretion) genes since a recent study demonstrated that PXR contributes to regulation of basal expression of a variety of human CYP genes in primary cultures of human hepatocytes [8]. These authors showed that after infection with Ad hPXR–siRNA, both basal and ligand-activated expression of CYP2A6, CYP2C8, CYP3A4, and CYP3A5 levels were significantly affected.

Besides PXR's well-known role in mediating ligand activation of ADME gene transcription, there are growing reports of additional biological functions ascribed to this receptor (Figure 9.1). For example, PXR ligand activation can cause hepatic steatosis [9], decrease hepatic fibrosis [10], and decrease the inflammatory response [11]. Additional reports have demonstrated that PXR can cross talk with VDR signaling [12]. PXR KO mice have revealed a role for PXR in glucocorticoid and mineralocorticoid homeostasis [13]. Finally, activated PXR appears to cross talk with both FOXO1 and FOXA2 altering glucose homeostasis [14, 15]. Hence, variation in PXR expression and ligand activation could affect a variety of biological pathways.

Since PXR expression in liver is highly correlated with CYP3A4 expression [16], a number of groups reasoned that the large population variation in basal and

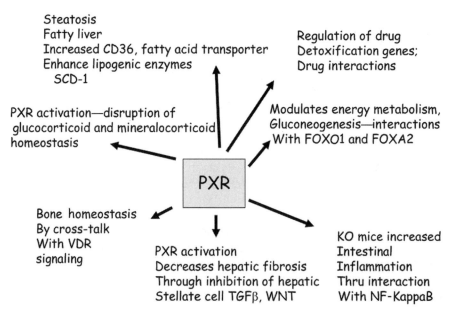

FIGURE 9.1   Biological functions associated with PXR ligand activation.

inducible CYP3A, and other ADME genes, might be explained by genetic variation in PXR [17–19]. This seemed like a reasonable hypothesis, particularly for CYP3A4, since cis-genetic variation in *CYP3A4*, the prototypical PXR target gene [20] cannot explain human variation in CYP3A4 expression. Moreover, there are no sequence variations in PXR binding sites in the *CYP3A4* gene [21]. In addition, because Blumberg et al. [2] proposed that PXR acts as a xenobiotic "sensor" that mediates the physiological response of multiple drug detoxification genes that have putative PXR binding elements, identification of functional polymorphisms in *PXR* might not only explain variation in induction of CYP3A4 and MDR1/P-gp, but might also be more generally applicable to variation in response to PXR ligands of many drug-metabolizing enzymes and transporters regulated by PXR [22].

Thus, several groups resequenced the *PXR* coding region and splice site junctions and subsequently determined whether common nucleotide variation could explain basal or inducible target gene expression. The idea that human PXR's LBD (ligand binding domain), in particular, might show sequence variation was an attractive one because there are significant differences in the PXR LBD sequence between species that is responsible for the well-known species differences in PXR ligand activation of target genes [1]. Indeed, it has been demonstrated that the LBD of PXR is under significant natural selection [23], further supporting the hypothesis that humans might show LBD genetic variation.

9.2 *PXR* GENE STRUCTURE

The *PXR* gene structure was first determined by Zhang et al. [17]. Zhang et al. used a mouse PXR mRNA (AF031814) sequence to search the human EST database for an *hPXR* clone (AA679591). This probe was used to screen a human genomics BAC library (Research Genetics). One BAC with a 150 kb insert was determined to contain the entire hPXR coding region (AF061056) by PCR amplification using primers for the 3′-UTR and the 5′-UTR of *hPXR*. The *hPXR* gene was sequenced by hybridizing plasmid subclone libraries of the *hPXR* BAC and sequencing the positive clones. However, since this approach would not guarantee isolation of all noncoding intronic and upstream regions of the gene, a BAC shotgun sequencing strategy was also utilized. Assembly of these sequences in combination with the raw shotgun sequences from Baylor Sequencing Center (AC069444, BAC clone RP11-169N13) generated a full sequence of the entire *PXR* gene. The human *PXR* gene is 38 kb, localized on chromosome (chr) 3q12q13.3 (Locus Link 8856, Golden Path, May 2004 Build), and comprises nine exons, with exons 2–9 encoding PXR.

9.3 PXR ALTERNATIVE mRNAs

Alignment of all ests in GenBank revealed that there are a number of est transcripts that differ in the length of the amino terminal end. The primary transcript, in terms

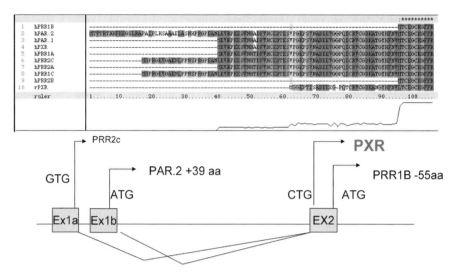

FIGURE 9.2   PXR has multiple isoforms with different amino terminal ends. (*a*) Amino acid alignment of PXR est isoforms. PXR is the most abundant isoform. (*b*) Generation of PXR, PAR.2, and PRR isoforms utilizing different initiation codons in exons 1a, 1b, and 2. See color insert.

of abundance (number of ests), is hPXR (Figure 9.2). Whereas hPXR contains exon 1a, and utilizes the initiator CTG in exon 2, hPAR.2 is created by alternative splicing with exon 1b, and uses an alternative ATG initiator methionine codon in exon 1b. Thus, compared to hPXR, hPAR.2 is a larger version with an additional 177 nt at the amino terminal end encoding an additional 39 amino acids. Bertilsson's group compared the ligand activation of hPXR.1 and hPAR.2 and found no significant difference in transactivation of a CYP3A4-luciferase reporter plasmid [3]. There are several additional PXR ests termed as hPRR that utilize an initiator GTG in exon 1a to generate a transcript that is 22 amino acids longer than PXR at the amino terminal end. There have been no reports on tissue distribution or functionality of these transcripts. Nonetheless, there is precedent for nuclear hormone receptors with longer amino terminal ends to have unique tissue distribution and/or function. For example, PPARγ2 differs from PPARγ1 in having 37 additional amino acids at the amino terminal end. This isoform also differs from PPARγ1 in its tissue distribution. In spite of the fact that PPARγ1 is much more highly and ubiquitously expressed, mice lacking PPARγ2 show impaired development of adipose tissue and insulin sensitivity [24].

Alternative splicing of hPXR (also referred to as hPXR.1) also gives rise to two additional hPXR transcripts in a variety of tissues [16, 25]; PXR.2 (splice variant 2, SV2) has a 111 bp deletion at the 5′ end of exon 5. This alternative PXR mRNA was originally detected in breast tissue and results in an in-frame deletion of 37 amino acids (174–210) from the LBD of PXR protein [26]. PXR.2 is produced by usage of a cryptic splice acceptor site within exon 5. Human PXR.3 (SV3) was identical to an alternative PXR transcript in mouse, mPXR.2 [27], with deletion of 123 bp from the 5' end of exon 5 and loss of 41 amino acids from the LBD of PXR.

Human variation in expression of PXR.1, PAR.2, and PXR.2 was reported in hepatic and intestinal tissues. Studies on the abundance of PXR transcripts revealed that in liver PXR.2 mRNAs represent 6.7% and PXR.3 represents 0.33% of total PXR transcripts [16]. Screening of a panel of 36 human tissues for PXR transcripts revealed that PXR, PXR.2, and PXR.3 were each present in heart, colon, stomach, adrenal, bone marrow, and fetal brain, but PXR is always the most abundant transcript [16]. PXR and PXR splice variants were also detected in human brain thalamus, spinal cord, pons, and medulla. Notably, the brain contains endogenous PXR ligands including dihydroepiandrosterone (DHEA), dehydroepiandrosterone sulfate (DHEAS), allopregnanolone, and pregnenolone. PXR in the brain could also be activated by drugs that cross the blood–brain barrier and brain acting drugs such as nicotine [16]. Notably, two comprehensive studies of PXR mRNA expression in mouse brain and various regions of mouse brain [28, 29] failed to detect mPXR. This stands in contrast to *in situ* hybridization results from the Allen Brain Institute, which found that PXR was highly expressed throughout the mouse brain (http://www.brainatlas.org/aba/). These discrepant results need to be clarified since brain PXR could alter the concentration of endogenous brain chemicals, protect from toxicity of PXR agonists that could cross the blood–brain barrier, and because activation of PXR by allopregnanolone has been suggested to play a protective role in Niemann-Pick C disease [30]. Importantly, the mouse and human PXR promoters are vastly different in size (mouse PXR, 2.9 kb; human PXR, 16 kb), and it will be important to determine whether the reported differences between labs in brain PXR expression are influenced at all by species differences in the PXR promoter.

In silico analysis of the splice variants to predict functional consequence has been done [16]. Alignment of PXR.1, PXR.2, and PXR.3 with other nuclear hormone receptor family members revealed that PXR.2 and PXR.3 lack portions of the LBD that are also absent in FXR, LXRβ, CAR, LXRα, RXRα, and ERα. This finding made it difficult to predict whether PXR.2 and PXR.3 would bind ligands because bile acids and oxysterols bind to the smaller LBDs in FXR and LXR, respectively. However, five of the amino acids missing from PXR.2 and PXR.3 are part of the PXR ligand binding pocket suggesting these alternative mRNAs might not bind ligands.

The functional consequence of PXR.2 has been briefly examined in cell-based assays. Compared to PXR, PXR.2 had marked reduction of basal and ligand (rifampin and corticosterone) induced transactivation activity of a CYP3A4 reporter in LS174 T cells [18], and, similarly, the orthologous mouse isoform mPXR.2 showed reduced response to ligands compared to mPXR.1 [27]. Likewise, some of the UGT1As (PXR targets) were induced by rifampin in Caco-2 cells transfected with PXR.1 and PAR.2, but not with hPXR.2 [31].

Fukuen et al. [25] identified seven additional splice variants of PXR in human liver. Seventy percent of the transcripts had an alteration in exon 5. However, they appear to be expressed at levels significantly lower than PXR or PXR.2 or PXR.3 and have not been functionally characterized.

9.4   GENETIC VARIANTS IN *PXR*'s EXONS AND THEIR FUNCTIONAL CONSEQUENCES

Several groups devised and implemented deep resequencing strategies for human *PXR*'s exons. Two hundred twenty-seven *PXR* SNPs (single nucleotide polymorphisms) in Nr1I2 have been deposited at dbSNP (http://www.ncbi.nlm.nih.gov/ SNP/snp_ref.cgi?chooseRs=all&locusId=8856&mrna=NM_003889.3&ctg=NT_ 005612.15&prot=NP_003880.3&orien). This compendium includes many, but not all, sequence variations reported.

Multiple SNPs in *PXR* exons have been reported encoding nonsynonymous changes (Table 9.1, Figure 9.3). A number of *PXR* SNPs are African American specific and were never found in European or Asian populations (E18K, P27S, D163G, and A370T). Once *PXR* variants had been discovered, some investigators used a diverse set of *in silico* approaches to analyze the resequencing data and identify, among all variants, those most likely to have an effect on function and to pursue in association studies and in biochemical tests. For amino acid substitutions, well-established online tools were used such as SIFT (sorting intolerant from tolerant) [34] and Polyphen (www.bork.embl-heidelberg.de/PolyPhen) that combine information on the biochemical and structural properties of amino acids and sequence conservation to estimate the probability of an effect on function. For noncoding regions, SNPs located at highly conserved positions in the sequence and/or in a predicted transcription binding site, which is part of a high probability cluster, were also assigned a higher priority for future genotype association studies. Evolutionary tests were also performed on the resequencing data mainly with the goal of investigating the role of natural selection in shaping the observed patterns of variation that would improve the assessment of the probability of an effect on function.

SNPs were picked for association studies on the basis of allele frequencies and results from *in silico* predictions, and association studies were carried out in various populations, including association of *PXR* variants with PXR and CYP3A4 expression in human livers and primary human hepatocytes. In addition, the *PXR* variants were cloned, expressed, and their functional consequence determined in cell-based assays.

P27S (*PXR**2) and G36R (*PXR**3) SNPs in exon 2 were observed by Zhang et al. [17] and Hustert et al. [18]. The proline to serine change could affect hydrophobicity at position 27 and analysis by protein phosphorylation prediction analysis indicated that this change produces a serine phosphorylation site having a high score. *PXR**2 (P27S) was population specific—it was absent in Caucasians but observed in African Americans with a frequency ranging from 15% to 20%. The *PXR**2 allele had no significant affect on the hepatic CYP3A4 content [17]. *PXR**3 (G36R) was present in a region of lower evolutionary conservation, suggesting it might be less functionally important to PXR function. Further gel shift assays using a CYP3A4–ER6 oligonucleotide demonstrated that the amount of PXRE–PXR–RXR complex formed when *PXR**3 and *PXR**2 proteins were used was slightly greater than that when *PXR**1 was used. However, transient transactivation assays revealed no difference between

TABLE 9.1 *PXR* Allele Nomenclature[a]

	Position in AF364606	Location	Effect	Functional Consequences	Population Allele Frequency	Reference
*PXR*2*	C79T	Exon 2	P27S, DBD	Hepatic CYP3A4 not different; no effect on basal or inducible CYP3A4	C = 0.0, AA = 0.20	[17]
*PXR*3*	G106A	Exon 2	G36R, DBD		C = 0.01, AA = 0.03	[17]
*PXR*4*	G4321A	Exon 4	R122Q, conserved	Reduced affinity in EMSA for PXR binding sequence, reduced ligand activation in transient transfection assays	C = 0.01, AA = 0.0	[17]
	T4448C	Exon 4	Syn		C = 0.01, AA = 0.05	[17]
	C5458T	Exon 5	Syn		C = 0.01, AA = 0.0	[17]
	A7635G	Intron 5		Rif induced intestinal CYP3A4 protein: GG>AA; IBD	C = 0.35, AA = 0.77	[17, 32]
	G7767A	Exon 6	Syn		C = 0.01, AA = 0.05	[17]
	C8055T	Intron 6		Rif induced intestinal CYP3A4 protein: TT, TC>CC; IBD	C = 0.15, AA = 0.18	[17, 32]
	C10620T	3′-UTR		Hepatocytes 6-β-testosterone hydroxylation following Rif TT, TC<CC; IBD	C = 0.11, AA = 0.14	[17, 32]
	G10799A	3′-UTR		Hepatocytes 6-β-testosterone hydroxylation following Rif AA, AG<GG	C = 0.13, AA = 0.14	[17]
	A11156C	3′-UTR		Intestinal P-gp protein CC, AC<AA	C = 0.16, AA = 0.33	[17]
	T11193C	3′-UTR		Intestinal P-gp protein CC, TC<TT	C = 0.16, AA = 0.30	[17]
*PXR*9*	G52A	Exon 2	E18K	Same as WT in transactivation assays, basal and Rif and corticosterone	C = 0.0, AA = 0.014	[18]

TABLE 9.1 *PXR* Allele Nomenclature[a]

	Position in AF364606	Location	Effect	Functional Consequences	Population Allele Frequency	Reference
PXR*2	C79T	Exon 2	P27S, DBD	Same as WT in transactivation assays, basal and Rif and corticosterone	C = 0.0, AA = 0.15	[18]
PXR*3	G106A	Exon 2	G36R, DBD	Same as WT in transactivation assays for basal and Rif but 40% increase with corticosterone	C = 0.03, AA = 0.0	[18]
PXR*10	G4374A	Exon 4	V140M	50% lower protein in LS174 T cells, increase in basal transactivation activity in LS174 T cells, reduced induction by Rif and corticosterone	C = 0.002, AA = 0.0	[18]
PXR*11	A4444G	Exon 4	D163G	Complete loss of basal transactivation activity in LS174 T cells, reduced induction by corticosterone, but promoter dependent enhanced induction by Rif	C = 0.0, AA = 0.014	[18]
PXR*12	G8528A	Exon 8	A370T	Increase in basal transactivation activity in LS174 T cells, reduced induction by Rif and corticosterone.	C = 0.0, AA = 0.016	[18]
PXR*5	C2904T	Exon 3	R98S	Loss of DNA binding and transactivation in HepG2 cells	J = 0.0024	[19]
PXR*6	G4399A	Exon 4	R148Q	Similar transactivation as WT	J = 0.0024	[19]
PXR*7	C8561T	Exon 8	R381W	Ligand-dependent reduced transactivation	J = 0.0024	[19]
PXR*8	A9863G	Exon 9	I403V	Ligand-dependent reduced transactivation	J = 0.0024	[19]
PAR.2*2	6bp del-24020	Intron 1a	HNF1 binding site	Complete loss of PAR.2 promoter activity in HepG2 cells	J = 0.27	[33]

Source: Adapted from [60].

[a]C, Caucasian; AA, African American; J, Japanese; IBD, inflammatory bowel disease.

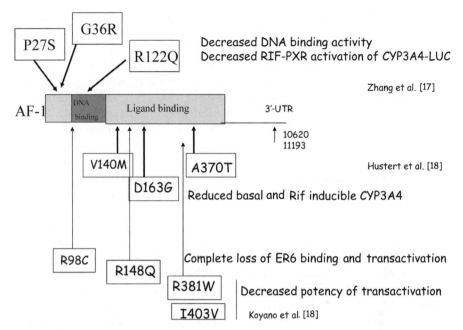

FIGURE 9.3 PXR coding variants. Amino acid changes encoded by PXR allelic variants are shown. 3'-UTR SNPs 10620 and 11193 were reported to influence survival in human cholestatic liver disease (primary sclerosing cholangitis).

*PXR*1*, *PXR*2*, and *PXR*3* in transactivation of a CYP3A4 reporter by two PXR ligands [17, 18].

Hustert et al. [18] reported that *PXR*9*, encoding a E18K amino acid change, when transfected into LS174 cells made less PXR protein, but demonstrated similar PXR ligand activation of transfected reporter plasmids containing PXR binding elements. The glutamic acid at position 18 is highly conserved in human, rabbit, rat, and mouse, suggesting it has an important function.

An exon 3, 2904C>T change (*PXR*5*) was identified by Koyano et al. [19] in a Japanese population. This variation resulted in an R98S change in the PXR*5 protein but was present in a Japanese population at a very low frequency. *PXR*5* failed to transactivate basal expression or ligand-mediated induction of a CYP3A4 reporter construct in Cos-7 and HepG2 cells. *PXR*5* also demonstrated compromised binding to an ER6 DNA element in gel shift assays [18, 35]. On the basis of the crystal structure and mutagenesis analysis, the DNA binding domain has been defined to consist of two zinc finger subdomains followed by a C-terminal extension [36]. *PXR*5* is located in this extension adjacent to the fourth Cys residue of the second zinc finger. This location may explain *PXR*5'*s complete loss of binding to DNA.

Exon 4, which encodes part of the LBD, harbors the greatest number of SNPs among the *PXR* coding exons. Four nonsynonymous SNPs identified in exon 4 are R122Q (*PXR*4*) [17], V140M (*PXR*10*) [18], R148Q (*PXR*6*) [19, 35], and D163G

(*PXR*11*) [18]. Unfortunately, only one person was heterozygous for *PXR*4* making it impossible to ascertain its function *in vivo*. The arginine (R) at position 122 was well conserved in 20 other orphan nuclear receptors, suggesting the 122R>Q change would have a functional consequence [17]. The R122Q substitution changes a basic amino acid to an uncharged one, thereby potentially altering DNA binding. Homology modeling suggested that the rare variant R122Q in *PXR*4* is in a site of direct DNA contact in the last α-helix of the DNA binding domain. Compared with *PXR*1* and variants *PXR*2* and *PXR*3*, only the *PXR*4* protein had significantly decreased affinity for the PXR binding site in electromobility shift assays and attenuated ligand activation of the *CYP3A4* reporter plasmids in transient transfection assays [17]. Thus, the *PXR*4* variant could affect the hepatic expression of CYP3A4 and could be contributing to existing variation in CYP3A4 in those rare individuals with this allele.

V140M (*PXR*10*) and D163G (*PXR*11*) exon 4 SNPs were identified in Caucasians and African Americans, respectively [18]. The Caucasian-specific allele, *PXR*10*, demonstrated enhanced basal but reduced rifampin and corticosterone transactivation of reporter plasmids harboring PXR binding elements in LS174 T cells. Interestingly, the *PXR*11* allele demonstrated complete loss of basal transactivation for CYP3A4 and DR3 reporter constructs. However, compared with *PXR*1*, rifampin-liganded *PXR*11* showed threefold higher activation of a CYP3A4 reporter [18]. The functional consequence of *PXR*11* was promoter specific, depending on the sequence context of the PXR binding element.

*PXR*6*, which results in a R148Q change in the PXR protein, was identified in a Japanese population. This amino acid change occurs in the LBD. At high concentrations of rifampin, *PXR*6* and *PXR*1* similarly activated a reporter plasmid, while the transactivation potential of *PXR*6* was reduced at lower rifampin concentrations. *PXR*6* was similar to *PXR*1* when paclitaxel was used as an activator; however, it had lower transactivation capability at higher levels of clotrimazole. It is of interest that amino acid 148 in rat PXR encodes a glutamine, suggesting the R148Q substitution is a tolerant change, or may have species-specific effects on PXR ligand binding.

A370T and R381W coding changes in exon 8 are encoded by *PXR*12* and *PXR*7*, respectively. *PXR*8* encodes an I403V change and is localized in exon 9. *PXR*7* and *PXR*8* were identified as rare alleles in the Japanese population. LS174 cells transiently transfected with PXR*12 (A370T) showed approximately twofold higher basal activation of DR3 and CYP3A4 reporter constructs, respectively, compared to PXR.1. However, upon ligand treatment, *PXR*12* showed slightly reduced induction of both reporter plasmids, compared to PXR.1 [18]. R381 and I403, the amino acids that are changed in *PXR*7* and *PXR*8*, respectively, are evolutionarily conserved in human, rat, mice, and rabbit, suggesting they are functionally important. Compared to PXR.1, both these variants demonstrated reduced transactivation of a CYP3A4 reporter upon ligand treatment (Rif, clotrimazole, and paclitaxel), but similar DNA binding and nuclear localization [35].

Notably, the functional consequences of PXR coding variants that lie in the LBD have been analyzed for their effects on binding of ligand to PXR or transactivation by PXR of reporter plasmids. Although the prototypical target tested in all of these studies has been CYP3A4, it is also possible that these SNPs could alter PXR function and affect drug–drug interactions in a ligand, promoter, and tissue-specific fashion, and this remains to be determined.

9.5   GENETIC VARIANTS IN INTRONS 2–8 AND THE 3'-UTR OF *PXR* AND THEIR FUNCTIONAL CONSEQUENCES

Numerous SNPs have been identified in introns of *PXR*. Zhang et al. [17] reported 14 SNPs in the introns and analyzed them for creation of cryptic splice sites or disruption of existing splice sites using a splice site prediction program (http://www.fruitfly.org/seq_tools/splice.html). Only C8357G in intron 7 created a cryptic splice site with a score of 0.98, but amplification of the *PXR* cDNA from livers with this intron 7 SNP did not reveal any aberrant splicing [17]. A7635G in intron 5 was associated with higher induction of intestinal CYP3A4 by rifampin. Homozygous 7635GG individuals demonstrated a twofold higher induction of intestinal CYP3A as compared to persons homozygous 7635AA. Individuals with at least one 8055C>T allele had twofold higher intestinal CYP3A following treatment with rifampicin as compared to persons with a CC genotype. However, the molecular basis of association of these SNPs with inducibility phenotype is yet to be explored. Possible explanations for the relationship of these genotypes to CYP3A phenotype are as follows: (a) the intronic SNPs can destabilize/stabilize the PXR pre-mRNA which in turn influences the level of PXR; or (b) there are yet unknown causal SNPs that may be in LD with these intronic SNPs. Recently, Karlsen et al. [37] reported a weak association of SNP7635 in intron 5 with onset of primary sclerosing cholangitis. SNP 252A>G in intron 2 was reported to be associated with oral midazolam clearance [38].

Out of nine SNPs reported in the 3'-UTR of PXR, four demonstrated association with expression of target genes. The 10620T and 10799A were associated with twofold lower testosterone 6β-hydroxylase activity following rifampin treatment of primary hepatocytes [17]. Persons with at least one 3'-UTR 11156C or one 11193G allele had 1.45-fold lower intestinal MDR levels as compared to persons homozygous for 1156AA and 11193TT alleles. Six additional intronic SNPs and one 3'-UTR SNP 26 bp downstream of the stop codon have been reported but not yet analyzed for any functional consequences [19]. Karlsen et al. [37] reported an association of Zhang's [17] 3'-UTR SNPs 10620 (rs1054190) and 11193 (rs3814058) and onset of primary sclerosing cholangitis. This was the first demonstration that PXR SNPs could influence survival in human cholestatic liver disease.

9.6   RESEQUENCING STRATEGY FOR THE *PXR* PROMOTER AND INTRON 1

The low number of detected *PXR* cSNPs relative to the extensive interindividual variability in PXR and CYP3A activity motivated an extensive survey of sequence variation in the promoter and intron 1 of *PXR*. In addition, these regions were targeted because some whole genome studies for sources of human variability in gene expression found that the majority of variability in gene expression is likely due to *trans*-acting factors or their cognate binding sequences. Comparative genomics coupled with computational prediction of transcription site clusters were used to guide selection of genomic targets for resequencing.

Comparative analysis of the orthologous PXR gene in multiple species using the UCSC (http://genome.ucsc.edu) and ECR [evolutionary conserved (http://ecrbrowser.dcode.org/)] browsers revealed regions of extensive sequence homology in *PXR* from various species. The genes flanking human *PXR*—telomeric 5′ (*AAT1*—testis-specific AMY-1 binding protein—also known as C3orf15) and centromeric 3′ (*GSK3β*, glycogen synthase kinase-3beta)—are conserved between human and rodents. However, the distance between *PXR* and the upstream gene AAT1 was 16 kb in humans versus only 2.9 kb in mice. Although both the UCSC and ECR browsers indicate a region with a high degree of conservation in intron 1, this conserved region represents a retroposed pseudogene (Prohibitin) present in *hPXR* intron 1, but not present in a syntenic region of mouse *PXR* (*mPXR* is on chr 6, while the pseudogene is on mouse chr 15). Local gene-to-gene alignment of human and mouse *PXR* using rVISTA (http://genome.lbl.gov/vista/rvista/submit.shtml) also indicated that the greatest evolutionary conservation was in the proximal promoter and intron 1. ClusterBuster online tool (zlab.bu.edu/cluster-buster), which utilizes TF (transcription factor) matrices to identify binding site clusters of high probability, was also used to identify *PXR* genomic regions with a high probability of being functional, since SNPs in those regions would be most likely to have a functional phenotype.

9.7   GENETIC VARIATION IN THE *PXR* PROMOTER AND 5′-UTR AND ITS FUNCTIONAL RELEVANCE

Deep resequencing of the *PXR* promoter and intron 1 was done targeting these regions. The DNA resequenced was from the polymorphism discovery resource (PDR). This is a useful resource for SNP discovery because the PDR panel was designed to represent the genetic diversity of the US residents who have ancestors from Europe, Africa, the Americas, and Asia. Eighty-nine SNPs were identified. Genotyping these SNPs in specific races revealed striking differences in allele frequency between populations. Zhang et al. [17] first reported on sequence variations in *PXR*'s promoter. Association between *PXR* genotype and CYP3A4 basal and inductive phenotype was used to analyze any affects of the *PXR* SNPs. Seven SNPs were identified in the 1 kb promoter region and two in the 5′-UTR of *PXR*. The promoter of PXR was analyzed for any potential cis-acting elements that might be influenced by these SNPs. It was observed that −1010G>A promoter SNP was present in a putative HSTF (heat shock transcription factor) binding site, −831T>C was within a NF-κB binding element, and −202G>A was present within a consensus binding element for c/EBP and HNF-1A (hepatic nuclear factor 1A) sites. The SNP at −165 was within a putative progesterone receptor binding site. Zhang et al. phenotyped approximately 100 persons for the expression and/or PXR-mediated inducibility of CYP3A4 and MDR1—both of which are PXR targets. The *PXR* genotype to CYP3A phenotype analysis demonstrated that the -831T was associated with a higher fold rifampin induction of the ERMBT (erythromycin breath test), a marker of CYP3A4 hepatic activity. Individuals with the -831TT genotype had a twofold higher ERMBT after treatment with rifampin as compared to subjects with a -831CC genotype. This SNP was in partial LD with the SNPs A-24381C in the 5′-UTR and G-24113A in intron 1, but no DNA response element was found at these positions. Another promoter SNP -25564G>A was identified in two subjects, one of whom was at the lower extreme of CYP3A4 content and another who had the lowest nifedipine clearance. However, due to an insufficient number of subjects (these were retrospective analyses), further studies with larger study populations are required in order to confirm the effect of this SNP.

A number of additional *PXR* promoter SNPs have been demonstrated to be associated with hepatic gene expression [39]. Basal hepatic CYP3A4 activity was related to several linked promoter SNPs including the 44477T>C located at -1359 bp in the *PXR* promoter. Thus, *SNP 44477 (-1359) was selected as the promoter LD tag* (in LD with a 6 bp deletion in intron 1a (SNP 46370 [33]) and a 46551C nucleotide insertion in intron 1a). Among females homozygous for the 44477T allele, testosterone 6β-hydroxylase activity was approximately threefold higher compared to persons homozygous for 44477C. Among males the 45005 C>T located at -831 bp in the PXR promoter was associated with significantly lower hepatic PXR levels.

The 5′-UTR SNP -24381 (called -566 in King et al. [40]) was associated with higher P-gp and CYP3A4 expression in colon tumors. This SNP has also been associated with risk of IBD (inflammatory bowel disease). This SNP was not found to be associated with hepatic PXR regulated traits [17]. It may be that the functional consequence of this SNP is tissue specific, or that it requires further studies in larger populations.

9.8 GENETIC VARIATION IN *PXR*'s INTRON 1 AND ITS FUNCTIONAL RELEVANCE

A number of SNPs in LD in PXR intron 1 were associated with CYP3A4 (Table 9.2). A 6 bp deletion in intron 1a (-24020) has been reported [39, 33]. Intron 1a also represents the promoter for the *PXR* variant PAR2 that uses exon 1b as the first coding exon. This 6 bp deletion occurred with a frequency of 27–28% in the Japanese population and was associated with complete loss of *PXR* promoter activity in HepG2 cells [33].

Although any of the SNPs alone or in combination with the intron 1 LD block could be affecting PXR activity, one LD TAG SNP was randomly chosen to represent the block of associated SNPs. For example, *63396C>T was chosen as an intron 1 LD Tag SNP* [it was in LD with intron 1 SNPs: 63704G>A and a polymorphic repeat at 63813 (CAAA)5/6(CA)12/13; and in partial LD with intron 1 SNPs 63877T>C

TABLE 9.2 *PXR* Promoter and Intron Variants Reported to Have a Functional Consequence[a]

Position in AF364606	Position from Translation Start Site (+1)[b]	Position from Transcription Start Site (+1)[c]	rs Number (HAPMAP)	Putative or In Vivo Functional Consequences	Population Allele Frequency (Caucasians)	Reference
44477T>C	−25913	−1359	rs1523130	STAT1, 3, 6, NFAT sites lost in T allele; hepatic CYP3A	T = 0.4581, C = 0.73	[39]
45005C>T	−25564G>A	−1010	rs3814055	HSTF, COMP1 sites	C = 0.01, AA = 0.09	[17]
	−25385	−831		SNP present in NF-κB, ISGF-3 sites; associated with IBD, CD, UC; hepatic CYP3A	C = 0.69, T = 0.31	[39, 17, 32]
45634G>A	−24756	−202		Rif induced ERMBT and fold induction after Rif: TT>CC	C = 0.39, AA = 0.32	[17]
	−24737A>G	−183		C/EBP site gained with A allele	C = 0.01, AA = 0.14	[39, 17]
	−24719C>G	−165		GR	C = 0.01, AA = 0	[17]
				PR	C = 0.01, AA = 0	[17]
46009	−24381A>C	Exon 1a	rs1523127	Associated with IBD, CD, UC; higher P-gp and CYP3A4 in colon tumors		[17, 32, 40]
46277	−24113G>A	Intron 1a		Rif induced ERMBT and fold induction after Rif: AA>GG	C = 0.39, AA = 0.32	[17]
46370 6 bp del	−24020	5′-UTR (Ex1b)		HNF1 site lost with deletion; hepatic CYP3A	WT = 0.71, Del = 0.29	[39]
46555 C ins	−23839	5′-UTR (Ex1b)	rs3842689	C insertion present in STRE site; hepatic CYP3A	5C = 0.71, 6C = 0.29	[17]
56348C>A	−14042	Intron 1b	rs11421631	SNP present in DR3 site; hepatic CYP3A	C = 0.71, A = 0.29	[39]
58188T>C	−12202		rs13085558	Hepatic CYP3A		[39]

63396C>T	−6994	Intron 1b	rs2472677	SNP present in HNF3β site; hepatic CYP3A	C = 0.38, T = 0.62	[39]
63704A>G	−6686	Intron 1b	rs12492296	Hepatic CYP3A	G = 0.38, A = 0.62	[39]
63813(CAAA)5/6, (CA)12/14	−6577	Intron 1b	rs4267673, rs7372335	HNF3β site gained by (CAAA)CA insertion; hepatic CYP3A	5/12 = 0.62, 6/12 = 0.38	[39]
63877T>C	−6513	Intron 1b	rs6438546	Hepatic CYP3A	T = 0.63, C = 0.37	[39]
66034T>C	−4356	Intron 1b	rs13059232	Hepatic CYP3A	T = 0.35, C = 0.65	[39]
68381G>T	−2009	Intron 1b	rs4688040	SNP present in HNF3α site; hepatic CYP3A	G = 0.79, T = 0.21	[39]
68456G>T	−1934	Intron 1b		Hepatic CYP3A	G = 0.80, T = 0.20	[39]
68740T>A	−1650	Intron 1b	rs2472679	Hepatic CYP3A	T = 0.96, A = 0.04	[39]
69245C>T	−1145	Intron 1b		CREB site lost in T allele; hepatic CYP3A	C = 0.95, T = 0.05	[39]
69789A>G	−601		rs7643645*	HNF4 site lost in G allele; hepatic CYP3A	A = 0.64, G = 0.36	[39]
	252A>G	Intron 2		Oral midazolam clearance	A = 0.36, G = 0.64	[38]
	7635A>G	Intron 5		Rif induction intestinal CYP3A; IBD and CD; PSC	A = 0.23, G = 0.77	[17, 32, 37]
	8055C>T	Intron 6		Rif induction intestinal CYP3A; IBD and CD	C = 0.82, T = 0.18	[17, 32, 37]
	10620C>T	3′-UTR		Rif induction of CYP3A in human hepatocytes; PSC	C = 0.86, T = 0.14	[17]
	11193T>C	3′-UTR		Intestine mdr1 mRNA; PSC	T = 0.70, C = 0.30	[17, 37]

[a] C, Caucasian; AA, African American; J, Japanese; IBD, inflammatory bowel disease; CD, Chron's disease; UC, ulcerative colitis; ERMBT, erythromycin breath test (hepatic CYP3A activity); PSC, primary sclerosing cholangitis.
[b] Position in AF364606, +1 the translation start site.
[c] Position in AF061056, +1 the transcription start site.

and 66034C>T]. Persons homozygous for the 63396C had threefold lower CYP3A4 activity compared to persons heterozygous or homozygous for the 63396T allele (Table 9.2). Females homozygous (TT) for the intron 1B 58188T>C SNP had significantly lower (2.3-fold) basal CYP3A4 activity versus 58188TC subjects. A number of *PXR* intron 1 SNPs associated with rifampin-induced CYP3A4 activity: Individuals homozygous for the intron 1 LD Tag SNP 63396C demonstrated significantly higher induction in CYP3A4 activity versus individuals homozygous for the major 63396T allele. Likewise, rifampicin-mediated fold induction in CYP3A4 activity was two times higher in subjects with 63877CC/66034TT genotypes as compared to subjects homozygous with 63877TT/66034CC genotypes. Similar effects were observed for the 68740T>A/69245C>T SNPs. The 69789A>G intron 1 genotype was associated with lower hepatic MDR1 mRNA, the effect being more pronounced in females. Persons homozygous for the 63396C intron 1 tag SNP had lower hepatic PAR.2 mRNA levels compared to subjects with at least one 63396T allele.

9.9 *IN SILICO* ANALYSIS FOR FUNCTIONAL EFFECT OF SNPs IN *PXR*'s PROMOTER, 5′-UTR, AND INTRON 1

A number of groups used Transfac to determine if any *PXR* SNPs, particularly those associated with any of the hepatic traits measured, were located in transcription factor binding sites or disrupting or creating any transcription factor sites. Interestingly, a number of SNPs in LD at the 3′ end of intron 1 were located in HNF3 binding sites (Figure 9.4): (a) the 63396 LD tag SNP is located in a putative HNF3β (also known as FOXA2, fork head transcription factor FoxA2) binding site; (b) the 63813 (CAAA)5/6 (CA)12/13 gain in polymorphic repeats increased the number of HNF3β binding sites; and (c) screening the HAPMAP data for any additional SNPs that were in LD revealed SNP 65104T>C that was also predicted to gain an HNF3β binding site. The 69245C>T change (which occurs in LD with the 68740 T>A that was associated with lower CYP3A4 expression in hepatocytes) disrupts a binding site for CREB. The 69789 A>G SNP, which was associated with lower expression of multiple PXR targets in liver, was present in a HNF4 binding site and the A>G change is predicted to result in the loss of the binding site.

The 44477T>C (−1359 promoter) SNP results in the loss of one binding site each for STAT1, STAT3, STAT6, HELIOSA, and NFAT. The 44477 SNP is in LD with the 6 bp deletion, which is located in an HNF1 binding site at position 46370, and a 46551 C ins, which is present in an STRE (stress response element) site.

9.10 SNPs IN *PXR*'s PROMOTER AND INTRON 1 AFFECT PUTATIVE HNF BINDING SITES

The results from several studies suggest that HNF3β (also known as fork head transcription factor FoxA2) and other hepatic transcription factors such as HNF4 and HNF1 may influence transcriptional activity of the *PXR* promoter and that SNPs

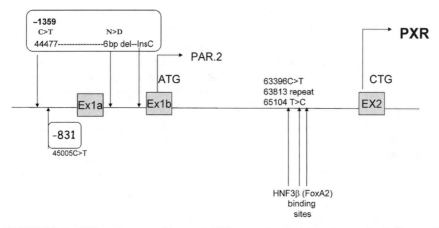

FIGURE 9.4   *PXR* promoter and intron 1 SNPs associated with various traits in liver and intestine.

in the transcription factor binding sites alter PXR expression/activity. For example, PXR SNPs associated with CYP3A4 expression were residing in and predicted to alter HNF binding sites. The 69789A>G SNP destroyed a potential HNF4 binding site and was associated with lower MDR1, CYP3A4, and PXR mRNA expression (Table 9.2). This is consistent with results [41–43] on the important role of HNF4 in regulating hepatic PXR expression. Most striking was the fact that the cluster of SNPs tagged with the 63396C>T SNP, in LD with the 63704G>A, 63813(CAAA)n/(CA)n, and 65104T>C, were each predicted to affect multiple HNF3β binding sites, and these SNPs were highly associated with basal and inducible CYP3A4 activity in liver and primary human hepatocytes. This result was interesting because of several reports that HNF3β can regulate PXR expression in liver [44] that HNF3β is recruited to the mouse *PXR* promoter (between −167 and −193 bp) during liver development [43] and that HNF3β cross talks with *PXR* to modulate its drug-induced activation of lipid metabolism (and glucose levels) in fasting mice [14]. Because HNF3β is an hepatic regulator of bile acid and glucose homeostasis [45], this regulation of PXR by HNF3β and cross talk with HNF3β suggests not only that HNF3β may be a master regulator of PXR but also that PXR, in addition to its known role in bile acid biology, may have a role in modulating hepatic glucose levels.

9.11   *PXR* SNPs HAVE BEEN ASSOCIATED WITH INTESTINAL AND HEPATIC INFLAMMATION AND DISEASES

The 45005T (−831 bp) promoter SNP, located in a putative NF-κB and ISFG-2 site, was associated with lower hepatic PXR mRNA expression in human liver. Another group recently reported that the same 45005 (23585)C allele was associated with increased susceptibility to IBD, Crohn's disease, and ulcerative colitis [32]. This result was intriguing because Zhou et al. [11] reported that PXR KO mice show significant inflammation within the intestine. Coupled with the fact that activated PXR suppresses the inflammatory response, these results suggested that persons with lower PXR expression (the 45005T allele in livers) might have a higher inflammatory response. Consistent with this notion, PXR expression was lower in patients with IBD [32]. However, paradoxically the 45005 (25385)C allele was significantly more frequent in IBD patients than the control population. There are several possible explanations for the discrepant results. The SNP, which effects a putative NF-κB site, could behave quite differently in the non inflammatory versus inflammatory state and there could be tissue-specific differences between liver and colon in the functional consequence of this SNP. A *PXR* 3′-UTR SNP (rs1054190) and an intron 5 SNP were shown to modify disease course for patients with primary sclerosing cholangitis, a hepatic disease where bile duct pathology leads to inflammation, fibrosis, and ultimately cirrhotic disease [37].

9.12   *PXR* STRUCTURAL VARIATION AND OTHER GENOMIC FEATURES

DNA structural variations including large deletions, duplications, inversions, and insertions are another source of genetic variation. Recent reports have shown that humans may differ to a greater degree in structural variations than in SNPs (taking into account the total number of nucleotides involved). Approximately 1300 structural variations were found in DNA from two subjects [46]. Iafrate [47] reported CNV (copy number variation) for *PXR* at chr 3q11 (one gain and one loss) using DNA from 39 healthy unrelated individuals. However, the authors used array-based comparative genomic hybridization (CGH) and the BAC clone showing *PXR* variation also contained the genes 5′ and 3′ of *PXR* (*AAT1* and *GSK3β*). DNA from the PDR24 subset was screened by quantitative real-time PCR for relative copy number (E. Schuetz and J. Lamba, unpublished data). Novel controls for chromosome copy number were diploid EBV-immortalized lymphoblasts that were converted by GMP conversion technology™ (GMP Genetics, Waltham, MA) to haploid cell lines, each monoallelic for one chromatid of chr 3, using Conversion technologies. Hybrid haploid cells were screened for each human chr 3 chromatid. Comparison of amplification of the DNA diploid and haploid for chr 3 demonstrated that CNV could be readily quantified. However, none of the PDR24 samples showed *PXR* CNV, suggesting it is not polymorphic (>1% frequency) in the population. Nevertheless, it was recently reported [48] that patients with bipolar disorder showed a significant increase in CNV of GSK3β and two other neighboring genes, including *PXR*, which is immediately 3′ to GSK3β. This data demonstrate that there may be a small number of individuals with bipolar disorder that show with *PXR* CNV.

   PXR was reported to be adjacent to a chromatin insulator element [49]. Insulators regulate gene activity by blocking enhancer–promoter interactions. However, although the authors indicated the AY457543 chromatin insulator binding site was in *PXR*, it is in fact in intron 1 of GSK3β, a gene 3′ of *PXR*. *PXR* is arranged head to tail, while GSK3β is arranged tail to head, which positions the insulator in a region most likely to be regulating GSK3β and not *PXR*.

9.13   PXR SUMMARY

Polymorphisms in nuclear hormone receptors are likely to contribute to the wide and inheritable variation in drug disposition, efficacy, and toxicity. In addition, they would contribute to human variation in the inductive drug response and the magnitude of inductive drug–drug interactions. In addition to PXR's well-known role in modifying expression of drug detoxification genes, it can significantly affect the inflammatory response in mouse intestine and liver, affect lipid deposition, glucose homeostasis, vitamin D levels, and glucocorticoid and mineralocorticoid homeostasis. The relationship of many newly identified *PXR* SNPs and risk of these and other diseases should help to strengthen the identity of which *PXR* SNPs are important to PXR expression/function in various tissues and disease states. Moreover, the recent finding that PXR −/− mice develop hepatic steatosis, and the growing body of work implying a connection between PXR, HNF3β, and glucose homeostasis, suggests that PXR may be a modulator gene in these diseases and that *PXR* genotyping in other disease settings is warranted.

9.14   *CAR* BACKGROUND

CAR belongs to NR1I subgroup of nuclear receptors and is also known as NR1I3. *CAR* was identified initially in 1994 while screening a human liver library with a degenerate oligonucleotide probe based on the P box region of receptor DBDs (DNA binding domains), it was called constitutive "active" receptor as it could transactivate genes in the absence of ligands [50, 51]. Later on, identification of CAR ligands, androstenol, and androstanol resulted in renaming it a constitutive "androstane" receptor [52]. As has been discussed in the previous chapters, CAR is one of the key nuclear hormone receptors involved in regulating the expression of multiple genes involved in drug disposition. Some of the CAR target genes involved in xenobiotic metabolism include cytochromes P450 (such as *CYP2Bs*, *CYP3As*, *CYP2Cs*) involved in the phase I of drug metabolism, conjugating enzymes (such as *GSTs*, *SULTs*, *UGTs*) involved in the phase II of drug metabolism, and drug transporters such as *MRPs*, *MDR1*, and *OATPs*. Biological roles for CAR have been extended beyond regulating detoxification genes. CAR has been shown to influence other physiological processes such as gluconeogenesis, bile acid homeostasis, and thyroid hormone levels. Thus, genetic variation influencing the expression and/or activity of CAR may influence the expression of its target genes that are involved not only in drug disposition but also in other biological processes.

Although most of the drug-metabolizing genes are polymorphic themselves, the influence of genetic polymorphisms in nuclear hormone receptors such as CAR and *PXR* will further contribute to the existing interindividual differences in the drug-metabolizing enzymes. The existence of wide range of interindividual variability in CAR mRNA expression (more than 200-fold) [53, 54] and its correlation with the mRNA expression of its target genes *CYP2B6* and *CYP3A4* [53–55] and other xenobiotic-metabolizing genes such as *CY2A6*, *CYP2B6*, *CYP2C*, and *UGT1A1* [56] further provides evidence for the influence of variable CAR expression in the altered expression or induction of the target genes. In this chapter, we will discuss the genetic polymorphisms as well as the alternatively spliced isoforms characterized in CAR so far.

9.15   *CAR* GENE STRUCTURE

The *CAR/NR1I3* gene is localized on chr 1q23 and has nine exons that code for 348 amino acids. Similar to other nuclear hormone receptors, CAR has an amino terminus DBD encoded by exons 2–3 and a C-terminal LBD encoded by exons 4–9, with an intervening hinge region. In contrast to other nuclear hormone receptors, *CAR* demonstrates lower homology in its LBD (74–79%) between human and rodents [57, 58]. This variation in the LBD is more striking at the DNA sequence level especially with respect to the rate of nonsynonymous and synonymous coding changes [59]. As is shown in Figure 9.5, *CAR* shares its 3′-UTR with the 3′-UTR of another gene known as *TOMM40L*.

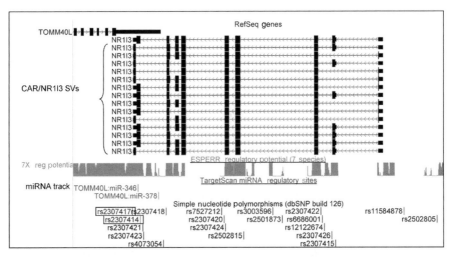

FIGURE 9.5   Snapshot of the human *CAR* gene locus from the UCSC genome browser. Top panel shows the RefSeq genes; the 3′-UTR of *TOMM40L* and *CAR* (NR1I3) overlap with each other. Some of the alternatively spliced isoforms of CAR are also shown in this panel. The ESPERR regulatory potential track displays regulatory potential scores computed from alignments of seven different species. The TargetScan miRNA track displays the potential miRNA binding sites in the *CAR* 3′-UTR. The SNPs from dbSNP occurring with the average heterozygosity of more than 0.005 are indicated in the bottom panel. The two SNPs in the 3′-UTR of *CAR* are boxed.

9.16   CAR ALTERNATIVELY SPLICED mRNAs

Alternative splicing is a mechanism that can generate multiple transcripts with similar or different functions from the same gene thereby resulting in profound functional diversity. There is precedent for alternate splicing among nuclear hormone receptors. The molecular complexity provided by alternate splicing in nuclear hormone receptors further complicates the signaling pathways. As has been observed with other nuclear hormone receptors, alternate splicing in CAR provides protein and functional diversity. Alternatively spliced isoforms of CAR demonstrate altered amino and C-terminal ends in addition to alterations in the DBD and LBD.

There have been extensive efforts to characterize and understand the functional consequences of alternatively spliced CAR isoforms. Approximately 26 alternatively spliced isoforms of CAR, originating from combinations of around eight splicing events, have been described for CAR (reviewed earlier [60]). In this section, we will discuss alternatively spliced isoforms of CAR that have been functionally characterized by various investigators.

The predominant isoforms of CAR are summarized in Figure 9.6. The splicing events resulting in deletions are indicated by upright triangles and insertions by inverted triangles, respectively. The potential effect of the alternative splicing on the CAR protein and the functional consequences as determined by *in vitro* studies are also summarized in Figure 9.6. For the sake of consistency, the SV numbers are kept same as indicated earlier [60]. The splicing event associated with deletion of exon 2 (indicated by SV22), which harbors the start site, results in use of an alternate start site in exon 1. The CAR isoforms skipping exon 2 lack the first zinc finger in the DBD and hence the ability to bind to CAR response elements and transactivate a CYP2B6-luciferase reporter [54]. Skipping of exons 4 and 5 results in premature termination codons due to change in the reading frame and hence loss of CAR expression.

The most common and extensively investigated CAR isoforms are characterized by a 12 bp insertion at the 5′ end of exon 7 (SV23) resulting in an in-frame insertion of four amino acids (SPTV) and a 15 bp insertion at the 5′ end of exon 8 (SV24) resulting in an in-frame insertion of five amino acids (APYLT), respectively. These two events can occur independent of each other or together.

SV23 characterized by a four amino acid insertion in the LBD adds an extra loop between helices 6 and 7. This isoform interacts weakly with CAR response elements and with various coactivators [61–65]. This variant was capable of transactivating *CYP2B6* and *MDR1* reporters in some studies but not *CYP3A4* and *UGT1A1* reporters [61–65]. Recently, it has been shown that upon RXRα overexpression SV23 shows RXR-dependent transactivation activity and interaction with SRC-1 as well as RXR-dependent response to clotrimazole and androstanol [66].

SV24, with a five amino acid insertion in the LBD, extends the loop between helices 8 and 9. This isoform has compromised CAR activity as assessed by its ability to bind CAR response elements, ability to transactivate of various reporter constructs, and ability to interact with the coactivators, as summarized in Figure 9.6 [61–65]. As shown by SV23, SV24 also demonstrates RXRα-dependent and

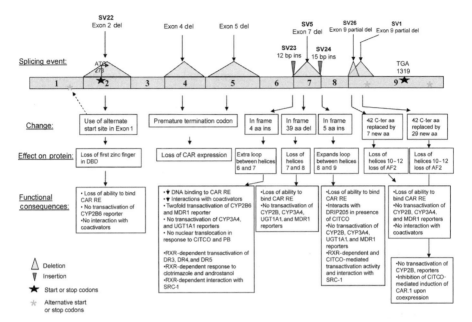

FIGURE 9.6 Summary of major alternatively spliced events in CAR and their impact on CAR mRNA and protein structure as well as their possible functional consequences (based on *in vitro* studies).

CITCO-induced transactivation of DR4x3, *CYP2B6*, and *CYP3A4* reporter constructs as well as RXRα-dependent and CITCO-mediated interaction with coactivator SRC-1 [67]. This indicates that SV24 is distinct from CAR.1 in regulation of target genes and demonstrates both RXRα and ligand dependence.

CAR.SV5, characterized by in-frame deletion of 39 amino acids due to skipping of exon 7, also demonstrates loss of CAR function [61–65].

Arnold et al. [61] identified CAR.SV26 with partial deletion of exon 9, which results in replacement of 42 C-terminal amino acids by seven new amino acids. This isoform has loss of CAR activity as indicated in Figure 9.6 and discussed earlier [60]. SV1 identified by Lamba et al. [64] uses a further downstream cryptic splice site and results in an altered C-terminal end of CAR protein along with the five amino acid insertion. Use of this cryptic splice site removes the 42 C-terminal amino acids coded by exon 9 and adds 29 new amino acids. Unpublished work from our lab has demonstrated that this isoform lacks the ability to transactivate a *CYP2B6* reporter (Figure 9.7*a*). However, upon coexpression with CAR.1 (wild-type), it suppresses CITCO-induced transactivation of a CYP2B6-luciferase reporter by CAR.1 (Figure 9.7*b*). Thus, the relative expression of this CAR isoform with respect to wild-type CAR (CAR.1) could influence the activation of target genes.

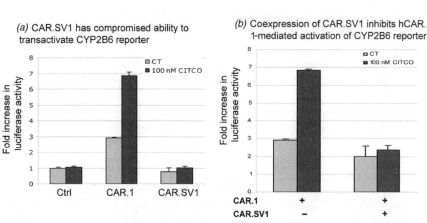

FIGURE 9.7   Functional characterization of CAR.SV1. CAR.SV1 is characterized by partial deletion of exon 9 as well as a 15 bp insertion as depicted in the schematic representation of CAR.1 and CAR.SV1 isoforms. (*a*) CAR.SV1 is unable to transactivate CYP2B6-luciferase reporter in presence or in absence of CITCO (a CAR activator) as compared to CAR.1 (WT). (*b*) Upon coexpression CAR.SV1 significantly suppresses the CITCO-mediated induction of CYP2B6-luciferase reporter transactivation by CAR.1.

Similar to human CAR, alternative splicing of CAR is evolutionarily conserved and has been reported in other species such as mouse [51], rat [68, 69], and chimp [60], indicating their role in providing protein diversity to CAR.

Some of the CAR isoforms demonstrate tissue-specific expression. This might influence tissue-specific expression of wild-type CAR transcript and would also influence tissue-specific activation of target genes as well as tissue-specific drug–drug interactions. In rat, a Car.SV splice variant (with retention of intron 6) expressed in lung has been implicated in loss of PB-induced expression of CYP2b in lung as compared to liver [68]. More recently a rat Car.SV isoform with retention of intron 7 has been shown to retain transactivation potential (for PBREM reporter); it suppresses cotransfected rat Car.1 (wild-type) mediated transactivation of a PBREM reporter [69].

Finally, the presence of different relative amounts of these alternatively spliced isoforms would contribute toward interindividual variability in level of the properly spliced CAR isoforms. Although preliminary, the relative expression of the CAR.SV24 as compared to CAR.1 has been shown to vary between 38% and 42% when compared in liver samples from four donors [63]. Interindividual variability in the relative expression of CAR isoforms could thus influence CAR-mediated transcriptional activation of target genes. Future studies in a greater number of subjects will elaborate on the contribution of the alternatively spliced isoforms in influencing the levels of normally spliced isoform and the expression of CAR target genes.

9.17   *CAR* GENETIC VARIANTS (SNPs) AND THEIR FUNCTIONAL CONSEQUENCES

To date, three reports focusing on the identification and/or functional characterization of SNPs in CAR have been published. Ikeda et al. [70, 71] identified 29 SNPs in a Japanese population: 16 were in the 5′ flanking region, 6 were exonic, and 10 were intronic. Four of the six exonic SNPs resulted in amino acid changes namely Val133Gly (exon 4), His246Arg (exon 7), Leu308Pro (exon 9), and Asn323Ser (exon 9) all within the LBD of CAR. The location in gene and protein and the frequency observed in different ethnic groups for the nonsynonymous and the synonymous coding changes are indicated in Table 9.3. As summarized in Table 9.3 functional studies carried out by coexpressing CAR variant protein with *CYP3A4* promoter/enhancer luciferase reporter revealed that CAR-246Arg variant and CAR-208Pro variant had substantially lower transactivation activity as compared to CAR wild-type (WT) protein [71]. Further CAR-246Arg variant was not responsive to CITCO, a CAR activator. This might be due to the fact that His246Arg change is present in helix 7, which is important for ligand selection [74]. Future studies are required to evaluate and understand the role of these variants in influencing the expression of CAR target genes.

Thompson et al. [59] sequenced all the coding exons and conserved intronic sequences (based on the sequence homology between six different species) in three different ethnic groups (European, African American, and Han People from Los Angeles). A nonsynonymous coding polymorphism resulting in Arg97Trp change was observed in the African American population but was absent in the other two ethnic groups tested. Using an *in silico* approach (SIFT), this polymorphism was predicted to be intolerant for the protein. However, there is no functional data to support this prediction at present. In addition to this polymorphism, eight SNPs (primarily intronic) that also altered transcription factor binding sites (namely, HNF1, CEBP, USF, and NF1) were identified in *CAR*. Future studies are required to further interrogate the functional consequences and regulatory effect of these polymorphisms.

Additional SNPs in *CAR* from different databases (HAPMAP, PERLEGEN, PharmGKB) as well as from some of the above-mentioned studies have been summarized earlier (reviewed earlier [60]). Further *in silico* analysis of the *CAR* gene using the UCSC genome browser (http://genome.ucsc.edu) revealed that *CAR* shares its 3′-UTR with the 3′-UTR of *TOMM40L* (translocase of outer mitochondrial membrane 40). As indicated in Figure 9.5 (a snapshot from USCS genome browser) this shared 3′-UTR demonstrates high regulatory potential compared to the *CAR* 5′-UTR as indicated by the 7X regulatory potential track, which displays regulatory potential (RP) scores computed from alignments of human, chimpanzee (panTro2), macaque (rheMac2), mouse (mm8), rat (rn4), dog (canFam2), and cow (bosTau2) (further information can be obtained from UCSC browser). Further, there are also two micro RNA binding sites in the shared 3′-UTR as shown by the TargetScan miRNA

TABLE 9.3 Coding Polymorphisms in *NR1I3* (*CAR*) Reported So Far[a]

Position wrt Translation Start Site as +1 (z30425)	Nucleotide Change	rs Number	Location	Amino Acid	Frequency				Functional Effects (Reference)
					CA [59]	AA [59]	Han [59]	J [71]	
249	A>G	rs2307425	Exon 4	Ser83Ser	0	0	0	0	
258	C>G		Exon 4	Ala86Ala	0	0.02	0	0	
289	C>T		Exon 4	**Arg97Trp**	0	0.02	0	0	Intolerant change as predicted by SIFT; disrupts predicted Exonic splicing enhancer sequence [59]
398	T>G		Exon 4 (Helix 2)	Val133gly	0	0	0	0.002	Not significantly different transactivation of *CYP3A4* promoter enhancer reporter; induced by CITCO [71]
540	C>T	rs2307424	Exon 5	Pro180Pro	0.42	0.16	0.45	0.521	No effect on docetaxel and doxorubin PK and PD [72, 73]
737	A>G		Exon 7 (Helix 7)	**His246Arg**	0	0	0	0.003	Significantly lower transactivation of *CYP3A4* promoter enhancer reporter, no response to CITCO [71]
923	T>C		Exon 9 (Helix 10)	**Leu308Pro**	0	0	0	0.0015	Significantly lower transactivation of *CYP3A4* promoter enhancer reporter, induced by CITCO [71]
930	G>A	rs45445294	Exon 9	Ala310Ala	0.02	0	0		
968	A>G		Exon 9 (Helix 10)	**Asn323Ser**	0	0	0	0.003	Not significantly different transactivation of *CYP3A4* promoter enhancer reporter; induced by CITCO [71]
1032	G>A		Exon 9	Gln344Gln	0	0	0	0.002	

[a]Nonsynonymous coding changes are in bold.

regulatory sites track, which shows conserved mammalian microRNA regulatory target sites for conserved microRNA families in the 3′-UTR regions of RefSeq genes, as predicted by the program TargetScanS. Future studies are required to evaluate the role of miRNAs in regulating the expression of CAR.

There are at least two SNPs (boxed in Figure 9.5) in the dbSNP track that are present in the 3′-UTR and could have functional significance. Since the coding polymorphisms in CAR are rare (Table 9.3), CAR hepatic expression shows ~200-fold interindividual variation; the variation in CAR may be due to either sequence variations 5′ or 3′ to the *CAR* coding region, or may be due to nucleotide variation in transcription factors regulating *CAR*.

Two recent reports have evaluated the relationship of a synonymous Pro180Pro polymorphism in CAR (Table 9.3) with respect to the pharmacokinetics and/or pharmacodynamics of docetaxel and doxorubicin in breast cancer patients [72, 73]. They did not find an association of individual variants in CAR or other genes with the PK and PD of these drugs. However, gene interactions were observed between the CAR Pro180Pro and HNF4α Met49Val variants, with the subjects having wild type genotypes for both the variants experiencing lower docetaxel-induced neutropenia versus subjects having other genotypes.

In addition to genetic polymorphisms, gender and ethnicity have been indicated to influence the expression of CAR and its target genes. Hepatic CAR mRNA levels in females are higher than in males and also correlate with the expression and activity of CYP2B6 [54]. It has also been observed that CYP2B6-mediated ifosfamide-*N* chlorethylation activity is higher in females than in males, which might be contributing to higher neurotoxic side effects in females [75]; however, additional studies are required to confirm and further explore the influence of gender in CAR-mediated regulation of drug disposition.

As regards influence of ethnicity, it has been reported that livers from Hispanic donors have higher activity of CAR target genes (*CYP2B6*, *CYP2A6*, and *CYP2C8*) [54, 76]. Using a genome wide gene expression approach it has been observed that, as compared to Whites, Hispanic donor livers have a significant overexpression of a significant number of CAR-activated genes [77]. However, further *in vivo* studies are required to confirm (a) whether Hispanics have higher basal and/or inducible CAR target gene expression in liver, and (b) whether this is attributable to sequence diversity in CAR or some other modulator gene.

9.18   CAR SUMMARY

CAR genetic variation is predicted to contribute to the wide and inheritable variation in drug disposition, efficacy, and toxicity as well as magnitude of drug–drug interactions. Polymorphisms in the CAR coding region appear to be rare. The *in vivo* functional consequences of *CAR* nucleotide variation remain to be tested but these association studies are complicated by the intrinsic polymorphic expression of *CAR* target genes such as CYP2B6, CYPA6, and CYP2C9. Nevertheless, the significant constitutive activity of CAR, even in the absence of ligands [78], predicts that *CAR* genetic variation will significantly contribute to basal expression of target genes. Importantly, since CAR not only regulates drug detoxification gene expression [53, 54, 56], but has also been implicated in the pathogenesis of nonalcoholic steatohepatitis (NASH) [79], in Fas-mediated hepatocyte injury [80], in maintaining thyroid hormone levels [81, 82], in the pathogenesis of primary biliary cirrhosis and cholestasis [83], and in promoting hepatocyte proliferation and tumorigenesis induced by CAR ligands PB and TCPOBOP [84], it will be important in the future to determine whether CAR genetic variation is associated with or modulates these disease phenotypes.

REFERENCES

1. Lehmann, J. M., McKee, D. D., Watson, M. A., Willson, T. M., Moore, J. T., and Kliewer, S. A. (1998) The human orphan nuclear receptor PXR is activated by compounds that regulate CYP3A4 gene expression and cause drug interactions. *J. Clin. Invest.* **102**, 1016–1023.

2. Blumberg, B., Sabbagh, W., Jr., Juguilon, H., Bolado, J., Jr., van Meter, C. M., Ong, E. S., and Evans, R. M. (1998) SXR, a novel steroid and xenobiotic-sensing nuclear receptor. *Genes Dev.* **12**, 3195–3205.

3. Bertilsson, G., Heidrich, J., Svensson, K., Asman, M., Jendeberg, L., Sydow-Backman, M., Ohlsson, R., Postlind, H., Blomquist, P., and Berkenstam, A. (1998) Identification of a human nuclear receptor defines a new signaling pathway for CYP3A induction. *Proc. Natl Acad. Sci. USA* **95**, 12208–12213.

4. Watkins, P. B., Murray, S. A., Winkelman, L. G., Heuman, D. M., Wrighton, S. A., and Guzelian, P. S. (1989) Erythromycin breath test as an assay of glucocorticoid-inducible liver cytochromes P-450. Studies in rats and patients. *J. Clin. Invest.* **83**, 688–697.

5. Kolars, J. C., Schmiedlin-Ren, P., Schuetz, J. D., Fang, C., and Watkins, P. B. (1992) Identification of rifampin-inducible P450IIIA4 (CYP3A4) in human small bowel enterocytes. *J. Clin. Invest.* **90**, 1871–1878.

6. Perry, W. and Jenkins, M. V. (1986) Hepatic mixed function oxidase induction during rifampicin/isoniazid therapy in Indian vegetarians. *Int. J. Clin. Pharmacol. Ther. Toxicol.* **24**, 344–348.

7. Lown, K. S., Mayo, R. R., Leichtman, A. B., Hsiao, H. L., Turgeon, D. K., Schmiedlin-Ren, P., Brown, M. B., Guo, W., Rossi, S. J., Benet, L. Z., et al. (1997) Role of intestinal P-glycoprotein (mdr1) in interpatient variation in the oral bioavailability of cyclosporine. *Clin. Pharmacol. Ther.* **62**, 248–260.

8. Kojima, K., Nagata, K., Matsubara, T., and Yamazoe, Y. (2007) Broad but distinct role of pregnane X receptor on the expression of individual cytochrome p450s in human hepatocytes. *Drug Metab. Pharmacokinet.* **22**, 276–286.

9. Zhou, J., Zhai, Y., Mu, Y., Gong, H., Uppal, H., Toma, D., Ren, S., Evans, R. M., and Xie, W. (2006) A novel pregnane X receptor-mediated and sterol regulatory element-binding protein-independent lipogenic pathway. *J. Biol. Chem.* **281**, 15013–15020.

10. Marek, C. J., Tucker, S. J., Konstantinou, D. K., Elrick, L. J., Haefner, D., Sigalas, C., Murray, G. I., Goodwin, B., and Wright, M. C. (2005) Pregnenolone-16alpha-carbonitrile inhibits rodent liver fibrogenesis via PXR (pregnane X receptor)-dependent and PXR-independent mechanisms. *Biochem. J.* **387**, 601–608.

11. Zhou, C., Tabb, M. M., Nelson, E. L., Grun, F., Verma, S., Sadatrafiei, A., Lin, M., Mallick, S., Forman, B. M., Thummel, K. E., et al. (2006) Mutual repression between steroid and xenobiotic receptor and NF-kappaB signaling pathways links xenobiotic metabolism and inflammation. *J. Clin. Invest.* **116**, 2280–2289.

12. Zhou, C., Assem, M., Tay, J. C., Watkins, P. B., Blumberg, B., Schuetz, E. G., and Thummel, K. E. (2006) Steroid and xenobiotic receptor and vitamin D receptor crosstalk mediates CYP24 expression and drug-induced osteomalacia. *J. Clin. Invest.* **116**, 1703–1712.

13. Zhai, Y., Pai, H. V., Zhou, J., Amico, J. A., Vollmer, R. R., and Xie, W. (2007) Activation of pregnane X receptor disrupts glucocorticoid and mineralocorticoid homeostasis. *Mol. Endocrinol.* **21**, 138–147.

14. Nakamura, K., Moore, R., Negishi, M., and Sueyoshi, T. (2007) Nuclear pregnane X receptor cross-talk with FoxA2 to mediate drug-induced regulation of lipid metabolism in fasting mouse liver. *J. Biol. Chem.* **282**, 9768–9776.

15. Kodama, S., Koike, C., Negishi, M., and Yamamoto, Y. (2004) Nuclear receptors CAR and PXR cross talk with FOXO1 to regulate genes that encode drug-metabolizing and gluconeogenic enzymes. *Mol. Cell. Biol.* **24**, 7931–7940.

16. Lamba, V., Yasuda, K., Lamba, J. K., Assem, M., Davila, J., Strom, S., and Schuetz, E. G. (2004) PXR (NR1I2): splice variants in human tissues, including brain, and identification of neurosteroids and nicotine as PXR activators. *Toxicol. Appl. Pharmacol.* **199**, 251–265.

17. Zhang, J., Kuehl, P., Green, E. D., Touchman, J. W., Watkins, P. B., Daly, A., Hall, S. D., Maurel, P., Relling, M., Brimer, C., et al. (2001) The human pregnane X receptor: genomic structure and identification and functional characterization of natural allelic variants. *Pharmacogenetics* **11**, 555–572.

18. Hustert, E., Zibat, A., Presecan-Siedel, E., Eiselt, R., Mueller, R., Fuss, C., Brehm, I., Brinkmann, U., Eichelbaum, M., Wojnowski, L., et al. (2001) Natural protein variants of pregnane X receptor with altered transactivation activity toward CYP3A4. *Drug Metab. Dispos.* **29**, 1454–1459.

19. Koyano, S., Kurose, K., Ozawa, S., Saeki, M., Nakajima, Y., Hasegawa, R., Komamura, K., Ueno, K., Kamakura, S., Nakajima, T., et al. (2002) Eleven novel single nucleotide polymorphisms in the NR1I2 (PXR) gene, four of which induce non-synonymous amino acid alterations. *Drug Metab. Pharmacokinet.* **17**, 561–565.

20. Sata, F., Sapone, A., Elizondo, G., Stocker, P., Miller, V. P., Zheng, W., Raunio, H., Crespi, C. L., and Gonzalez, F. J. (2000) CYP3A4 allelic variants with amino acid substitutions in exons 7 and 12: evidence for an allelic variant with altered catalytic activity. *Clin. Pharmacol. Ther.* **67**, 48–56.

21. Kuehl, P., Zhang, J., Lin, Y., Lamba, J., Assem, M., Schuetz, J., Watkins, P. B., Daly, A., Wrighton, S. A., Hall, S. D., et al. (2001) Sequence diversity in CYP3A promoters and characterization of the genetic basis of polymorphic CYP3A5 expression. *Nat. Genet.* **27**, 383–391.

22. Rosenfeld, J. M., Vargas, R., Jr., Xie, W., and Evans, R. M. (2003) Genetic profiling defines the xenobiotic gene network controlled by the nuclear receptor pregnane X receptor. *Mol. Endocrinol.* **17**, 1268–1282.

23. Krasowski, M. D., Yasuda, K., Hagey, L. R., and Schuetz, E. G. (2005) Evolutionary selection across the nuclear hormone receptor superfamily with a focus on the NR1I subfamily (vitamin D, pregnane X, and constitutive androstane receptors). *Nucl. Recept.* **3**, 2.

24. Zhang, J., Fu, M., Cui, T., Xiong, C., Xu, K., Zhong, W., Xiao, Y., Floyd, D., Liang, J., Li, E., et al. (2004) Selective disruption of PPARgamma 2 impairs the development of adipose tissue and insulin sensitivity. *Proc. Natl Acad. Sci. USA* **101**, 10703–10708.

25. Fukuen, S., Fukuda, T., Matsuda, H., Sumida, A., Yamamoto, I., Inaba, T., and Azuma, J. (2002) Identification of the novel splicing variants for the hPXR in human livers. *Biochem. Biophys. Res. Commun.* **298**, 433–438.

26. Dotzlaw, H., Leygue, E., Watson, P., and Murphy, L. C. (1999) The human orphan receptor PXR messenger RNA is expressed in both normal and neoplastic breast tissue. *Clin. Cancer Res.* **5**, 2103–2107.

27. Kliewer, S. A., Moore, J. T., Wade, L., Staudinger, J. L., Watson, M. A., Jones, S. A., McKee, D. D., Oliver, B. B., Willson, T. M., Zetterstrom, R. H., et al. (1998) An orphan nuclear receptor activated by pregnanes defines a novel steroid signaling pathway. *Cell* **92**, 73–82.

28. Bookout, A. L., Jeong, Y., Downes, M., Yu, R. T., Evans, R. M., and Mangelsdorf, D. J. (2006) Anatomical profiling of nuclear receptor expression reveals a hierarchical transcriptional network. *Cell* **126**, 789–799.

29. Gofflot, F., Chartoire, N., Vasseur, L., Heikkinen, S., Dembele, D., Le, M. J., and Auwerx, J. (2007) Systematic gene expression mapping clusters nuclear receptors according to their function in the brain. *Cell* **131**, 405–418.

30. Langmade, S. J., Gale, S. E., Frolov, A., Mohri, I., Suzuki, K., Mellon, S. H., Walkley, S. U., Covey, D. F., Schaffer, J. E., and Ory, D. S. (2006) Pregnane X receptor (PXR) activation: a mechanism for neuroprotection in a mouse model of Niemann-Pick C disease. *Proc. Natl Acad. Sci. USA* **103**, 13807–13812.

31. Gardner-Stephen, D., Heydel, J. M., Goyal, A., Lu, Y., Xie, W., Lindblom, T., Mackenzie, P., and Radominska-Pandya, A. (2004) Human PXR variants and their differential effects on the regulation of human UDP-glucuronosyltransferase gene expression. *Drug Metab. Dispos.* **32**, 340–347.

32. Dring, M. M., Goulding, C. A., Trimble, V. I., Keegan, D., Ryan, A. W., Brophy, K. M., Smyth, C. M., Keeling, P. W., O'Donoghue, D., O'Sullivan, M., et al. (2006) The pregnane X receptor locus is associated with susceptibility to inflammatory bowel disease. *Gastroenterology* **130**, 341–348.

33. Uno, Y., Sakamoto, Y., Yoshida, K., Hasegawa, T., Hasegawa, Y., Koshino, T., and Inoue, I. (2003) Characterization of six base pair deletion in the putative HNF1-binding site of human PXR promoter. *J. Hum. Genet.* **48**, 594–597.

34. Ng, P. C. and Henikoff, S. (2002) Accounting for human polymorphisms predicted to affect protein function. *Genome Res.* **12**, 436–446.

35. Koyano, S., Kurose, K., Saito, Y., Ozawa, S., Hasegawa, R., Komamura, K., Ueno, K., Kamakura, S., Kitakaze, M., Nakajima, T., et al. (2004) Functional characterization of four naturally occurring variants of human pregnane X receptor (PXR): one variant causes dramatic loss of both DNA binding activity and the transactivation of the CYP3A4 promoter/enhancer region. *Drug Metab. Dispos.* **32**, 149–154.

36. Watkins, R. E., Wisely, G. B., Moore, L. B., Collins, J. L., Lambert, M. H., Williams, S. P., Willson, T. M., Kliewer, S. A., and Redinbo, M. R. (2001) The human nuclear xenobiotic receptor PXR: structural determinants of directed promiscuity. *Science* **292**, 2329–2333.

37. Karlsen, T. H., Lie, B. A., Frey, F. K., Thorsby, E., Broome, U., Schrumpf, E., and Boberg, K. M. (2006) Polymorphisms in the steroid and xenobiotic receptor gene influence survival in primary sclerosing cholangitis. *Gastroenterology* **131**, 781–787.

38. He, P., Court, M. H., Greenblatt, D. J., and von Moltke, L. L. (2006) Human pregnane X receptor: genetic polymorphisms, alternative mRNA splice variants, and cytochrome P450 3A metabolic activity. *J. Clin. Pharmacol.* **46**, 1356–1369.

39. Lamba, J., Lamba, V., Strom, S., Venkataramanan, R., and Schuetz, E. (2007) Novel SNPs in the promoter and intron 1 of human PXR/NR1I2 and their association with CYP3A4 expression. *Drug Metab. Dispos.* **36**, 169–181.

40. King, C. R., Xiao, M., Yu, J., Minton, M. R., Addleman, N. J., Van Booven, D. J., Kwok, P. Y., McLeod, H. L., and Marsh, S. (2007) Identification of NR1I2 genetic variation using resequencing. *Eur. J. Clin. Pharmacol.* **63**, 547–554.

41. Tirona, R. G., Lee, W., Leake, B. F., Lan, L. B., Cline, C. B., Lamba, V., Parviz, F., Duncan, S. A., Inoue, Y., Gonzalez, F. J., et al. (2003) The orphan nuclear receptor HNF4alpha determines PXR- and CAR-mediated xenobiotic induction of CYP3A4. *Nat. Med.* **9**, 220–224.

42. Kamiya, A., Inoue, Y., and Gonzalez, F. J. (2003) Role of the hepatocyte nuclear factor 4alpha in control of the pregnane X receptor during fetal liver development. *Hepatology* **37**, 1375–1384.

43. Kyrmizi, I., Hatzis, P., Katrakili, N., Tronche, F., Gonzalez, F. J., and Talianidis, I. (2006) Plasticity and expanding complexity of the hepatic transcription factor network during liver development. *Genes Dev.* **20**, 2293–2305.

44. Gibson, G. G., Phillips, A., Aouabdi, S., Plant, K., and Plant, N. (2006) Transcriptional regulation of the human pregnane-X receptor. *Drug Metab. Rev.* **38**, 31–49.

45. Rausa, F. M., Tan, Y., Zhou, H., Yoo, K. W., Stolz, D. B., Watkins, S. C., Franks, R. R., Unterman, T. G., and Costa, R. H. (2000) Elevated levels of hepatocyte nuclear factor 3beta in mouse hepatocytes influence expression of genes involved in bile acid and glucose homeostasis 1. *Mol. Cell. Biol.* **20**, 8264–8282.

46. Korbel, J. O., Urban, A. E., Affourtit, J. P., Godwin, B., Grubert, F., Simons, J. F., Kim, P. M., Palejev, D., Carriero, N. J., Du, L., et al. (2007) Paired-end mapping reveals extensive structural variation in the human genome. *Science* **318**, 420–426.

47. Iafrate, A. J., Feuk, L., Rivera, M. N., Listewnik, M. L., Donahoe, P. K., Qi, Y., Scherer, S. W., and Lee, C. (2004) Detection of large-scale variation in the human genome. *Nat. Genet.* **36**, 949–951.

48. Lachman, H. M., Pedrosa, E., Petruolo, O. A., Cockerham, M., Papolos, A., Novak, T., Papolos, D. F., and Stopkova, P. (2007) Increase in GSK3beta gene copy number variation in bipolar disorder. *Am. J. Med. Genet. B Neuropsychiatr. Genet.* **144**, 259–265.

49. Mukhopadhyay, R., Yu, W., Whitehead, J., Xu, J., Lezcano, M., Pack, S., Kanduri, C., Kanduri, M., Ginjala, V., Vostrov, A., et al. (2004) The binding sites for the chromatin insulator protein CTCF map to DNA methylation-free domains genome-wide. *Genome Res.* **14**, 1594–1602.

50. Baes, M., Gulick, T., and Choi, H. S. (1994) A new orphan member of the nuclear hormone receptor superfamily that interacts with a subset of retinoic acid response elements. *Mol. Cell. Biol.* **14**, 1544–1551.

51. Choi, H. S., Chung, M., and Tzameli, I. (1997) Differential transactivation by two isoforms of the orphan nuclear hormone receptor CAR. *J. Biol. Chem.* **272**, 23565–23571.

52. Forman, B. M., Tzameli, I., and Choi, H. S. (1998) Androstane metabolites bind to and deactivate the nuclear receptor CAR-beta. *Nature* **395**, 612–615.

53. Chang, T. K. H., Bandiera, S. M., and Chen, J. (2003) Constitutive androstane receptor and pregnane X receptor gene expression in human liver: interindividual variability and correlation with CYP2B6 mRNA levels. *Drug Metab. Dispos.* **31**, 7–10.

54. Lamba, V., Yasuda, K., Davila, J., Lamba, J., Strom, S., Hancock, M. L., Fackenthal, J. D., Rogan, P. K., Ring, B., Wrighton, S. A., et al. (2003) Hepatic CYP2B6 expression: gender and ethnic differences and relationship to CYP2B6 genotype and CAR (constitutive androstane receptor) expression. *J. Pharmacol. Exp. Ther.* **307**, 906–922.

55. Pascussi, J. M., Drocourt, L., Gerbal-Chaloin, S., Fabre, J. M., Maurel, P., and Vilarem, M. J. (2001) Dual effect of dexamethasone on CYP3A4 gene expression in human hepatocytes. Sequential role of glucocorticoid receptor and pregnane X receptor. *Eur. J. Biochem.* **268**, 6346–6358.

56. Wortham, M., Czerwinski, M., He, L., Parkinson, A., and Wan, Y. J. (2007) Expression of constitutive androstane receptor, hepatic nuclear factor 4 alpha, and P450 oxidoreductase genes determines interindividual variability in basal expression and activity of a broad scope of xenobiotic metabolism genes in the human liver. *Drug Metab. Dispos.* **35**, 1700–1710.

57. Moore, L. B., Maglich, J. M., Mckee, D. D., Wisely, B., Willson, T. M., Kliewer, S. A., Lambert, M. H., and Moore, J. T. (2002) Pregnane X receptor (PXR), constitutive androstane receptor (CAR), and benzoate X receptor (BXR) define three pharmacologically distinct classes of nuclear receptors. *Mol. Endocrinol.* **16**, 977–986.

58. Zhang, Z., Burch, P. E., Cooney, A. J., Lanz, R. B., Pereira, F. A., Wu, J., Gibbs, R. A., Weinstock, G., and Wheeler, D. A. (2004) Genomic analysis of the nuclear receptor family: new insights into structure, regulation, and evolution from the rat genome. *Genome Res.* **14**, 580–590.

59. Thompson, E. E., Kuttab-Boulos, H., Krasowski, M. D., and Di, R. A. (2005) Functional constraints on the constitutive androstane receptor inferred from human sequence variation and cross-species comparisons. *Hum. Genomics* **2**, 168–178.

60. Lamba, J., Lamba, V., and Schuetz, E. (2005) Genetic variants of PXR (NR1I2) and CAR (NR1I3) and their implications in drug metabolism and pharmacogenetics. *Curr. Drug Metab.* **6**, 369–383.

61. Arnold, K. A., Eichelbaum, M., and Burk, O. (2004) Alternative splicing affects the function and tissue-specific expression of the human constitutive androstane receptor. *Nucl. Recept.* **2**, 1.

62. Auerbach, S. S., Ramsden, R., Stoner, M. A., Verlinde, C., Hassett, C., and Omiecinski, C. J. (2003) Alternatively spliced isoforms of the human constitutive androstane receptor. *Nucl. Acids Res.* **31**, 3194–3207.

63. Jinno, H., Tanaka-Kagawa, T., Hanioka, N., Ishida, S., Saeki, M., Soyama, A., Itoda, M., Nishimura, T., Saito, Y., Ozawa, S., et al. (2004) Identification of novel alternative splice variants of human constitutive androstane receptor and characterization of their expression in the liver. *Mol. Pharmacol.* **65**, 496–502.

64. Lamba, J. K., Lamba, V., Yasuda, K., Lin, Y. S., Assem, M., Thompson, E., Strom, S., and Schuetz, E. (2004) Expression of constitutive androstane receptor splice variants in human tissues and their functional consequences. *J. Pharmacol. Exp. Ther.* **311**, 811–821.

65. Savkur, R. S., Wu, Y. F., Bramlett, K. S., Wang, M. M., Yao, S. F., Perkins, D., Totten, M., Searfoss, G., Ryan, T. P., Su, E. W., et al. (2003) Alternative splicing within the ligand binding domain of the human constitutive androstane receptor. *Mol. Genet. Metab.* **80**, 216–226.

66. Auerbach, S. S., Dekeyser, J. G., Stoner, M. A., and Omiecinski, C. J. (2007) CAR2 displays unique ligand binding and RXRalpha heterodimerization characteristics. *Drug Metab. Dispos.* **35**, 428–439.

67. Auerbach, S. S., Stoner, M. A., Su, S., and Omiecinski, C. J. (2005) Retinoid X receptor-alpha-dependent transactivation by a naturally occurring structural variant of human constitutive androstane receptor (NR1I3). *Mol. Pharmacol.* **68**, 1239–1253.

68. Kanno, Y., Aoki, A., Nakahama, T., and Inouye, Y. (2003) Role of defective splicing of mRNA in the lack of pulmonary expression of constitutively active receptor in rat. *J. Health Sci.* **49**, 541–546.

69. Kanno, Y., Aoki, S., Mochizuki, M., Mori, E., Nakahama, T., and Inouye, Y. (2005) Expression of constitutive androstane receptor splice variants in rat liver and lung and their functional properties. *Biol. Pharm. Bull.* **28**, 2058–2062.

70. Ikeda, S., Kurose, K., Ozawa, S., Sai, K., Hasegawa, R., Komamura, K., Ueno, K., Kamakura, S., Kitakaze, M., Tomoike, H., et al. (2003) Twenty-six novel single nucleotide polymorphisms and their frequencies of the NR1I3 (CAR) gene in a Japanese population. *Drug Metab. Pharmacokinet.* **18**, 413–418.

71. Ikeda, S., Kurose, K., Jinno, H., Sai, K., Ozawa, S., Hasegawa, R., Komamura, K., Kotake, T., Morishita, H., Kamakura, S., et al. (2005) Functional analysis of four naturally occurring variants of human constitutive androstane receptor. *Mol. Genet. Metab.* **86**, 314–319.

72. Hor, S. Y., Lee, S. C., Wong, C. I., Lim, Y. W., Lim, R. C., Wang, L. Z., Fan, L., Guo, J. Y., Lee, H. S., Goh, B. C., et al. (2007) PXR, CAR and HNF4alpha genotypes and their association with pharmacokinetics and pharmacodynamics of docetaxel and doxorubicin in Asian patients. *Pharmacogenom. J.* **8**, 139–146.

73. Tham, L.-S., Holford, N. H. G., Hor, S.-Y., Tan, T., Wang, L., Lim, R.-C., Lee, H.-S., Lee, S.-C., and Goh, B.-C. (2007) Lack of association of single-nucleotide polymorphisms in pregnane X receptor, hepatic nuclear factor 4{alpha}, and constitutive androstane receptor with docetaxel pharmacokinetics. *Clin. Cancer Res.* **13**, 7126–7132.

74. Jyrkkarinne, J., Windshugel, B., Makinen, J., Ylisirnio, M., Perakyla, M., Poso, A., Sippl, W., and Honkakoski, P. (2005) Amino acids important for ligand specificity of the human constitutive androstane receptor. *J. Biol. Chem.* **280**, 5960–5971.

75. Schmidt, R., Baumann, F., Hanschmann, H., Geissler, F., and Preiss, R. (2001) Gender difference in ifosfamide metabolism by human liver microsomes. *Eur. J. Drug Metab. Pharmacokinet.* **26**, 193–200.

76. Parkinson, A., Mudra, D. R., Johnson, C., Dwyer, A., and Carroll, K. M. (2004) The effects of gender, age, ethnicity, and liver cirrhosis on cytochrome P450 enzyme activity in human liver microsomes and inducibility in cultured human hepatocytes. *Toxicol. Appl. Pharmacol.* **199**, 193–209.

77. Finkelstein, D., Lamba, V., Assem, M., Rengelshausen, J., Yasuda, K., Strom, S., and Schuetz, E. (2006) ADME transcriptome in hispanic versus white donor livers: evidence of a globally enhanced NR1I3 (CAR, constitutive androstane receptor) gene signature in hispanics. *Xenobiotica* **36**, 989–1012.

78. Honkakoski, P., Zelko, I., Sueyoshi, T., and Negishi, M. (1998) The nuclear orphan receptor CAR-retinoid X receptor heterodimer activates the phenobarbital-responsive enhancer module of the CYP2B gene. *Mol. Cell. Biol.* **18**, 5652–5658.

79. Yamazaki, Y., Kakizaki, S., Horiguchi, N., Sohara, N., Sato, K., Takagi, H., Mori, M., and Negishi, M. (2007) The role of the nuclear receptor constitutive androstane receptor in the pathogenesis of non-alcoholic steatohepatitis. *Gut* **56**, 565–574.

80. Baskin-Bey, E. S., Huang, W., Ishimura, N., Isomoto, H., Bronk, S. F., Braley, K., Craig, R. W., Moore, D. D., and Gores, G. J. (2006) Constitutive androstane receptor (CAR) ligand, TCPOBOP, attenuates Fas-induced murine liver injury by altering Bcl-2 proteins. *Hepatology* **44**, 252–262.

81. Maglich, J. M., Watson, J., McMillen, P. J., Goodwin, B., Willson, T. M., and Moore, J. T. (2004) The nuclear receptor CAR is a regulator of thyroid hormone metabolism during caloric restriction. *J. Biol. Chem.* **279**, 19832–19838.

82. Qatanani, M., Zhang, J., and Moore, D. D. (2005) Role of the constitutive androstane receptor in xenobiotic-induced thyroid hormone metabolism. *Endocrinology* **146**, 995–1002.

83. Rutherford, A. E. and Pratt, D. S. (2006) Cholestasis and cholestatic syndromes. *Curr. Opin. Gastroenterol.* **22**, 209–214.

84. Qatanani, M. and Moore, D. D. (2005) CAR, the continuously advancing receptor, in drug metabolism and disease. *Curr. Drug Metab.* **6**, 329–339.

10

BEYOND PXR AND CAR, REGULATION OF XENOBIOTIC METABOLISM BY OTHER NUCLEAR RECEPTORS

MARTIN WAGNER, GERNOT ZOLLNER, AND MICHAEL TRAUNER

Division of Gastroenterology and Hepatology, Medical University of Graz, Graz, Austria

10.1 INTRODUCTION

The expression level of drug transporters and metabolizing enzymes determines the systemic and local drug exposure in a variety of tissues and the resulting pharmacological as well as toxicological effects. Most endogenous or exogenous compounds, including drugs, are eliminated from the body via a hepatic-biliary or renal-urinary elimination route or directly at the intestinal level as first line of defense, placing liver, kidney, and gut into the center of drug metabolism. Most drugs that enter the body are lipophilic and rendered more hydrophilic for their final elimination by biotransformation reactions in the liver or kidney. These enzymatic and transport steps can roughly be grouped as phase I reactions (hydroxylation), phase II reactions (conjugation), and phase III reactions (transport) [1].

Phase I reactions are mainly performed by cytochrome P450 enzymes belonging to a gene family consisting of about 300 members [2]. Cytochrome P450 enzyme reactions (oxidation, reduction, or hydrolysis) result in aliphatic and aromatic hydroxylation; O-, N-, or S-dealkylation; or dehalogenation. Typically, a hydroxyl group is generated which then can participate in phase II conjugation reactions (mainly glucuronidation, sulfation, and conjugation with glutathione). Finally, the intracellular levels of native or by phase I and II reactions modified drugs are then regulated via specific transport proteins localized on the basolateral and apical membrane of

Nuclear Receptors in Drug Metabolism   Edited by Wen Xie
Copyright © 2009 John Wiley & Sons, Inc.

hepatocytes, the renal tubular epithelial cells, and enterocytes, which facilitate export into bile, urine, or directly into the intestinal lumen (phase III), respectively. Transport proteins for xenobiotics are either ATP binding cassette (ABC) proteins, that is, ABCB (multidrug resistance proteins, MDR1 and MDR3), ABCC (multidrug resistance associated proteins, MRP1–4; putatively, MRP5–9), and ABCG transporters, organic anion transporting polypeptides (OATPs), and organic anion transporters (OATs) [3–5] (Figure 10.1).

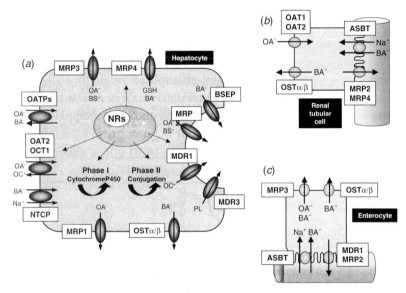

FIGURE 10.1 Drug metabolism and transport systems. (*a*) Hepatocyte. Hepatic drug uptake and export are mediated by specific transport systems on the basolateral (sinusoidal) and canalicular (apical) membrane, which are able to transport drugs in addition to their physiological endogenous substrates. Major drug uptake systems include the organic anion transporting polypeptides (OATPs), organic anion transporter 2 (OAT2), and organic cation transporter 1 (OCT1). Hepatic drug export is mediated mainly via multidrug resistance associated proteins (MRP1, MRP3, MRP4) at the basolateral membrane and via the bile salt export pump (BSEP), conjugate export pump MRP2, and the multidrug resistance protein 1 (MDR1) and the phospholipid flippase MDR3 at the canalicular membrane. The Na^+/taurocholate cotransporter NTCP and the organic solute transporter (OSTα/β) are involved in bile acid transport. Biotransformation from a nonpolar to a polar compound takes place in phase I (via cytochrome p450 enzymes) and phase II reactions (conjugation reactions). Distinct, partially overlapping nuclear receptor (NR) pathways are critically involved in the regulation of phase I, phase II, and transport processes. (*b*) Renal tubular cell. OAT1, OAT2, and potentially OSTα/β on the basolateral as well as MRP2 and MRP4 on the apical membrane play an important role in renal drug excretion. The apical sodium-dependent bile acid transporter (ASBT), OSTα/β together with MRP2 and MRP4 plays a critical role in bile acid homeostasis. (*c*) Enterocyte. Key drug transporters are MDR1, MRP2, MRP3, and potentially OSTα/β, which—together with CYP3A4 (not shown)—are responsible for first-pass intestinal drug metabolism. ASBT and OSTα/β are responsible for bile acid absorption.

Expression of genes involved in drug disposition and metabolism is largely under transcriptional control of ligand-activated nuclear receptors (NRs). In addition to the "classical" xenobiotic sensing NRs such as the pregnane X receptor (PXR) and constitutive androstane receptor (CAR), many other NRs have recently been identified as regulators of drug disposition genes. In contrast to PXR and CAR with their broad range of overlapping ligand specificities, these NRs are receptors for endogenous compounds (e.g., bile acids, fatty acids, vitamin D) and primarily regulate the metabolism of their activating ligands and intermediate metabolic products. However, farnesoid X receptor (FXR), peroxisome proliferator-activated receptors (PPARs), vitamin D receptor (VDR), hepatocyte nuclear factor (HNF) 4α, and the glucocorticoid receptor (GR) regulate multiple genes not only involved in the metabolism of their specific natural ligands but also genes involved in drug disposition and metabolism [1–5] (Table 10.1). This chapter focuses on NRs affecting drug metabolism beyond PXR and CAR.

10.2 FARNESOID X RECEPTOR

The farnesoid X receptor (FXR/NR1H4) was primarily identified as global regulator of bile acid synthesis, metabolism, and transport. FXR was the first nuclear orphan receptor shown to be activated by bile acids leading to a breakthrough in the field of bile acid research [6–8]. The most potent natural activator of FXR is chenodeoxycholic acid (CDCA), while therapeutically used ursodeoxycholic acid (UDCA) and other more hydrophilic bile acids are poor activators [6, 7, 9]. Lithocholic acid (LCA), which was also shown to activate PXR and VDR, seems to be an inverse agonist of FXR [10]. FXR is abundantly expressed in tissues belonging to the enterohepatic circulation (i.e., liver, gut) and in kidney [11]. However, FXR is also expressed in several tissues not belonging to typical bile acid target tissues (e.g., adrenal gland, heart, thymus, ovary, eye, spleen, testis, vasculature and vascular smooth muscle cells, white adipose tissue) [12–15]. FXR typically forms a heterodimer with the retinoid X receptor (RXR) and binds with high affinity to inverted repeat (IR) 1 response elements [16, 17], although binding to IR0 (for dehydroepiandrosterone sulfotransferase, *SULT2A1*) and everted repeats (ER) 8 (for *MRP2*) have also been reported [18, 19]. UDP-glucuronosyltransferase *UGT2B4* is the only gene so far identified which is activated by an FXR monomer [20].

Since bile acids represent the natural ligands for FXR activation, it may not be surprising that this receptor is the key regulator of bile acid homeostasis. FXR activates and represses genes involved in bile acid transport, synthesis, and metabolism, thus maintaining bile acid homeostasis [21, 22]. However, many of these transporters and phase I and II metabolizing enzymes are not only involved in bile acid transport but also in drug transport and metabolism, rendering FXR also a regulator of drug disposition.

10.2.1 FXR-Dependent Regulation of Basolateral Drug and Bile Acid Uptake

OATPs mediate transport of a wide range of amphiphatic substances including drugs and bile acids (Figure 10.1). The only member of the OATP/Oatp family positively

TABLE 10.1 Key Nuclear Receptors and Transcription Factors Beyond PXR and CAR Involved in Regulation of Drug Disposition

Nuclear Receptor	Expression	Ligands	Target Genes	General Effects/Functional Impact
FXR (NR1H4)	Liver, kidney, small intestine, adrenal glands, white adipose tissue, vascular smooth muscle	Bile acids, farnesol	*CYP3A4; UGT2B4, UGT2B7, SULT2A1; I-BABP, MRP2, BSEP, OATP1B3, MDR3, OSTα/β, Mdr-1; SHP, PPARα, PXR, FGF19*	Main function: bile acid, lipid, and glucose homeostasis; induction of phase I, II, and III drug (and bile acid) metabolism; induction of cellular bile acid export and modulation of bile formation
SHP (NR0B2)	Liver, small intestine, spleen, heart, pancreas, white and brown adipose tissue, skeletal muscle	— (So far no known ligand)	*SHP, HNF4, PXR, CAR, LRH-1, GR; Ntcp, ASBT, OATP1B1, OCT, OAT2; CYP7A1, CYP8B1, CYP27A*	Interference with coactivators of several NR (RAR:RXR, LXR:RXR, PPAR:RXR) and repression of respective target genes; repression of bile acid synthesis
VDR (NR1I1)	Intestine, bone, kidney, parathyroid glands, low in liver	1α,25-dihydroxyvitamin D₃, lithocholic acid	*CYP2B6, CYP2C9, CYP3A4; Sult2A1; Mrp3, ASBT; FXR,CYP27B1*	Main function: calcium homeostasis, cell proliferation and differentiation; induction of phase I, II, and III drug (and bile acid) metabolism; interference with FXR-dependent Bsep induction; negative regulation of calcitriol hydroxylation

Receptor	Tissue distribution	Ligand	Target genes	Function
RXRα (NR2B1)	Ubiquitous	All *trans*-retinoic acid	*Ntcp, Mrp2, ASBT*	Heterodimeric binding partner of other NRs; induction of drug and bile acid transporters
HNF4α (NR2A1)	Liver, kidney, small intestine, pancreas	Potentially fatty acids	*CYP2A6, CYP2B6, CYP2C8, CYP2C9, CYP2D6, CYP3A4, CYP3A5; SULT2A1; Ntcp, Oatp1a1, MDR1, OCT1, OAT1, OAT2; PXR, CAR, PPARα, FXR, HNF1α; Cyp7A1, CYP8B1, CYP27A1*	Main function: liver development and differentiation, lipid and bile metabolism; induction of phase I, II, and III drug (and bile acid) metabolism; stimulation of bile acid synthesis
HNF1α (TCF1)	Liver, pancreas, kidney, small intestine	– (So far no known ligand)	*Cyp1a2, Cyp2a5, Cyp2b10, Cyp2c29, Cyp2b9, Cyp2e1, Cyp3a11; UGT1A family; ASBT, OATP1B1, OATP1B3, Oatp1a1, Oatp1b2, Oatp1a4, Ntcp; FXR, HNF4, PXR, CYP7A1*	Main function: hepatocyte differentiation; induction of phase I, II, and III drug (and bile acid) metabolism; required for constitutive expression of various basolateral uptake systems
PPARα (NR1C1)	Liver, heart, kidney, muscle, brown adipose tissue	Fatty acids, fibrates, statins, eicosanoids, leukotrienes, NSAIDs	*CYP1A, CYP2A, CYP2C, CYP2E, CYP4A; UGT1A1, UGT1A4, UGT1A9, UGT2B4, GSTA1, GSTM2, SULT2A1; ASBT, Mdr2, Bsep, Oatp1a1, Ntcp, OCT1, OCT2, ABCA1; ABCD2, ABCD3; FXR, CYP7A1, CYP8B1, CYP27*	Main function: fatty acid homeostasis; induction of phase I, II, and III drug (and bile acid) metabolism; induction of biliary phospholipid secretion, repression of bile acid synthesis

(Continued)

TABLE 10.1 (*Continued*)

Nuclear Receptor	Expression	Ligands	Target Genes	General Effects/Functional Impact
PPARγ (NR1C3)	White and brown adipose tissue, large intestine, macrophages	Fatty acids, thiazolidinediones	*CYP4A1; UGT1A9; ABCA1, ABCG1, ABCG2; SHP*	Main function: adipocyte differentiation, insulin sensitivity, inflammation; induction of phase II and III drug metabolism
GR (NR3C1)	Ubiquitous	Glucocorticoids	*CYP2B6, CYP2C8, CYP2C9, CYP2C19, CYP3A4; ASBT, NTCP,* potentially *AE2, BSEP, MRP2; CAR, PXR, RXRα*	Main function: control of development, metabolism (of what) and immune response; induction of phase I drug (and bile acid) metabolism; stimulation of cellular bile acid uptake and export
AhR	Lung, thymus, kidney, liver	Xenobiotics (e.g., halogenated and polycyclic aromatic hydrocarbons including dioxin)	*CYP1A family, CYP1B1; GSTA2, UGT1A1, UGT1A6; ABCG2,* potentially *MRP2, Mrp3, Mrp5, Mrp6*	Main function: xenobiotic phase I and II drug metabolism
Nrf2	Ubiquitous	Reactive oxygene species (ROS), electrophiles	Cyp1a2; Cyp2a, Cyp4a, Cyp2c13; GSTalpha, GSTmu, GSTpi, NAD(P)H dehydrogenase quinone (NQO)1; MRP1–3, ABCG2	Main function: protection against oxidative stress; induction of phase I, II, and III drug metabolism

regulated by FXR is the human *OATP1B3/SLCO1B3* (formerly known as *OATP8*) [23]. Many bile acid and drug uptake systems at the basolateral membrane are negatively regulated by FXR. From a teleological point of view, bile acid-activated FXR may serve to limit accumulation of toxic bile acids in the hepatocyte via multiple mechanisms including repression of bile acid uptake systems. These repressive effects of FXR are mediated via the short heterodimer partner (SHP/NR0B2) and/or HNF4α and HNF1α and involve bile acid as well as cytokine signals [4, 24]. The promoter of human and rodent *SHP* contains an FXR response element and is transactivated by FXR [25]. SHP is an atypical member of the NR superfamily since it lacks the DNA binding domain but still contains its dimerization domain allowing to interact and negatively affect several other NRs [among them SHP itself, HNF4α, liver receptor homologue-1 (LRH-1), CAR:RXR, retinoic acid receptor (RAR):RXR, PXR:RXR, liver X receptor (LXR):RXR] [4, 26]. SHP-mediated gene repression involves competition with coactivators and its strong transcriptional repressor domain [4, 27, 28]. So far no ligand has been reported for SHP.

FXR/SHP-dependent repression has been reported for the main bile acid uptake system, the Na$^+$/taurocholate cotransporter, Ntcp (*Slc10A1*) (Figure 10.1). Rat *Ntcp* is transactivated by RXRα:RARα [28–30] and induction of SHP negatively interferes with RXRα:RARα mediated activation of the rat *Ntcp* promoter [31]. Other targets of bile acid-induced FXR/SHP effects on rodent *Ntcp* include HNF4α [32–34] and HNF1α [30, 35], a downstream target of HNF4α [36–39]. Bile acid-induced SHP inhibits the transcriptional activity of HNF4α [27, 36]. Thus, RXRα:RARα and HNF4α are both targets for SHP-mediated repression of the rat *Ntcp* promoter. However, regulation of NTCP/Ntcp by bile acids is complex and differs considerably among humans, mice, and rats [32]. The human *NTCP* promoter does not contain the rat RXRα:RARα and HNF4α response elements [32]. However, SHP suppresses GR-mediated activation of human *NTCP* independent of RXRα:RARα [40].

HNF1α, which itself is highly dependent upon HNF4α, is a key regulator of basolateral OATP/Oatp expression [36–39]. Human *OATP1B1/SLCO1B3* (formerly known as *OATP-C*) promoter activity critically depends on an intact HNF1α binding site [41]. FXR-activated SHP directly inhibits HNF4α-mediated transactivation of the *HNF1α* promoter explaining the suppressive effect of bile acids on the *OATP1B1* gene promoter [36]. HNF1α is also a potent transactivator of the mouse *Oatp1b2/Slco1b2* (formerly known as Oatp4) promoter and Oatp1b2 mRNA levels are markedly decreased in HNF1α knock-out mice [42]. The central role of HNF1α and HNF4α as key regulators of basolateral uptake systems is further underlined by reduced Ntcp, Oatp1a1/Slco1a1 (Oatp1) and Oatp1a4/Slco1a4 (Oatp2) expression in HNF1α and HNF4α knock-out mice [35, 43]. Thus, FXR regulates basolateral uptake systems in a complex manner involving SHP, HNF4α, and HNF1α. In addition, multiple FXR/SHP-independent pathways also account for Oatp/OATP and Ntcp/NTCP regulation [4, 32, 36, 44]. Other FXR-regulated genes involved in bile acid transport are the apical sodium-dependent bile acid transporter Asbt/ASBT (Slc10A2/SLC10A2) [45, 46] and the ileal bile acid binding protein (I-babp/I-BABP) [7, 47, 48].

10.2.2   FXR-Dependent Regulation of Phase I And II Metabolism

CYP3A4 is the most abundant cytochrome P450 in human liver and responsible for phase I metabolism of numerous xenobiotics and endobiotics. Its expression is regulated by several NRs including PXR and CAR [2]. In addition, the *CYP3A4* promoter harbors two functional FXR recognition sites, which mediate FXR-dependent CYP3A4 induction of primary bile acids [49]. These findings may explain elevated CYP3A expression in models of cholestasis [50, 51] and increased polyhydroxylated bile acid levels in serum and urine in human cholestatic liver disease [52–55]. However, in primary biliary cirrhosis (PBC) CYP3A4 expression is not induced, suggesting that post-transcriptional mechanism also plays a role [56]. FXR knock-out mice challenged with bile acids also display increased levels of Cyp3a11, the murine homologue of CYP3A4. This may be attributed to alternative activation of PXR by bile acids [51, 57, 58].

Among uridine 5'-diphosphate-glucuronosyltransferase (UGT) enzymes, UGT2B4 conjugates a large variety of endogenous and exogenous molecules and is considered to be the major bile acid conjugating UGT enzyme in human liver. The human UGT2B4 converts hydrophobic bile acids into more hydrophilic glucuronide derivatives. A bile acid response element in the UGT2B4 promoter to which FXR, but not RXR, binds was identified. Thus, *UGT2B4* is the only gene described so far to be activated by FXR through binding of an FXR monomer to a single hexameric DNA motif. RXR activation abolished the induction of UGT2B4 expression and inhibited binding of FXR, suggesting that RXR modulates FXR target gene activation [20]. This suggests that UGT2B4 gene induction by bile acids contributes to reduction of bile acid toxicity and metabolism of endobiotics and xenobiotics. In contrast to UGT2B4, UGT2B7 seems to be repressed by bile acids such as LCA and CDCA *in vitro*. In transfection experiments, *UGT2B7* promoter activity was decreased via a negative FXR response element [59].

Dehydroepiandrosterone sulfotransferase SULT2A1 is a cytosolic enzyme that mediates sulfoconjugation of endogenous hydroxysteroids (dehydroepiandrosterone, testosterone, bile acids), and diverse xenobiotic compounds. Upon sulfation, SULT2A1 substrates become polar, water soluble, and less toxic. The *SULT2A1* gene promoter contains an FXR response element and is activated by CDCA [19]. This is in contrast to repression of SULT2A1 by FXR agonists in a recent work, suggesting that FXR-mediated SULT2A1 regulation requires further investigation [60]. Similar to other phase I and phase II enzymes, SULT2A1 is also regulated by multiple NRs [61].

FXR has directly been linked to induction of multiple detoxification enzymes such as Mdr1a, Cyp2b10, Cyp2c38, Cyp4a10, Sult1d1, glutathione *S*-transferases Gsto2 and Gstt2 and Ugt1a1 in a recent study [62]. In addition, the xenobiotic sensor PXR is a target of FXR, suggesting that FXR activation might induce drug and bile acid breakdown indirectly via activation of PXR target genes [63].

10.2.3   FXR-Dependent Regulation of Canalicular and Basolateral Drug and Bile Acid Excretion

At the canalicular membrane the multispecific organic anion transporter *MRP2/Mrp2* (*ABCC2/Abcc2*) (which is also involved in export of divalent bile acids, as well as bilirubin and glutathione conjugates) and the bile salt export pump *BSEP/Bsep* (*ABCB11/Abcb11*) are directly transactivated by FXR [18, 64–66]. MDR3, the canalicular phospholipid flippase for biliary phospholipid excretion, is also directly positively regulated by FXR [67]. Phospholipids together with bile acids and cholesterol form mixed micelles in bile, thus protecting the biliary epithelium from the detergent effects of bile acids. Tight control of the phospholipid to bile acid ratio is therefore critical and—at least in part—controlled by FXR-regulated MDR3 and BSEP expression. MDR3 defects in humans result in a variety of cholestatic syndromes ranging from progressive familial intrahepatic cholestasis in neonates to biliary cirrhosis in adults as well as cholestasis of pregnancy and drug-induced cholestasis [68]. In addition, indirect evidence suggests that the multidrug transporter Mdr1 may in part be regulated by FXR [69].

The bile acid, steroids, and organic solutes transporting heterodimer OSTα/β, which is located at the basolateral membrane of all key tissues involved in bile acid and drug metabolism (e.g., liver, kidney, intestine, cholangiocytes), are directly activated via an FXR response element [70–74]. Expression of other basolateral export systems (Mrp3, Mrp4, Mrp5, and Mrp6) seems to be independent of FXR as demonstrated in bile acid-fed FXR knock-out mice [75].

10.2.4 FXR-Dependent Effects on Sterol and Lipid Metabolism

It should be mentioned that enzymes involved in bile acid syntheses such as CYP7A1, CYP8B1, and CYP27 are regulated by FXR/SHP-dependent and FXR/SHP-independent pathways and the reader is referred to detailed recent reviews [4, 21, 76, 77]. FXR activation greatly impacts on cholesterol levels. Bile acid sequestrants such as cholestyramine and colesevelam lower LDL-cholesterol by reducing bile acid-dependent FXR activation leading to increased cholesterol catabolism for bile acid synthesis [78]. In addition to LDL-cholesterol, FXR also impacts on HDL metabolism either via reduction of *ApoA1* gene or by modulating the plasma HDL-cholesterol clearance via scavenger receptor B-1 (SR-B1) expression [79, 80]. FXR is also involved in the regulation of triglyceride metabolism. This is mediated mainly via repression of sterol response element binding protein (SREBP) 1c expression and its lipogenic target genes, fatty acid synthetase (FAS) and acetyl CoA carboxylase (ACC) [81]. In addition, FXR effects on serum triglyceride levels may be explained not only by the repression of de novo lipogenesis but also by their effects on activators (e.g., apolipoprotein C-II) and inhibitors (e.g., apolipoprotein C-II) of lipoprotein lipase activity and VLDL clearance [82, 83]. Moreover, FXR participates in the control of hepatic glucose metabolism, particularly in gluconeogenesis, via modulating expression levels of phosphoenolpyruvate carboxykinase (PEPCK) and glucose-6-phosphatase (G6Pase) [84].

Taken together, FXR not only regulates bile acid homeostasis but also greatly impacts on transporters and enzymes involved in drug metabolism. Effects on phase I–III genes by bile acid-activated FXR may therefore at least in part explain alterations of drug metabolism in cholestatic liver disease [62, 85, 86].

10.3   HEPATOCYTE NUCLEAR FACTOR 4α

Hepatocyte nuclear factor 4α (HNF4α/NR2A1) is a highly conserved member of the NR superfamily and is expressed at highest levels, in liver, intestine, kidney, and pancreas [87, 88]. HNF4α has an essential role in development, oncogenesis and maintenance of organ function. As such, HNF4α contributes to regulation of a large fraction of the liver transcriptomes by binding directly to almost half of the actively transcribed genes [89]. HNF4α functions as a homodimer and can activate gene transcription in the absence of exogenous ligands [90, 91]. However, HNF4α binding activity may be modulated by fatty acyl-coenzyme A (CoA) thioesters [92], suggesting an important role in the control of metabolic status. Furthermore, mutations in the HNF4α gene cause maturity onset diabetes of the young (MODY1), a rare form of non-insulin-dependent diabetes mellitus inherited in an autosomal dominant pattern and characterized by defective secretion of insulin [93]. A large number of putative HNF4α target genes have been identified, including those encoding several apolipoproteins, blood coagulation factors, and enzymes involved in lipid, amino acid, and glucose metabolism [88]). Conditional HNF4α knock-out mice accumulate lipid in the liver and exhibit greatly reduced serum cholesterol and triglyceride levels and increased serum bile acid concentrations [43]. This was explained by a disruption of very–low density lipoprotein (VLDL) secretion, an increase in hepatic cholesterol uptake, and a decrease in bile acid uptake into hepatocytes due to down-regulation of the major basolateral bile acid transporters Ntcp and Oatp1a1.

HNF4α also plays an important role in regulating drug metabolism and transport. The human organic cation transporter 1 (OCT1) and the organic anion transporter 2 (OAT2/SLC22A7), which both mediate the hepatocellular uptake of numerous drugs and endobiotics from sinusoidal blood into hepatocytes, are transactivated by HNF4α and suppressed by bile acids via FXR and SHP [94, 95]. HNF4α also regulates the human organic anion transporter 1 (OAT1, SLC22A6), which is localized to the basolateral membranes of renal tubular epithelial cells, where it plays a critical role in the excretion of anionic compounds [96].

A number of CYP genes harbor putative HNF4α binding sites in their promoter and enhancer sequences [97]. Several studies using mobility shift and recombinant promoter analysis have shown that HNF4α plays a positive role in the regulation of rodent Cyp2c12 [98], Cyp3a [99, 100], Cyp2a4 [101], Cyp2c1/2 [102] and human CYP3A4 [103, 104], CYP2C8 [105], CYP2C9 [106], and CYP2D6 [107]. In addition to these promoter studies, blockage of HNF4α translation in human hepatocytes showed a dose-dependent down-regulation of CYP3A4, CYP3A5, and CYP2A6 and to a lesser degree of CYP2B6, CYP2C9, and CYP2D6 expression [108]. Furthermore, HNF4α regulates the basal and CAR/PXR-induced expression of human *SULT2A1* [61].

HNF4α is also an important regulator of bile acid synthesis. CYP7A1 (the key enzyme in bile acid synthesis), CYP27A1 (catalyzing the first step in the alternative pathway of bile acid synthesis), and CYP8B1 (controlling the ratio of CDCA to CA) harbor an HNF4α binding site in their gene promoters [109–111].

In addition to direct target gene regulation, HNF4α may also indirectly act via activation of other NRs. An HNF4α binding site was characterized in the *PXR* promoter, thereby regulating responses to xenobiotics through activation of the *PXR* gene during fetal liver development [112, 113]. The human CAR promoter is regulated by HNF4α [114] and expression of CAR is reduced in mice lacking HNF4α indicating a role for HNF4α also in the regulation of the second classical xenobiotic NR [103]. Moreover, FXR and PPARα gene transcription is activated by HNF4α [115–117]. HNF4α is a critical regulator of the liver-enriched transcription factor HNF1α [36–39], which itself plays a key role in the regulation of drug and bile acid transport and metabolism [35, 118]. Taken together, these data indicate that HNF4α is a major *in vivo* regulator of genes involved in the control of lipid and bile acid homeostasis as well as drug transport and metabolism either directly or indirectly via interactions with other NRs and transcription factors.

10.4   VITAMIN D RECEPTOR

The vitamin D receptor (VDR/NR1I1) is a member of the superfamily of steroid hormone receptors. It regulates calcium homeostasis, cell proliferation, and differentiation, and exerts immunomodulatory and antimicrobial functions [119]. VDR binds to and mediates the calcemic effects of calcitriol (1α,25-dihydroxyvitamin D3) after forming an heterodimer with RXR. 1α,25-dihydroxyvitamin D3 negatively regulates its own synthesis by repressing the 25-hydroxyvitamin D_3 1α-hydroxylase (*CYP27B1*) in a cell-type selective event that involves different combinations of multiple VDR response elements [120, 121].

VDR has also been demonstrated to be an intestinal receptor for the toxic secondary bile acid LCA [122]. Activation of VDR by vitamin D or LCA induces expression of CYP3A4 that can detoxify LCA via phase I hydroxylation [122]. Low expression of VDR in hepatocytes [123] and the coexpression of VDR and CYP3A4 in enterocytes indicate that dietary vitamin D may modulate first-pass drug metabolism in the intestine. The murine multispecific anion transporter Mrp3, which is also expressed in intestine, harbors a VDR response element in its promoter region and is transactivated upon calcitriol and LCA treatment [124]. VDR-stimulated Mrp3 expression could thus further contribute to intestinal first-pass drug metabolism.

In addition to CYP3A4, ligand-activated VDR has been shown to induce the expression of CYP2B6 and CYP2C9 [122, 125, 126]. VDR—together with PXR and CAR—controls the basal and inducible expression of these CYP genes through competitive interaction with the same battery of responsive elements [125]. *SULT2A1/Sult2a1* is another target for VDR [127]. However, VDR is not the only NR involved in the regulation of *Sult2a1* expression. FXR, PXR, and CAR bind to the same inverse repeat (IR) 0 element within the rodent *Sult2a1* gene promoter [19, 128–131].

VDR also plays an important role in bile acid homeostasis, since ASBT mRNA and promoter activity are increased by calcitriol [132]. In addition, VDR negatively interacts with FXR and calcitriol inhibits FXR transactivation *in vitro* [133]. Thus VDR is an important regulator of drug metabolism and disposition and—due to its high expression in enterocytes—a potent modulator of first-pass drug metabolism in intestine.

10.5 GLUCOCORTICOID RECEPTOR

The glucocorticoid receptor (GR/NR3C1) is ubiquitously expressed in the body and regulates numerous functions including repression of transcriptional responses to inflammatory signals. The natural ligands for GR are glucocorticoids but UDCA used for the treatment of cholestatic disorder was also reported to activate GR [134, 135]. GR plays an important role in the regulation of drug disposition genes. However, the role of the GR on the mechanism of CYP induction by glucocorticoids has not yet been fully resolved since glucocorticoids may not only directly increase target gene transcription via activation of GR but may also modulate the expression of other NRs. As such, CAR has been identified as a primary GR response gene with a glucocorticoid-responsive element in its promoter region. In addition, glucocorticoids increase the levels of PXR and RXRα mRNAs and proteins, leading to the potentiation of xenobiotic-mediated induction of CYP2B6, CYP2C8/9, and CYP3A4 [136]. This effect is likely to be mediated through the direct transcriptional activation of these NRs by GR [137]. In addition to its transcriptional induction, dexamethasone also increases translocation of CAR protein into the nucleus [136].

CYP2C9 appears to be a primary glucocorticoid-responsive gene containing a GR in addition to a CAR-responsive element in its gene promoter [138]). Glucocorticoid-responsive elements were also identified in CYP2C8 and in CYP2C19 that mediate dexamethasone induction via GR [105, 139]. Mice lacking GR show a decrease in the level of CYP2B and, when challenged with dexamethasone, fail to induce CYP2B proteins in contrast to their wild-type littermates [140].

GR is also involved in the regulation of bile acid transport and has been shown to directly transactivate human *NTCP* [40] and *ASBT* [141]. GR also appears to modulate anion exchanger AE2 expression [142]. Taken together, GR not only modulates and potentiates action of PXR and CAR target genes, but also directly regulates gene expression of various genes involved in drug metabolism.

10.6 PEROXISOME PROLIFERATOR-ACTIVATED RECEPTORS

PPARs are ligand-activated NRs and have originally been named after their peroxisome proliferation-inducing properties in rodents. However, PPAR agonists do not induce peroxisome proliferation in primates or human, [143]. PPARα, PPARβ, and PPARγ are dietary lipid sensors, which control lipid homeostasis and cellular differentiation from adipocytes. As such, almost all occurring natural fatty acids and eicosanoids are natural ligands for PPARs. PPARα (NR1C1) is highly expressed in heart, liver, kidney and brown fat, tissues with a high rate of β-oxidation of fatty acids, while PPARγ is mainly expressed in white adipose tissue [143]. Upon activation, PPARs heterodimerize with RXR and bind to DR1 response elements [144]. PPARs regulate the expression of various genes crucial for lipid and glucose metabolism [145, 146]. The main tasks of PPARα are regulation of fatty acid uptake, intracellular fatty acid transport, and β-oxidation of fatty acid. In addition, PPARα regulates expression of apoAI and apoAIII, major HDL lipoproteins [143]. Thus, therapeutically targeting PPARα with fibrates reduces fatty acids and hypertriglyceridemia and raises HDL.

Drug and bile acid metabolism are also regulated by PPARα. PPARα induces gene transcription of human and rodent hepatic organic cation transporters OCT1/Oct1 and OCT2/Oct2 [147, 148]. Both the *SULT2A1* and *UGT2B4* gene promoters contain PPAR response elements and are activated by fibrates [149, 150]. Molecular evidence for a cross talk between the PPARα and FXR transcriptional pathways has been provided by identification of PPARα as an FXR target gene [151]. Thus, UGT2B4 expression can be directly induced via activation of PPARα (and of FXR) and indirectly via FXR-dependent induction of PPARα, which then activates *UGT2B4* transcription. *UGT1A9* is also a PPARα (and PPARγ) target gene [152, 153].

Fibrates and other PPARα activators directly induce canalicular Mdr2 thereby inducing biliary phospholipid output [154–156]. In addition, ASBT/Asbt expression in liver (cholangiocytes) and intestine is induced by targeting PPARα [157, 158], resulting in increased bile acid absorption from the intestine and bile ducts. These effects on biliary phospholipid, bile acid levels, and bile acid metabolism may reduce the toxicity/aggressiveness of bile composition. Indeed, clinical trials using fibrates in patients with primary biliary cirrhosis, a chronic cholestatic disorder involving the small bile ducts, showed beneficial effects on biochemical parameters and in part also on histological findings [159–162]. In addition, PPARα plays a role in the regulation of the bile acid synthesis enzymes CYP8B1 [163] and CYP7A1 [164]. Collectively, the transcriptional effects on transport systems phase II metabolizing enzymes indicate that PPARα may be an important modulator of the metabolism of endobiotics (e.g., bile acids) and xenobiotics.

PPARγ is therapeutically targeted by thiazolidinediones and is a key regulator of adipogenesis and insulin sensitivity. PPARγ induces expression of the cholesterol transporter ABCA1 and ABCG1 [165]. Recently, a response element in ABCG2, a protective pump against endogenous and exogenous toxic agents and highly expressed in stem cells monocyte-derived dendritic cells, has been reported [166]. Similar to PPARα, PPARγ also regulates expression of UGT1A9 [152]. This suggests that PPARγ and respective agonists may also modulate drug disposition.

10.7 ARYL HYDROCARBON RECEPTOR (AhR)

The AhR plays an important role in normal physiology (e.g., cell cycle control, female reproduction), in metabolic adaption to xenobiotics, and in dioxin toxicity [167, 168]. AhR is a high affinity receptor for halogenated and polycyclic aromatic hydrocarbons including dioxin. Among natural ligands are tryptamine, bilirubin and biliverdin, and lipoxin A4 [169]. AhR binds together with its aryl hydrocarbon receptor nuclear translocator (Arnt) to xenobiotic response elements and regulates mainly phase I (CYP1A family) and phase II (GSTA2 and UGT1A1 and UGT1A6) enzymes [170]. Experiments in rodents and Caco-2 cell lines with inducers and antagonists of AhR also linked AhR activation to the induction of xenobiotic transport proteins such as Mrp2, Mrp3, Mrp5, and Mrp6 and the breast cancer resistance protein ABCG2/BCRP

[171, 172]. However, the responsible xenobiotic response elements in the promoter region still remain to be elucidated. Note that the typical CAR ligands phenobarbital and TCPOBOP increase ligand binding and mRNA levels of AhR [173], suggesting complex NR cross talk.

10.8 CONCLUSIONS

In addition to the classical xenosensing receptors such as PXR and CAR, other NRs (i.e., FXR, VDR, LXR, GR, or HNF4α), which have mainly been linked to bile acid and lipid metabolism, are also important modulators of a number of genes involved in drug metabolism and transport. The overlapping range of target genes of these receptors may serve as a redundant safety mechanism to achieve adequate protection against potentially toxic xenobiotics, even if one pathway is compromised due to a disease state. Moreover, NR interactions and cross talk can enhance and fine-tune the metabolic answers.

Our increasing knowledge on NR activation will lead to devolvement of novel drugs. Stimulation of NRs has become a highly interesting approach to target lipid, bile acid, and glucose homeostasis. Given the complex regulatory mechanisms and NR cross talk, however, targeting a specific gene or pathway with a NR ligand without interfering with other pathways is a challenging task. On the other hand, our knowledge on NR activation will allow us to predict potential drug side effects and adverse drug interactions. In addition, identifying variations and polymorphisms in the NR environment may be of importance since this also affects individual drug deposition, effects, and toxicity.

However, our current understanding of these regulatory pathways is largely based on molecular mechanisms, and data from *in vitro* studies may only be extrapolated with caution to complex situations in humans *in vivo*. Especially, NR interaction and cross talk make the prediction of *in vivo* responses difficult. Further research is required for a better understanding of the detailed transcriptional mechanism for proper drug and therapeutics design.

REFERENCES

1. Xu, C., Li, C. Y., and Kong, A. N. (2005) Induction of phase I, II and III drug metabolism/transport by xenobiotics. *Arch. Pharm. Res.* **28**, 249–268.
2. Handschin, C. and Meyer, U. A. (2003) Induction of drug metabolism: the role of nuclear receptors. *Pharmacol. Rev.* **55**, 649–673.
3. Tirona, R. G. and Kim, R. B. (2005) Nuclear receptors and drug disposition gene regulation. *J. Pharm. Sci.* **94**, 1169–1186.
4. Eloranta, J. J. and Kullak-Ublick, G. A. (2005) Coordinate transcriptional regulation of bile acid homeostasis and drug metabolism. *Arch. Biochem. Biophys.* **433**, 397–412.
5. Urquhart, B. L., Tirona, R. G., and Kim, R. B. (2007) Nuclear receptors and the regulation of drug-metabolizing enzymes and drug transporters: implications for interindividual variability in response to drugs. *J. Clin. Pharmacol.* **47**, 566–578.

6. Parks, D. J., Blanchard, S. G., Bledsoe, R. K., Chandra, G., Consler, T. G., Kliewer, S. A., Stimmel, J. B., Willson, T. M., Zavacki, A. M., Moore, D. D., et al. (1999) Bile acids: natural ligands for an orphan nuclear receptor. *Science* **284**, 1365–1368.

7. Makishima, M., Okamoto, A. Y., Repa, J. J., Tu, H., Learned, R. M., Luk, A., Hull, M. V., Lustig, K. D., Mangelsdorf, D. J., and Shan, B. (1999) Identification of a nuclear receptor for bile acids. *Science* **284**, 1362–1365.

8. Wang, H., Chen, J., Hollister, K., Sowers, L. C., and Forman, B. M. (1999) Endogenous bile acids are ligands for the nuclear receptor FXR/BAR. *Mol. Cell* **3**, 543–553.

9. Lew, J. L., Zhao, A., Yu, J., Huang, L., De Pedro, N., Pelaez, F., Wright, S. D., and Cui, J. (2004) The farnesoid X receptor controls gene expression in a ligand- and promoter-selective fashion. *J. Biol. Chem.* **279**, 8856–8861.

10. Yu, J., Lo, J. L., Huang, L., Zhao, A., Metzger, E., Adams, A., Meinke, P. T., Wright, S. D., and Cui, J. (2002) Lithocholic acid decreases expression of bile salt export pump through farnesoid X receptor antagonist activity. *J. Biol. Chem.* **277**, 31441–31447.

11. Lu, T. T., Repa, J. J., and Mangelsdorf, D. J. (2001) Orphan nuclear receptors as eLiXiRs and FiXeRs of sterol metabolism. *J. Biol. Chem.* **276**, 37735–37738.

12. Otte, K., Kranz, H., Kober, I., Thompson, P., Hoefer, M., Haubold, B., Remmel, B., Voss, H., Kaiser, C., Albers, M., et al. (2003) Identification of farnesoid X receptor beta as a novel mammalian nuclear receptor sensing lanosterol. *Mol. Cell. Biol.* **23**, 864–872.

13. Huber, R. M., Murphy, K., Miao, B., Link, J. R., Cunningham, M. R., Rupar, M. J., Gunyuzlu, P. L., Haws, T. F., Kassam, A., Powell, F., et al. (2002) Generation of multiple farnesoid-X-receptor isoforms through the use of alternative promoters. *Gene* **290**, 35–43.

14. Bishop-Bailey, D., Walsh, D. T., and Warner, T. D. (2004) Expression and activation of the farnesoid X receptor in the vasculature. *Proc. Natl Acad. Sci. USA* **101**, 3668–3673.

15. Rizzo, G., Disante, M., Mencarelli, A., Renga, B., Gioiello, A., Pellicciari, R., and Fiorucci, S. (2006) The farnesoid X receptor promotes adipocyte differentiation and regulates adipose cell function in vivo. *Mol. Pharmacol.* **70**, 1164–1173.

16. Laffitte, B. A., Kast, H. R., Nguyen, C. M., Zavacki, A. M., Moore, D. D., and Edwards, P. A. (2000) Identification of the DNA binding specificity and potential target genes for the farnesoid X-activated receptor. *J. Biol. Chem.* **275**, 10638–10647.

17. Urizar, N. L., Dowhan, D. H., and Moore, D. D. (2000) The farnesoid X-activated receptor mediates bile acid activation of phospholipid transfer protein gene expression. *J. Biol. Chem.* **275**, 39313–39317.

18. Kast, H. R., Goodwin, B., Tarr, P. T., Jones, S. A., Anisfeld, A. M., Stoltz, C. M., Tontonoz, P., Kliewer, S., Willson, T. M., and Edwards, P. A. (2002) Regulation of multidrug resistance-associated protein 2 (ABCC2) by the nuclear receptors pregnane X receptor, farnesoid X-activated receptor, and constitutive androstane receptor. *J. Biol. Chem.* **277**, 2908–2915.

19. Song, C. S., Echchgadda, I., Baek, B. S., Ahn, S. C., Oh, T., Roy, A. K., and Chatterjee, B. (2001) Dehydroepiandrosterone sulfotransferase gene induction by bile acid activated farnesoid X receptor. *J. Biol. Chem.* **276**, 42549–42556.

20. Barbier, O., Torra, I. P., Sirvent, A., Claudel, T., Blanquart, C., Duran-Sandoval, D., Kuipers, F., Kosykh, V., Fruchart, J. C., and Staels, B. (2003) FXR induces the UGT2B4 enzyme in hepatocytes: a potential mechanism of negative feedback control of FXR activity. *Gastroenterology* **124**, 1926–1940.

21. Trauner, M. and Boyer, J. L. (2003) Bile salt transporters: molecular characterization, function, and regulation. *Physiol. Rev.* **83**, 633–671.

22. Kullak-Ublick, G. A., Stieger, B., and Meier, P. J. (2004) Enterohepatic bile salt transporters in normal physiology and liver disease. *Gastroenterology* **126**, 322–342.

23. Jung, D., Podvinec, M., Meyer, U. A., Mangelsdorf, D. J., Fried, M., Meier, P. J., and Kullak-Ublick, G. A. (2002) Human organic anion transporting polypeptide 8 promoter is transactivated by the farnesoid X receptor/bile acid receptor. *Gastroenterology* **122**, 1954–1966.

24. Zollner, G., Marschall, H. U., Wagner, M., and Trauner, M. (2006) Role of nuclear receptors in the adaptive response to bile acids and cholestasis: pathogenetic and therapeutic considerations. *Mol. Pharm.* **3**, 231–251.

25. Goodwin, B., Jones, S. A., Price, R. R., Watson, M. A., McKee, D. D., Moore, L. B., Galardi, C., Wilson, J. G., Lewis, M. C., Roth, M. E., et al. (2000) A regulatory cascade of the nuclear receptors FXR, SHP-1, and LRH-1 represses bile acid biosynthesis. *Mol. Cell* **6**, 517–526.

26. Kanaya, E., Shiraki, T., and Jingami, H. (2004) The nuclear bile acid receptor FXR is activated by PGC-1alpha in a ligand-dependent manner. *Biochem. J.* **382**, 913–921.

27. Lee, Y. K., Dell, H., Dowhan, D. H., Hadzopoulou-Cladaras, M., and Moore, D. D. (2000) The orphan nuclear receptor SHP inhibits hepatocyte nuclear factor 4 and retinoid X receptor transactivation: two mechanisms for repression. *Mol. Cell. Biol.* **20**, 187–195.

28. Li, D., Zimmerman, T. L., Thevananther, S., Lee, H. Y., Kurie, J. M., and Karpen, S. J. (2002) Interleukin-1 beta-mediated suppression of RXR:RAR transactivation of the Ntcp promoter is JNK-dependent. *J. Biol. Chem.* **277**, 31416–31422.

29. Denson, L. A., Auld, K. L., Schiek, D. S., McClure, M. H., Mangelsdorf, D. J., and Karpen, S. J. (2000) Interleukin-1beta suppresses retinoid transactivation of two hepatic transporter genes involved in bile formation. *J. Biol. Chem.* **275**, 8835–8843.

30. Karpen, S. J., Sun, A. Q., Kudish, B., Hagenbuch, B., Meier, P. J., Ananthanarayanan, M., and Suchy, F. J. (1996) Multiple factors regulate the rat liver basolateral sodium-dependent bile acid cotransporter gene promoter. *J. Biol. Chem.* **271**, 15211–15221.

31. Denson, L. A., Sturm, E., Echevarria, W., Zimmerman, T. L., Makishima, M., Mangelsdorf, D. J., and Karpen, S. J. (2001) The orphan nuclear receptor, shp, mediates bile acid-induced inhibition of the rat bile acid transporter, ntcp. *Gastroenterology* **121**, 140–147.

32. Jung, D., Hagenbuch, B., Fried, M., Meier, P. J., and Kullak-Ublick, G. A. (2004) Role of liver-enriched transcription factors and nuclear receptors in regulating the human, mouse, and rat NTCP gene. *Am. J. Physiol. Gastrointest. Liver Physiol.* **286**, G752–G761.

33. Dietrich, C. G., Martin, I. V., Porn, A. C., Voigt, S., Gartung, C., Trautwein, C., and Geier, A. (2007) Fasting induces basolateral uptake transporters of the SLC family in the liver via HNF4{alpha} and PGC1{alpha}. *Am. J. Physiol. Gastrointest. Liver Physiol.* **293**, G585–G590.

34. Geier, A., Dietrich, C. G., Balasubramanian, N., Suchy, F. J., Gartung, C., Matern, S., and Ananthanarayanan, M. (2006) Mouse and rat Ntcp genes are directly transactivated via a conserved HNF-4alpha element in the proximal promoter region. A previously unidentified role [Abstract]. *Hepatology* **42**, 459A.

35. Shih, D. Q., Bussen, M., Sehayek, E., Ananthanarayanan, M., Shneider, B. L., Suchy, F. J., Shefer, S., Bollileni, J. S., Gonzalez, F. J., Breslow, J. L., et al. (2001) Hepatocyte nuclear factor-1alpha is an essential regulator of bile acid and plasma cholesterol metabolism. *Nat. Genet.* **27**, 375–382.

36. Jung, D. and Kullak-Ublick, G. A. (2003) Hepatocyte nuclear factor 1 alpha: a key mediator of the effect of bile acids on gene expression. *Hepatology* **37**, 622–631.

37. Wang, B., Cai, S. R., Gao, C., Sladek, F. M., and Ponder, K. P. (2001) Lipopolysaccharide results in a marked decrease in hepatocyte nuclear factor 4 alpha in rat liver. *Hepatology* **34**, 979–989.

38. Tian, J. M. and Schibler, U. (1991) Tissue-specific expression of the gene encoding hepatocyte nuclear factor 1 may involve hepatocyte nuclear factor 4. *Genes Dev.* **5**, 2225–2234.

39. Miura, N. and Tanaka, K. (1993) Analysis of the rat hepatocyte nuclear factor (HNF) 1 gene promoter: synergistic activation by HNF4 and HNF1 proteins. *Nucleic Acids Res.* **21**, 3731–3736.

40. Eloranta, J. J., Jung, D., and Kullak-Ublick, G. A. (2006) The human Na+-taurocholate cotransporting polypeptide gene is activated by glucocorticoid receptor and peroxisome proliferator-activated receptor-{gamma} coactivator-1{alpha}, and suppressed by bile acids via a small heterodimer partner-dependent mechanism. *Mol. Endocrinol.* **20**, 65–79.

41. Jung, D., Hagenbuch, B., Gresh, L., Pontoglio, M., Meier, P. J., and Kullak-Ublick, G. A. (2001) Characterization of the human OATP-C (SLC21A6) gene promoter and regulation of liver-specific OATP genes by hepatocyte nuclear factor 1 alpha. *J. Biol. Chem.* **276**, 37206–37214.

42. Li, N. and Klaassen, C. D. (2004) Role of liver-enriched transcription factors in the down-regulation of organic anion transporting polypeptide 4 (oatp4; oatplb2; slc21a10) by lipopolysaccharide. *Mol. Pharmacol.* **66**, 694–701.

43. Hayhurst, G. P., Lee, Y. H., Lambert, G., Ward, J. M., and Gonzalez, F. J. (2001) Hepatocyte nuclear factor 4alpha (nuclear receptor 2A1) is essential for maintenance of hepatic gene expression and lipid homeostasis. *Mol. Cell. Biol.* **21**, 1393–1403.

44. Wang, L., Lee, Y. K., Bundman, D., Han, Y., Thevananther, S., Kim, C. S., Chua, S. S., Wei, P., Heyman, R. A., Karin, M., et al. (2002) Redundant pathways for negative feedback regulation of bile acid production. *Dev. Cell* **2**, 721–731.

45. Chen, F., Ma, L., Dawson, P. A., Sinal, C. J., Sehayek, E., Gonzalez, F. J., Breslow, J., Ananthanarayanan, M., and Shneider, B. L. (2003) Liver receptor homologue-1 mediates species- and cell line-specific bile acid-dependent negative feedback regulation of the apical sodium-dependent bile acid transporter. *J. Biol. Chem.* **278**, 19909–19916.

46. Neimark, E., Chen, F., Li, X., and Shneider, B. L. (2004) Bile acid-induced negative feedback regulation of the human ileal bile acid transporter. *Hepatology* **40**, 149–156.

47. Grober, J., Zaghini, I., Fujii, H., Jones, S. A., Kliewer, S. A., Willson, T. M., Ono, T., and Besnard, P. (1999) Identification of a bile acid-responsive element in the human ileal bile acid-binding protein gene. Involvement of the farnesoid X receptor/9-cis-retinoic acid receptor heterodimer. *J. Biol. Chem.* **274**, 29749–29754.

48. Sinal, C. J., Tohkin, M., Miyata, M., Ward, J. M., Lambert, G., and Gonzalez, F. J. (2000) Targeted disruption of the nuclear receptor FXR/BAR impairs bile acid and lipid homeostasis. *Cell* **102**, 731–744.

49. Gnerre, C., Blattler, S., Kaufmann, M. R., Looser, R., and Meyer, U. A. (2004) Regulation of CYP3A4 by the bile acid receptor FXR: evidence for functional binding sites in the CYP3A4 gene. *Pharmacogenetics* **14**, 635–645.

50. Stedman, C., Robertson, G., Coulter, S., and Liddle, C. (2004) Feed-forward regulation of bile acid detoxification by CYP3A4: studies in humanized transgenic mice. *J. Biol. Chem.* **279**, 11336–11343.

51. Marschall, H. U., Wagner, M., Bodin, K., Zollner, G., Fickert, P., Gumhold, J., Silbert, D., Fuchsbichler, A., Sjovall, J., and Trauner, M. (2006) Fxr(–/–) mice adapt to biliary obstruction by enhanced phase I detoxification and renal elimination of bile acids. *J. Lipid Res.* **47**, 582–592.

52. Alme, B., Bremmelgaard, A., Sjovall, J., and Thomassen, P. (1977) Analysis of metabolic profiles of bile acids in urine using a lipophilic anion exchanger and computerized gas–liquid chromatography-mass spectrometry. *J. Lipid Res.* **18**, 339–362.

53. Bremmelgaard, A. and Sjovall, J. (1979) Bile acid profiles in urine of patients with liver diseases. *Eur. J. Clin. Invest.* **9**, 341–348.

54. Bremmelgaard, A. and Sjovall, J. (1980) Hydroxylation of cholic, chenodeoxycholic, and deoxycholic acids in patients with intrahepatic cholestasis. *J. Lipid Res.* **21**, 1072–1081.

55. Shoda, J., Tanaka, N., Osuga, T., Matsuura, K., and Miyazaki, H. (1990) Altered bile acid metabolism in liver disease: concurrent occurrence of C-1 and C-6 hydroxylated bile acid metabolites and their preferential excretion into urine. *J. Lipid Res.* **31**, 249–259.

56. Zollner, G., Wagner, M., Fickert, P., Silbert, D., Gumhold, J., Zatloukal, K., Denk, H., and Trauner, M. (2007) Expression of bile acid synthesis and detoxification enzymes and the alternative bile acid efflux pump MRP4 in patients with primary biliary cirrhosis. *Liver Int.* **27**, 920–929.

57. Guo, G. L., Lambert, G., Negishi, M., Ward, J. M., Brewer, H. B., Jr., Kliewer, S. A., Gonzalez, F. J., and Sinal, C. J. (2003) Complementary roles of farnesoid X receptor, pregnane X receptor, and constitutive androstane receptor in protection against bile acid toxicity. *J. Biol. Chem.* **278**, 45062–45071.

58. Schuetz, E. G., Strom, S., Yasuda, K., Lecureur, V., Assem, M., Brimer, C., Lamba, J., Kim, R. B., Ramachandran, V., Komoroski, B. J., et al. (2001) Disrupted bile acid homeostasis reveals an unexpected interaction among nuclear hormone receptors, transporters, and cytochrome P450. *J. Biol. Chem.* **276**, 39411–39418.

59. Lu, Y., Heydel, J. M., Li, X., Bratton, S., Lindblom, T., and Radominska-Pandya, A. (2005) Lithocholic acid decreases expression of UGT2B7 in Caco-2 cells: a potential role for a negative farnesoid X receptor response element. *Drug Metab. Dispos.* **33**, 937–946.

60. Miyata, M., Matsuda, Y., Tsuchiya, H., Kitada, H., Akase, T., Shimada, M., Nagata, K., Gonzalez, F. J., and Yamazoe, Y. (2006) Chenodeoxycholic acid-mediated activation of the farnesoid X receptor negatively regulates hydroxysteroid sulfotransferase. *Drug Metab. Pharmacokinet.* **21**, 315–323.

61. Echchgadda, I., Song, C. S., Oh, T., Ahmed, M., De La Cruz, I. J., and Chatterjee, B. (2007) The xenobiotic-sensing nuclear receptors PXR, CAR and orphan nuclear receptor HNF-4{alpha} in the regulation of human steroid-/bile acid-sulfotransferase. *Mol. Endocrinol.* **21**, 2099–2111.

62. Amador-Noguez, D., Dean, A., Huang, W., Setchell, K., Moore, D., and Darlington, G. (2007) Alterations in xenobiotic metabolism in the long-lived little mice. *Aging Cell* **6**, 453–470.

63. Jung, D., Mangelsdorf, D. J., and Meyer, U. A. (2006) Pregnane X receptor is a target of farnesoid X receptor. *J. Biol. Chem.* **281**, 19081–19091.

64. Ananthanarayanan, M., Balasubramanian, N., Makishima, M., Mangelsdorf, D. J., and Suchy, F. J. (2001) Human bile salt export pump promoter is transactivated by the farnesoid X receptor/bile acid receptor. *J. Biol. Chem.* **276**, 28857–28865.

65. Gerloff, T., Geier, A., Roots, I., Meier, P. J., and Gartung, C. (2002) Functional analysis of the rat bile salt export pump gene promoter. *Eur. J. Biochem.* **269**, 3495–3503.

66. Plass, J. R., Mol, O., Heegsma, J., Geuken, M., Faber, K. N., Jansen, P. L., and Muller, M. (2002) Farnesoid X receptor and bile salts are involved in transcriptional regulation of the gene encoding the human bile salt export pump. *Hepatology* **35**, 589–596.

67. Huang, L., Zhao, A., Lew, J. L., Zhang, T., Hrywna, Y., Thompson, J. R., De Pedro, N., Royo, I., Blevins, R. A., Pelaez, F., et al. (2003) Farnesoid X receptor activates transcription of the phospholipid pump MDR3. *J. Biol. Chem.* **278**, 51085–51090.

68. Trauner, M., Fickert, P., and Wagner, M. (2007) MDR3 (ABCB4) defects: a paradigm for the genetics of adult cholestatic syndromes. *Semin. Liver Dis.* **27**, 77–98.

69. Stedman, C., Liddle, C., Coulter, S., Sonoda, J., Alvarez, J. G., Evans, R. M., and Downes, M. (2006) Benefit of farnesoid X receptor inhibition in obstructive cholestasis. *Proc. Natl Acad. Sci. USA* **103**, 11323–11328.

70. Dawson, P. A., Hubbert, M., Haywood, J., Craddock, A. L., Zerangue, N., Christian, W. V., and Ballatori, N. (2005) The heteromeric organic solute transporter alpha-beta, Ostalpha-Ostbeta, is an ileal basolateral bile acid transporter. *J. Biol. Chem.* **280**, 6960–6968.

71. Seward, D. J., Koh, A. S., Boyer, J. L., and Ballatori, N. (2003) Functional complementation between a novel mammalian polygenic transport complex and an evolutionarily ancient organic solute transporter, OSTalpha-OSTbeta. *J. Biol. Chem.* **278**, 27473–27482.

72. Wang, W., Seward, D. J., Li, L., Boyer, J. L., and Ballatori, N. (2001) Expression cloning of two genes that together mediate organic solute and steroid transport in the liver of a marine vertebrate. *Proc. Natl Acad. Sci. USA* **98**, 9431–9436.

73. Landrier, J. F., Eloranta, J. J., Vavricka, S. R., and Kullak-Ublick, G. A. (2006) The nuclear receptor for bile acids, FXR, transactivates human organic solute transporter-alpha and -beta genes. *Am. J. Physiol. Gastrointest. Liver Physiol.* **290**, G476–G485.

74. Lee, H., Zhang, Y., Lee, F. Y., Nelson, S. F., Gonzalez, F. J., and Edwards, P. A. (2006) FXR regulates organic solute transporter alpha and beta in the adrenal gland, kidney and intestine. *J. Lipid Res.* **47**, 201–214.

75. Zollner, G., Wagner, M., Moustafa, T., Fickert, P., Silbert, D., Gumhold, J., Fuchsbichler, A., Halilbasic, E., Denk, H., Marschall, H. U., et al. (2006) Coordinated induction of bile acid detoxification and alternative elimination in mice: role of FXR-regulated organic solute transporter-{alpha}/beta in the adaptive response to bile acids. *Am. J. Physiol. Gastrointest. Liver Physiol.* **290**, G923–G932.

76. Chiang, J. Y. (2004) Regulation of bile acid synthesis: pathways, nuclear receptors, and mechanisms. *J. Hepatol.* **40**, 539–551.

77. Zollner, G. and Trauner, M. (2006) Molecular mechanisms of cholestasis. *Wien. Med. Wochenschr.* **156**, 380–385.

78. Kuipers, F., Stroeve, J. H., Caron, S., and Staels, B. (2007) Bile acids, farnesoid X receptor, atherosclerosis and metabolic control. *Curr. Opin. Lipidol.* **18**, 289–297.

79. Claudel, T., Sturm, E., Duez, H., Torra, I. P., Sirvent, A., Kosykh, V., Fruchart, J. C., Dallongeville, J., Hum, D. W., Kuipers, F., et al. (2002) Bile acid-activated nuclear receptor FXR suppresses apolipoprotein A-I transcription via a negative FXR response element. *J. Clin. Invest.* **109**, 961–971.

80. Lambert, G., Amar, M. J., Guo, G., Brewer, H. B., Jr., Gonzalez, F. J., and Sinal, C. J. (2003) The farnesoid X-receptor is an essential regulator of cholesterol homeostasis. *J. Biol. Chem.* **278**, 2563–2570.

81. Watanabe, M., Houten, S. M., Wang, L., Moschetta, A., Mangelsdorf, D. J., Heyman, R. A., Moore, D. D., and Auwerx, J. (2004) Bile acids lower triglyceride levels via a pathway involving FXR, SHP, and SREBP-1c. *J. Clin. Invest.* **113**, 1408–1418.

82. Trauner, M. (2007) A little orphan runs to fat: the orphan receptor small heterodimer partner as a key player in the regulation of hepatic lipid metabolism. *Hepatology* **46**, 1–5.

83. Claudel, T., Staels, B., and Kuipers, F. (2005) The farnesoid X receptor: a molecular link between bile acid and lipid and glucose metabolism. *Arterioscler. Thromb. Vasc. Biol.* **25**, 2020–2030.

84. Stayrook, K. R., Bramlett, K. S., Savkur, R. S., Ficorilli, J., Cook, T., Christe, M. E., Michael, L. F., and Burris, T. P. (2005) Regulation of carbohydrate metabolism by the farnesoid X receptor. *Endocrinology* **146**, 984–991.

85. Olomu, A. B., Vickers, C. R., Waring, R. H., Clements, D., Babbs, C., Warnes, T. W., and Elias, E. (1988) High incidence of poor sulfoxidation in patients with primary biliary cirrhosis. *N. Engl. J. Med.* **318**, 1089–1092.

86. Davies, M. H., Ngong, J. M., Pean, A., Vickers, C. R., Waring, R. H., and Elias, E. (1995) Sulphoxidation and sulphation capacity in patients with primary biliary cirrhosis. *J. Hepatol.* **22**, 551–560.

87. Miquerol, L., Lopez, S., Cartier, N., Tulliez, M., Raymondjean, M., and Kahn, A. (1994) Expression of the L-type pyruvate kinase gene and the hepatocyte nuclear factor 4 transcription factor in exocrine and endocrine pancreas. *J. Biol. Chem.* **269**, 8944–8951.

88. Sladek, F. M. (1994) *Hepatocyte Nuclear Factor 4*, R. G. Landes Company, Austin, pp. 207–230.

89. Odom, D. T., Zizlsperger, N., Gordon, D. B., Bell, G. W., Rinaldi, N. J., Murray, H. L., Volkert, T. L., Schreiber, J., Rolfe, P. A., Gifford, D. K., et al. (2004) Control of pancreas and liver gene expression by HNF transcription factors. *Science* **303**, 1378–1381.

90. Ladias, J. A., Hadzopoulou-Cladaras, M., Kardassis, D., Cardot, P., Cheng, J., Zannis, V., and Cladaras, C. (1992) Transcriptional regulation of human apolipoprotein genes ApoB, ApoCIII, and ApoAII by members of the steroid hormone receptor superfamily HNF-4, ARP-1, EAR-2, and EAR-3. *J. Biol. Chem.* **267**, 15849–15860.

91. Sladek, F. M., Zhong, W. M., Lai, E., and Darnell, J. E., Jr. (1990) Liver-enriched transcription factor HNF-4 is a novel member of the steroid hormone receptor superfamily. *Genes Dev.* **4**, 2353–2365.

92. Hertz, R., Magenheim, J., Berman, I., and Bar-Tana, J. (1998) Fatty acyl-CoA thioesters are ligands of hepatic nuclear factor-4alpha. *Nature* **392**, 512–516.

93. Yamagata, K., Furuta, H., Oda, N., Kaisaki, P. J., Menzel, S., Cox, N. J., Fajans, S. S., Signorini, S., Stoffel, M., and Bell, G. I. (1996) Mutations in the hepatocyte nuclear factor-4alpha gene in maturity-onset diabetes of the young (MODY1). *Nature* **384**, 458–460.

94. Saborowski, M., Kullak-Ublick, G. A., and Eloranta, J. J. (2006) The human organic cation transporter-1 gene is transactivated by hepatocyte nuclear factor-4alpha. *J. Pharmacol. Exp. Ther.* **317**, 778–785.

95. Popowski, K., Eloranta, J. J., Saborowski, M., Fried, M., Meier, P. J., and Kullak-Ublick, G. A. (2005) The human organic anion transporter 2 gene is transactivated by hepatocyte nuclear factor-4 alpha and suppressed by bile acids. *Mol. Pharmacol.* **67**, 1629–1638.

96. Ogasawara, K., Terada, T., Asaka, J., Katsura, T., and Inui, K. (2007) Hepatocyte nuclear factor-4{alpha} regulates the human organic anion transporter 1 gene in the kidney. *Am. J. Physiol. Renal Physiol.* **292**, F1819–F1826.

97. Venepally, P., Chen, D., and Kemper, B. (1992) Transcriptional regulatory elements for basal expression of cytochrome P450IIC genes. *J. Biol. Chem.* **267**, 17333–17338.

98. Sasaki, Y., Takahashi, Y., Nakayama, K., and Kamataki, T. (1999) Cooperative regulation of CYP2C12 gene expression by STAT5 and liver-specific factors in female rats. *J. Biol. Chem.* **274**, 37117–37124.

99. Ogino, M., Nagata, K., Miyata, M., and Yamazoe, Y. (1999) Hepatocyte nuclear factor 4-mediated activation of rat CYP3A1 gene and its modes of modulation by apolipoprotein AI regulatory protein I and v-ErbA-related protein 3. *Arch. Biochem. Biophys.* **362**, 32–37.

100. Huss, J. M. and Kasper, C. B. (1998) Nuclear receptor involvement in the regulation of rat cytochrome P450 3A23 expression. *J. Biol. Chem.* **273**, 16155–16162.

101. Yokomori, N., Nishio, K., Aida, K., and Negishi, M. (1997) Transcriptional regulation by HNF-4 of the steroid 15alpha-hydroxylase P450 (Cyp2a-4) gene in mouse liver. *J. Steroid Biochem. Mol. Biol.* **62**, 307–314.

102. Chen, D., Lepar, G., and Kemper, B. (1994) A transcriptional regulatory element common to a large family of hepatic cytochrome P450 genes is a functional binding site of the orphan receptor HNF-4. *J. Biol. Chem.* **269**, 5420–5427.

103. Tirona, R. G., Lee, W., Leake, B. F., Lan, L. B., Cline, C. B., Lamba, V., Parviz, F., Duncan, S. A., Inoue, Y., Gonzalez, F. J., et al. (2003) The orphan nuclear receptor HNF4alpha determines PXR- and CAR-mediated xenobiotic induction of CYP3A4. *Nat. Med.* **9**, 220–224.

104. Matsumura, K., Saito, T., Takahashi, Y., Ozeki, T., Kiyotani, K., Fujieda, M., Yamazaki, H., Kunitoh, H., and Kamataki, T. (2004) Identification of a novel polymorphic enhancer of the human CYP3A4 gene. *Mol. Pharmacol.* **65**, 326–334.

105. Ferguson, S. S., Chen, Y., LeCluyse, E. L., Negishi, M., and Goldstein, J. A. (2005) Human CYP2C8 is transcriptionally regulated by the nuclear receptors constitutive androstane receptor, pregnane X receptor, glucocorticoid receptor, and hepatic nuclear factor 4alpha. *Mol. Pharmacol.* **68**, 747–757.

106. Ibeanu, G. C. and Goldstein, J. A. (1995) Transcriptional regulation of human CYP2C genes: functional comparison of CYP2C9 and CYP2C18 promoter regions. *Biochemistry* **34**, 8028–8036.

107. Cairns, W., Smith, C. A., McLaren, A. W., and Wolf, C. R. (1996) Characterization of the human cytochrome P4502D6 promoter. A potential role for antagonistic interactions between members of the nuclear receptor family. *J. Biol. Chem.* **271**, 25269–25276.

108. Jover, R., Bort, R., Gomez-Lechon, M. J., and Castell, J. V. (2001) Cytochrome P450 regulation by hepatocyte nuclear factor 4 in human hepatocytes: a study using adenovirus-mediated antisense targeting. *Hepatology* **33**, 668–675.

109. Yang, Y., Zhang, M., Eggertsen, G., and Chiang, J. Y. (2002) On the mechanism of bile acid inhibition of rat sterol 12alpha-hydroxylase gene (CYP8B1) transcription: roles of alpha-fetoprotein transcription factor and hepatocyte nuclear factor 4alpha. *Biochim. Biophys. Acta* **1583**, 63–73.

110. Zhang, M. and Chiang, J. Y. (2001) Transcriptional regulation of the human sterol 12alpha-hydroxylase gene (CYP8B1): roles of hepatocyte nuclear factor 4alpha in mediating bile acid repression. *J. Biol. Chem.* **276**, 41690–41699.

111. Chen, W. and Chiang, J. Y. (2003) Regulation of human sterol 27-hydroxylase gene (CYP27A1) by bile acids and hepatocyte nuclear factor 4alpha (HNF4alpha). *Gene* **313**, 71–82.

112. Li, J., Ning, G., and Duncan, S. A. (2000) Mammalian hepatocyte differentiation requires the transcription factor HNF-4alpha. *Genes Dev.* **14**, 464–474.

113. Kamiya, A., Inoue, Y., and Gonzalez, F. J. (2003) Role of the hepatocyte nuclear factor 4alpha in control of the pregnane X receptor during fetal liver development. *Hepatology* **37**, 1375–1384.

114. Ding, X., Lichti, K., Kim, I., Gonzalez, F. J., and Staudinger, J. L. (2006) Regulation of constitutive androstane receptor and its target genes by fasting, cAMP, hepatocyte nuclear factor alpha, and the coactivator peroxisome proliferator-activated receptor gamma coactivator-1alpha. *J. Biol. Chem.* **281**, 26540–26551.

115. Zhang, Y., Castellani, L. W., Sinal, C. J., Gonzalez, F. J., and Edwards, P. A. (2004) Peroxisome proliferator-activated receptor-gamma coactivator 1alpha (PGC-1alpha) regulates triglyceride metabolism by activation of the nuclear receptor FXR. *Genes Dev.* **18**, 157–169.

116. Pineda Torra, I., Jamshidi, Y., Flavell, D. M., Fruchart, J. C., and Staels, B. (2002) Characterization of the human PPARalpha promoter: identification of a functional nuclear receptor response element. *Mol. Endocrinol.* **16**, 1013–1028.

117. Lu, T. T., Makishima, M., Repa, J. J., Schoonjans, K., Kerr, T. A., Auwerx, J., and Mangelsdorf, D. J. (2000) Molecular basis for feedback regulation of bile acid synthesis by nuclear receptors. *Mol. Cell* **6**, 507–515.

118. Arrese, M. and Karpen, S. J. (2002) HNF-1 alpha: have bile acid transport genes found their "master"? *J. Hepatol.* **36**, 142–145.

119. Campbell, M. J. and Adorini, L. (2006) The vitamin D receptor as a therapeutic target. *Expert Opin. Ther. Targets* **10**, 735–748.

120. Turunen, M. M., Dunlop, T. W., Carlberg, C., and Vaisanen, S. (2007) Selective use of multiple vitamin D response elements underlies the 1 alpha,25-dihydroxyvitamin D3-mediated negative regulation of the human CYP27B1 gene. *Nucleic Acids Res.* **35**, 2734–2747.

121. Murayama, A., Kim, M. S., Yanagisawa, J., Takeyama, K., and Kato, S. (2004) Transrepression by a liganded nuclear receptor via a bHLH activator through co-regulator switching. *EMBO J.* **23**, 1598–1608.

122. Makishima, M., Lu, T. T., Xie, W., Whitfield, G. K., Domoto, H., Evans, R. M., Haussler, M. R., and Mangelsdorf, D. J. (2002) Vitamin D receptor as an intestinal bile acid sensor. *Science* **296**, 1313–1316.

123. Gascon-Barre, M., Demers, C., Mirshahi, A., Neron, S., Zalzal, S., and Nanci, A. (2003) The normal liver harbors the vitamin D nuclear receptor in nonparenchymal and biliary epithelial cells. *Hepatology* **37**, 1034–1042.

124. McCarthy, T. C., Li, X., and Sinal, C. J. (2005) Vitamin D receptor-dependent regulation of colon multidrug resistance-associated protein 3 gene expression by bile acids. *J. Biol. Chem.* **280**, 23232–23242.

125. Drocourt, L., Ourlin, J. C., Pascussi, J. M., Maurel, P., and Vilarem, M. J. (2002) Expression of CYP3A4, CYP2B6, and CYP2C9 is regulated by the vitamin D receptor pathway in primary human hepatocytes. *J. Biol. Chem.* **277**, 25125–25132.

126. Schmiedlin-Ren, P., Thummel, K. E., Fisher, J. M., Paine, M. F., and Watkins, P. B. (2001) Induction of CYP3A4 by 1 alpha,25-dihydroxyvitamin D3 is human cell line-specific and is unlikely to involve pregnane X receptor. *Drug Metab. Dispos.* **29**, 1446–1453.

127. Echchgadda, I., Song, C. S., Roy, A. K., and Chatterjee, B. (2004) Dehydroepiandrosterone sulfotransferase is a target for transcriptional induction by the vitamin D receptor. *Mol. Pharmacol.* **65**, 720–729.

128. Runge-Morris, M., Wu, W., and Kocarek, T. A. (1999) Regulation of rat hepatic hydroxysteroid sulfotransferase (SULT2-40/41) gene expression by glucocorticoids: evidence for a dual mechanism of transcriptional control. *Mol. Pharmacol.* **56**, 1198–1206.

129. Sonoda, J., Xie, W., Rosenfeld, J. M., Barwick, J. L., Guzelian, P. S., and Evans, R. M. (2002) Regulation of a xenobiotic sulfonation cascade by nuclear pregnane X receptor (PXR). *Proc. Natl Acad. Sci. USA* **99**, 13801–13806.

130. Saini, S. P., Sonoda, J., Xu, L., Toma, D., Uppal, H., Mu, Y., Ren, S., Moore, D. D., Evans, R. M., and Xie, W. (2004) A novel constitutive androstane receptor-mediated and CYP3A-independent pathway of bile acid detoxification. *Mol. Pharmacol.* **65**, 292–300.

131. Assem, M., Schuetz, E. G., Leggas, M., Sun, D., Yasuda, K., Reid, G., Zelcer, N., Adachi, M., Strom, S., Evans, R. M., et al. (2004) Interactions between hepatic Mrp4 and Sult2a as revealed by the constitutive androstane receptor and Mrp4 knockout mice. *J. Biol. Chem.* **279**, 22250–22257.

132. Chen, X., Chen, F., Liu, S., Glaeser, H., Dawson, P. A., Hofmann, A. F., Kim, R. B., Shneider, B. L., and Pang, K. S. (2006) Transactivation of rat apical sodium-dependent bile acid transporter and increased bile acid transport by 1alpha,25-dihydroxyvitamin D3 via the vitamin D receptor. *Mol. Pharmacol.* **69**, 1913–1923.

133. Honjo, Y., Sasaki, S., Kobayashi, Y., Misawa, H., and Nakamura, H. (2006) 1,25-dihydroxyvitamin D3 and its receptor inhibit the chenodeoxycholic acid-dependent transactivation by farnesoid X receptor. *J. Endocrinol.* **188**, 635–643.

134. Tanaka, H. and Makino, I. (1992) Ursodeoxycholic acid-dependent activation of the glucocorticoid receptor. *Biochem. Biophys. Res. Commun.* **188**, 942–948.

135. Miura, T., Ouchida, R., Yoshikawa, N., Okamoto, K., Makino, Y., Nakamura, T., Morimoto, C., Makino, I., and Tanaka, H. (2001) Functional modulation of the glucocorticoid receptor and suppression of NF-kappaB-dependent transcription by ursodeoxycholic acid. *J. Biol. Chem.* **276**, 47371–47378.

136. Pascussi, J. M., Drocourt, L., Fabre, J. M., Maurel, P., and Vilarem, M. J. (2000) Dexamethasone induces pregnane X receptor and retinoid X receptor-alpha expression in human hepatocytes: synergistic increase of CYP3A4 induction by pregnane X receptor activators. *Mol. Pharmacol.* **58**, 361–372.

137. Pascussi, J. M., Gerbal-Chaloin, S., Drocourt, L., Maurel, P., and Vilarem, M. J. (2003) The expression of CYP2B6, CYP2C9 and CYP3A4 genes: a tangle of networks of nuclear and steroid receptors. *Biochim. Biophys. Acta.* **1619**, 243–253.

138. Gerbal-Chaloin, S., Daujat, M., Pascussi, J. M., Pichard-Garcia, L., Vilarem, M. J., and Maurel, P. (2002) Transcriptional regulation of CYP2C9 gene. Role of glucocorticoid receptor and constitutive androstane receptor. *J. Biol. Chem.* **277**, 209–217.

139. Chen, Y., Ferguson, S. S., Negishi, M., and Goldstein, J. A. (2003) Identification of constitutive androstane receptor and glucocorticoid receptor binding sites in the CYP2C19 promoter. *Mol. Pharmacol.* **64**, 316–324.

140. Schuetz, E. G., Schmid, W., Schutz, G., Brimer, C., Yasuda, K., Kamataki, T., Bornheim, L., Myles, K., and Cole, T. J. (2000) The glucocorticoid receptor is essential for induction of cytochrome P-4502B by steroids but not for drug or steroid induction of CYP3A or P-450 reductase in mouse liver. *Drug Metab. Dispos.* **28**, 268–278.

141. Jung, D., Fantin, A. C., Scheurer, U., Fried, M., and Kullak-Ublick, G. A. (2004) Human ileal bile acid transporter gene ASBT (SLC10A2) is transactivated by the glucocorticoid receptor. *Gut* **53**, 78–84.

142. Alvaro, D., Gigliozzi, A., Marucci, L., Alpini, G., Barbaro, B., Monterubbianesi, R., Minetola, L., Mancino, M. G., Medina, J. F., Attili, A. F., et al. (2002) Corticosteroids modulate the secretory processes of the rat intrahepatic biliary epithelium. *Gastroenterology* **122**, 1058–1069.

143. Brown, J. D. and Plutzky, J. (2007) Peroxisome proliferator-activated receptors as transcriptional nodal points and therapeutic targets. *Circulation* **115**, 518–533.

144. Willson, T. M., Brown, P. J., Sternbach, D. D., and Henke, B. R. (2000) The PPARs: from orphan receptors to drug discovery. *J. Med. Chem.* **43**, 527–550.

145. Kota, B. P., Huang, T. H., and Roufogalis, B. D. (2005) An overview on biological mechanisms of PPARs. *Pharmacol. Res.* **51**, 85–94.

146. Nakata, K., Tanaka, Y., Nakano, T., Adachi, T., Tanaka, H., Kaminuma, T., and Ishikawa, T. (2006) Nuclear receptor-mediated transcriptional regulation in Phase I, II, and III xenobiotic metabolizing systems. *Drug Metab. Pharmacokinet.* **21**, 437–457.

147. Luci, S., Geissler, S., Konig, B., Koch, A., Stangl, G. I., Hirche, F., and Eder, K. (2006) PPARalpha agonists up-regulate organic cation transporters in rat liver cells. *Biochem. Biophys. Res. Commun.* **350**, 704–708.

148. Nie, W., Sweetser, S., Rinella, M., and Green, R. M. (2005) Transcriptional regulation of murine Slc22a1 (Oct1) by peroxisome proliferator agonist receptor-alpha and -gamma. *Am. J. Physiol. Gastrointest. Liver Physiol.* **288**, G207–G212.

149. Barbier, O., Duran-Sandoval, D., Pineda-Torra, I., Kosykh, V., Fruchart, J. C., and Staels, B. (2003) Peroxisome proliferator-activated receptor alpha induces hepatic expression of the human bile acid glucuronidating UDP-glucuronosyltransferase 2B4 enzyme. *J. Biol. Chem.* **278**, 32852–32860.

150. Fang, H. L., Strom, S. C., Cai, H., Falany, C. N., Kocarek, T. A., and Runge-Morris, M. (2005) Regulation of human hepatic hydroxysteroid sulfotransferase gene expression by the peroxisome proliferator-activated receptor alpha transcription factor. *Mol. Pharmacol.* **67**, 1257–1267.

151. Pineda, T. I., Claudel, T., Duval, C., Kosykh, V., Fruchart, J. C., and Staels, B. (2003) Bile acids induce the expression of the human peroxisome proliferator-activated receptor alpha gene via activation of the farnesoid X receptor. *Mol. Endocrinol.* **17**, 259–272.

152. Barbier, O., Villeneuve, L., Bocher, V., Fontaine, C., Torra, I. P., Duhem, C., Kosykh, V., Fruchart, J. C., Guillemette, C., and Staels, B. (2003) The UDP-glucuronosyltransferase

1A9 enzyme is a peroxisome proliferator-activated receptor alpha and gamma target gene. *J. Biol. Chem.* **278**, 13975–13983.

153. Barbier, O., Girard, H., Inoue, Y., Duez, H., Villeneuve, L., Kamiya, A., Fruchart, J. C., Guillemette, C., Gonzalez, F. J., and Staels, B. (2005) Hepatic expression of the UGT1A9 gene is governed by hepatocyte nuclear factor 4alpha. *Mol. Pharmacol.* **67**, 241–249.

154. Chianale, J., Vollrath, V., Wielandt, A. M., Amigo, L., Rigotti, A., Nervi, F., Gonzalez, S., Andrade, L., Pizarro, M., and Accatino, L. (1996) Fibrates induce mdr2 gene expression and biliary phospholipid secretion in the mouse. *Biochem. J.* **314** (Pt. 3), 781–786.

155. Miranda, S., Vollrath, V., Wielandt, A. M., Loyola, G., Bronfman, M., and Chianale, J. (1997) Overexpression of mdr2 gene by peroxisome proliferators in the mouse liver. *J. Hepatol.*, **26**, 1331–1339.

156. Kok, T., Bloks, V. W., Wolters, H., Havinga, R., Jansen, P. L., Staels, B., and Kuipers, F. (2003) Peroxisome proliferator-activated receptor alpha (PPARalpha)-mediated regulation of multidrug resistance 2 (Mdr2) expression and function in mice. *Biochem. J.* **369**, 539–547.

157. Jung, D., Fried, M., and Kullak-Ublick, G. A. (2002) Human apical sodium-dependent bile salt transporter gene (SLC10A2) is regulated by the peroxisome proliferator-activated receptor alpha. *J. Biol. Chem.* **277**, 30559–30566.

158. Wagner, M., Halilbasic, E., Marschall, H. U., Zollner, G., Fickert, P., Langner, C., Zatloukal, K., Denk, H., and Trauner, M. (2005) CAR and PXR agonists stimulate hepatic bile acid and bilirubin detoxification and elimination pathways in mice. *Hepatology* **42**, 420–430.

159. Ritzel, U., Leonhardt, U., Nather, M., Schafer, G., Armstrong, V. W., and Ramadori, G. (2002) Simvastatin in primary biliary cirrhosis: effects on serum lipids and distinct disease markers. *J. Hepatol.* **36**, 454–458.

160. Kurihara, T., Maeda, A., Shigemoto, M., Yamashita, K., and Hashimoto, E. (2002) Investigation into the efficacy of bezafibrate against primary biliary cirrhosis, with histological references from cases receiving long term monotherapy. *Am. J. Gastroenterol.* **97**, 212–214.

161. Kamisako, T. and Adachi, Y. (1995) Marked improvement in cholestasis and hypercholesterolemia with simvastatin in a patient with primary biliary cirrhosis. *Am. J. Gastroenterol.* **90**, 1187–1188.

162. Kanda, T., Yokosuka, O., Imazeki, F., and Saisho, H. (2003) Bezafibrate treatment: a new medical approach for PBC patients? *J. Gastroenterol.* **38**, 573–578.

163. Hunt, M. C., Yang, Y. Z., Eggertsen, G., Carneheim, C. M., Gafvels, M., Einarsson, C., and Alexson, S. E. (2000) The peroxisome proliferator-activated receptor alpha (PPARalpha) regulates bile acid biosynthesis. *J. Biol. Chem.* **275**, 28947–28953.

164. Patel, D. D., Knight, B. L., Soutar, A. K., Gibbons, G. F., and Wade, D. P. (2000) The effect of peroxisome-proliferator-activated receptor-alpha on the activity of the cholesterol 7 alpha-hydroxylase gene. *Biochem. J.* **351** (Pt. 3), 747–753.

165. Takeda, K., Ichiki, T., Tokunou, T., Funakoshi, Y., Iino, N., Hirano, K., Kanaide, H., and Takeshita, A. (2000) Peroxisome proliferator-activated receptor gamma activators downregulate angiotensin II type 1 receptor in vascular smooth muscle cells. *Circulation* **102**, 1834–1839.

166. Szatmari, I., Vamosi, G., Brazda, P., Balint, B. L., Benko, S., Szeles, L., Jeney, V., Ozvegy-Laczka, C., Szanto, A., Barta, E., et al. (2006) Peroxisome proliferator-activated

receptor gamma-regulated ABCG2 expression confers cytoprotection to human dendritic cells. *J. Biol. Chem.* **281**, 23812–23823.

167. Baba, T., Mimura, J., Nakamura, N., Harada, N., Yamamoto, M., Morohashi, K., and Fujii-Kuriyama, Y. (2005) Intrinsic function of the aryl hydrocarbon (dioxin) receptor as a key factor in female reproduction. *Mol. Cell. Biol.* **25**, 10040–10051.

168. Elferink, C. J. (2003) Aryl hydrocarbon receptor-mediated cell cycle control. *Prog. Cell Cycle Res.* **5**, 261–267.

169. Fujii-Kuriyama, Y. and Mimura, J. (2005) Molecular mechanisms of AhR functions in the regulation of cytochrome P450 genes. *Biochem. Biophys. Res. Commun.* **338**, 311–317.

170. Kohle, C. and Bock, K. W. (2007) Coordinate regulation of Phase I and II xenobiotic metabolisms by the Ah receptor and Nrf2. *Biochem. Pharmacol.* **73**, 1853–1862.

171. Maher, J. M., Cheng, X., Slitt, A. L., Dieter, M. Z., and Klaassen, C. D. (2005) Induction of the multidrug resistance-associated protein family of transporters by chemical activators of receptor-mediated pathways in mouse liver. *Drug Metab. Dispos.* **33**, 956–962.

172. Ebert, B., Seidel, A., and Lampen, A. (2005) Identification of BCRP as transporter of benzo[a]pyrene conjugates metabolically formed in Caco-2 cells and its induction by Ah-receptor agonists. *Carcinogenesis* **26**, 1754–1763.

173. Harper, P. A., Riddick, D. S., and Okey, A. B. (2006) Regulating the regulator: factors that control levels and activity of the aryl hydrocarbon receptor. *Biochem. Pharmacol.* **72**, 267–279.

11

EMERGING ROLE OF RETINOID-RELATED ORPHAN RECEPTOR (ROR) AND ITS CROSS TALK WITH LXR (LIVER X RECEPTOR) IN THE REGULATION OF DRUG-METABOLIZING ENZYMES

TAIRA WADA AND WEN XIE

Center for Pharmacogenetics, Department of Pharmaceutical Sciences, University of Pittsburgh, Pittsburgh, PA, USA

11.1   INTRODUCTION

The metabolic homeostasis of foreign chemicals (xenobiotics, including drugs) and endogenous compounds (endobiotics) is essential for the survival of mammals. Xenobiotic metabolism is facilitated by the phase I and phase II drug-metabolizing enzymes, as well as drug transporters [1, 2]. The phase I reactions include oxidation, reduction, hydrolysis, and hydration. The cytochrome P450 (CYP) enzymes play a critical and dominating role in phase I reactions. Phase II metabolism includes sulfation (mediated by sulfotransferases or SULTs), glucuronidation (mediated by UDP-glucuronosyltransferases or UGTs), or glutathione conjugation (mediated by glutathione *S*-transferases or GSTs). Products generated by phase I metabolism are generally more polar and can be better substrates for phase II conjugations. The conjugation by phase II enzymes results in increased polarity and solubility of xenobiotics and promotes xenobiotic excretion. Drug transporters, which are not the focus of this review, are responsible for the uptake and efflux of parent or biotransformed xenobiotics.

Nuclear Receptors in Drug Metabolism   Edited by Wen Xie
Copyright © 2009 John Wiley & Sons, Inc.

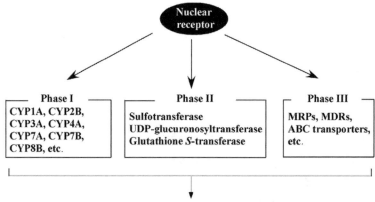

FIGURE 11.1   Proposed model of nuclear receptors as master xenobiotic receptors. Activation of nuclear receptors, such as PXR and CAR, may result in the concerted regulation of phase I and phase II enzymes and the "phase III" drug transporters.

Although the structure and function of phase I and phase II enzymes have been the subjects of extensive studies for several decades and it has been known that the expression of many drug-metabolizing enzymes is inducible, little is known about the transcriptional regulators that control the production of these enzymes. In the late 1990s, the role of nuclear receptors in the transcriptional regulation of drug-metabolizing enzymes began to emerge, which was highlighted by the cloning of the xenobiotic receptor pregnane X receptor (PXR) [3–6] and characterization of the constitutive androstane receptor (CAR) as another important xenobiotic nuclear receptor [7, 8]. In the past 10 years, considerable advances have been made in establishing nuclear receptors as master regulators of the expression of phase I and phase II drug-metabolizing enzymes and transporters (Figure 11.1) [1, 2].

In addition to PXR and CAR, several other nuclear receptors, such as vitamin D receptor (VDR), hepatocyte nuclear factor 4α (HNF4α), peroxisome proliferator-activated receptors (PPARs), and farnesoid X receptor (FXR), have also been implicated in the regulation of xenobiotic enzymes and transporters [9–12]. More recently, results from our lab have shown that the liver X receptor (LXR), a previously known sterol sensor, can also regulate the expression of phase II SULTs, such as the bile acid detoxifying hydroxysteroid sulfotransferase Sult2a9/2a1 [13] and the estrogen sulfotransferase (Est/Sult1e1) [14].

It has also been appreciated that nuclear receptors cross talk in regulating drug-metabolizing enzymes and transporters [1, 2]. There are several mechanisms by which nuclear receptors may cross talk. These include (1) share of DNA binding sites. Examples include shared regulation of CYP3A and CYP2B by PXR and CAR; shared regulation of CYP3A by PXR and VDR; shared regulation of MRP2 by PXR, CAR, and FXR; and shared regulation of Sult2a9/2a1 by PXR, FXR, CAR, and LXR. (2) Presence of multiple NR binding sites on the same target gene promoter. We have recently shown that PXR, LXR, and PPARγ cooperate to regulate the free fatty acid transporter CD36. In this case, three distinct NR response elements, a DR3/PXRE, a DR7/LXRE (LXR response element), and a DR1/PPRE, were found to cluster in 500 bp sequences of the mouse Cd36 gene promoter [75]. (3) Receptor mutual suppression. Having known that activation of PXR promotes bilirubin detoxification [17], we were surprised to find that the PXR null mice also showed an increased bilirubin clearance [18]. We proposed that the ligand-free PXR suppresses the constitutive activity of CAR. The increased bilirubin clearance in PXR null mice was the result of CAR derepression as a consequence of loss of PXR. Further studies suggested that the suppression was, at least in part, due to the disruption of ligand-independent recruitment of coactivator by CAR. Nuclear receptor cross talk has not only expanded the function of individual receptors but also plays an important role in forming the fail-safe metabolic safety net to protect against xeno- and endobiotic toxicants.

It is conceivable that nuclear receptor-mediated regulation of drug-metabolizing enzymes has implications in drug metabolism, drug toxicity, and drug–drug interactions. Interestingly, the same metabolic enzyme and transporter systems are responsible for the metabolism of numerous endobiotics, such as steroid hormones, bile acids, and lipid metabolites [15]. Consistent with this notion, studies have implicated nuclear receptor-mediated enzyme and transporter regulation in many pathophysiological conditions by impacting the homeostasis of endobiotics. Examples include the role of PXR and CAR in hyperbilirubinemia [16–18] and cholestasis [6, 19–22], and the role of LXR in cholestasis [13] and estrogen homeostasis [14].

11.2   ORPHAN NUCLEAR RECEPTOR RORα

RORs are members of the nuclear receptors that include three isoforms: RORα (NR1F1), RORβ (NR1F2), and RORγ (NR1F3) [23]. Each ROR isoform has a distinct pattern of tissue distribution [24]. RORα is expressed in many tissues, including cerebella Purkinje cells, liver, thymus, skeletal muscle, skin, lung, and kidney [25, 26]. RORβ exhibits a more restricted pattern of expression and is expressed in several regions of the central nervous system, retina, and pineal gland. RORγ is most highly expressed in thymus, but it is also detectable in many other tissues, including liver, kidney, and muscle [23, 27, 28]. The functional ligands of RORs remain elusive. It was suggested that cholesterol and its sulfonated derivatives might function as RORα ligands [29]. Other ligands suggested to bind to RORα are thiazolidinediones [30] and melatonin [31]. To our knowledge, none of these have been established as functional ligands for RORα. Evidence has been provided indicating that certain retinoids, including all-trans retinoic acid, can function as partial antagonists for RORβ and RORγ [32].

RORs bind either as monomers to the ROR response elements (RORE) within the target gene promoter regions. ROREs are composed of 6 bp A/T-rich region immediately preceding a consensus AGGTCA motif [28, 33]. However, RORα has been demonstrated to be able to bind as a homodimer to direct repeats of the consensus AGGTCA motif separated by two base pairs (DR2) [34]. Like many other nuclear

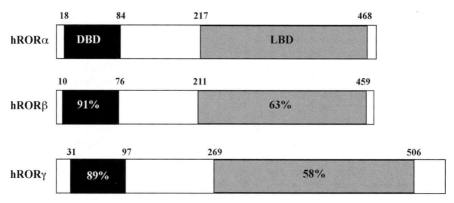

FIGURE 11.2 Comparison of the homology among three ROR isoforms. Percentages indicate amino acid identities in the DNA binding domain (DBD) and ligand binding domain (LBD) compared to RORα. The positions of amino acids are also labeled.

receptors, RORs are composed of an N-terminal DBD and an LBD located at the C-terminal. ROR isoforms share a highly conserved DBD but they are more diverse in their LBDs (Figure 11.2). The transcriptional activities of RORs are negatively and positively regulated through the recruitment of nuclear receptor corepressors and coactivators, respectively. It has been reported that corepressors (N-CoR, RIP140, and SMRT) and coactivators (GRIP, PBP, SRC-1, CBP, and PGC1α) can interact with RORα [34–38]. It was proposed that cell-specific interactions with specific coregulators may contribute to the molecular mechanism for distinct physiological functions of RORα [34].

Characterization of ROR null mice has revealed a number of important physiological functions of RORs. RORα$^{-/-}$ mice show an ataxic phenotype similar to that observed in the staggerer (sg/sg) mutant mice, which carry a natural deletion in the LBD, causing a frame shift and a truncated RORα protein [25, 26]. The RORα$^{sg/sg}$ mice exhibit a variety of phenotypes, including cerebella degeneration, abnormal circadian behavior, vascular dysfunction, muscular irregularities, osteoporosis, atherosclerosis, and altered immune responses [39–45], as summarized in Table 11.1. The atherosclerotic phenotype in the RORα$^{sg/sg}$ mice was associated with a marked hypo-α-lipoproteinemia due to a decreased expression of ApoAI, a transcriptional target of RORα in the intestine [49]. In the vascular system, RORα is involved in postischemic angiogenesis and differentiation and contractile function of smooth muscle cells [48]. The RORα$^{sg/sg}$ mice exhibit more extensive angiogenesis, leading to an increase of inflammatory cytokine production and eNOS protein level [47], suggesting RORα as a potent negative regulator of ischemia-induced angiogenesis. Inflammation is associated with both atherosclerosis and angiogenesis. The RORα$^{sg/sg}$ mice also have a delayed lymphocyte development, which may be accounted for by RORα-mediated transcriptional activation of IκκB, the inhibitor of NF-κB [46].

Among other ROR isoforms, RORβ is believed to be involved in the processing of sensory information, as RORβ$^{-/-}$ mice showed significant phenotypes in circadian

TABLE 11.1 Summary of Known Functions of RORα

Physiological Functions	References
Inflammation: Anti-inflammation (suppression of NF-κB activity by activating IκκB)	[34, 39, 40, 41, 46]
Circadian rhythm: A crucial component of the molecular circadian clock (regulating the clock gene BMAL1)	[42]
Angiogenesis: Promote ischemia-induced angiogenesis	[44, 47, 48]
Osteoporosis: Protection against osteoporosis (regulating bone sialoglycoprotein and osteocalcin)	[43, 45]
Metabolic homeostasis: Regulation of triglyceride and lipoprotein metabolism and protection against atherosclerosis (regulating ApoAI and ApoCIII)	[49, 50]
Xenosensor: Positive and negative regulation of phase I and phase II drug-metabolizing enzymes	[51, 52]

behaviors and retinal degeneration [53]. RORβ has been reported to regulate the blue opsin gene in cone photoreceptor development [54]. RORγ $^{-/-}$ mice lacked all lymph nodes and Peyer's patches, and contained reduced number of thymocytes [37, 55, 56], suggesting that RORγ plays an essential role in lymphoid organogenesis and thymopoiesis. Recent studies have demonstrated an important role for both RORα and RORγ in the differentiation of naive T cells into Th17 cells [57, 58].

11.3 A POTENTIAL ROLE OF RORs IN XENO- AND ENDOBIOTIC GENE REGULATION

Both RORα and RORγ have been shown to be expressed in the liver [51]. However, the hepatic function of these two receptors remains unknown until recently. To determine the role of RORα in the liver, we examined the gene expression profiles in livers of the RORα$^{sg/sg}$, RORγ null, and RORα/RORγ double knock-out (DKO) mice by microarray analysis [51]. To our surprise, the microarray results suggested that loss of RORα and/or RORγ had a major effect on the expression of multiple drug-metabolizing enzymes and transporters, as summarized in Table 11.2. In the ROR DKO mice, major regulations include the activation of Cyp2b9/10, Cyp4a10 and Cyp4a14, Est/Sult1e1 and Sult2a1/2a1, and suppression of Cyp7b1, Cyp8b1, Hsd3b4, and Hsd3b5. Interestingly, some genes are selectively controlled by RORα or RORγ, whereas some other genes are affected by both receptors. For example, Hsd3b4 and Hsd3b5, two enzymes involved in the deactivation of steroid hormone, were dramatically decreased in ROR DKO mice, but not in single knock-out mice [51].

Some of the gene regulations observed in the ROR loss of function knock-out mouse models have been further supported by studies using ROR gain of function experiments. For example, we showed that overexpression of RORα by transfection in primary mouse hepatocytes or mouse liver *in vivo* suppressed the expression of

TABLE 11.2   Summary of Major Regulation of Drug-Metabolizing Enzymes in the RORα Null Mice

		Microarray Regulation[a]	Confirmed by Northern Blot	Confirmed Real-Time PCR
Phase I enzymes	Cyp2b9	↑	+	−
	Cyp2b10	→	+	−
	Cyp2b13	↑	−	−
	Cyp2e1	→	−	−
	Cyp3a11	→	−	−
	Cyp3a25	→	−	−
	Cyp4a1	↑	−	+
	Cyp4a10	↑	−	+
	Cyp7b1	↓	+	+
	Cyp8b1	→	−	+
Phase II enzymes	Sult1e1	↑	+	+
	Sult2a1	→	+	+
	Sult1d1	→	−	−
	Sult1a1	→	−	−
	Gstt3	→	−	−
	Gstm4	→	−	−
	Gsta1	→	−	−
	Gstp1	→	−	−
	Hsd3b4	↓	+	−
	Hsd3b5	↓	+	+

[a] ↑, Up-regulated; ↓ down-regulated; →, no change.

Est/Sult1e1 and Sult2a9/2a1, consistent with the activation of the same genes in the RORα$^{sg/sg}$ mice [51].

To study the mechanism by which RORs affect the expression of CYP genes, we focused on the regulation of oxysterol 7α-hydroxylase (Cyp7b1), an enzyme that plays an important role in the homeostasis of cholesterol, oxysterols, and bile acids [59, 60]. The expression of Cyp7b1 gene was suppressed in the RORα$^{sg/sg}$ mice [51, 52], suggesting RORα as a positive regulator of Cyp7b1. Promoter analysis established Cyp7b1 as a transcriptional target of RORα and a functional RORE was identified in the Cyp7b1 gene promoter. Moreover, transfection of RORα induced the expression of endogenous Cyp7b1 in the mouse liver. Interestingly, Cyp7b1 regulation appeared to be RORα specific, as RORγ had little effect, consistent with the notion that RORα, but not RORγ, may play a dominating role in phase I and phase II enzyme regulation. Our microarray analysis also suggests that, although loss of RORγ alone affected the expression of some enzymes, the overall effect of RORγ null on metabolic enzyme regulation was not as dramatic as that observed in the RORα null or RORα/RORγ DKO mice [51]. The molecular mechanism by which RORs affect the expression of other drug-metabolizing enzymes remains to be established.

Although the primary focus of this article is the regulation of xeno- and endobiotic gene expression, it is worthwhile to mention that in addition to the regulation of drug-metabolizing enzyme and transporter genes, loss of RORα and/or RORγ also affect the expression of many other genes in the liver. Moreover, RORα$^{sg/sg}$ mice, but not the RORγ null mice, showed lower plasma triglyceride and cholesterol levels than the control mice. In contrast, RORγ null mice, but not RORα$^{sg/sg}$ mice, exhibited lower blood glucose levels. These results suggested RORα and RORγ may play distinct roles in the regulation of triglyceride and glucose homeostasis, respectively [51].

11.4   LXR AND ITS REGULATION OF DRUG-METABOLIZING ENZYMES

LXRs, both the α and β isoforms (NR1H2, 3), were cloned and initially defined as sterol sensors that can be activated by the endogenous cholesterol derivative hydroxycholesterols, as well as synthetic LXR agonists, such as T0901317 (TO1317) and GW3965 [61, 62]. Upon ligand activation, LXRs regulate gene expression by heterodimerization with the retinoid X receptor (RXR) and subsequent binding of LXR–RXR heterodimers to LXREs present in target gene promoters. LXRα is highly expressed in the liver and is also found in adipose, intestine, kidney, and macrophages, whereas LXRβ is ubiquitously expressed [63, 64].

LXR was first shown to have an antiatherosclerogenic effect by favoring an overall increase in cholesterol removal, while decreasing endogenous cholesterol synthesis and dietary absorption [64–66]. Despite their promises as antiatherosclerogenic targets, LXRs were linked to prolipogenic effects by activating the sterol regulatory element binding protein 1c (SREBP-1c), a transcriptional factor that regulates the expression of a battery of lipogenic enzymes, including stearoyl CoA desaturase-1 (SCD-1), acetyl CoA carboxylase (ACC), and fatty acid synthase (FAS) [65, 67–69]. In macrophages, activation of LXRs results in the efflux of cholesterol via the up-regulation of LXR target genes ABCA1, ABCG1, and ApoE [70]. Treatment with LXR agonists in macrophages prevented bacterial or LPS-triggered induction of inflammatory signals. LXR signaling can impact antimicrobial responses by regulating macrophage gene expression and apoptosis [71, 72].

LXRs have been recently shown to regulate drug-metabolizing enzymes, including SULTs, suggesting a broader function of LXR beyond being sterol sensors. We have recently reported that LXRs can regulate Sult2a9/2a1 and impact sensitivity to bile acid toxicity and cholestasis [13]. Genetic (using VP-LXRα transgene) or pharmacological (using a LXR agonist) activation of LXR in mice conferred a resistance to lithocholic acid (LCA)-induced hepatotoxicity and bile duct ligation (BDL)-induced cholestasis in female mice. In contrast, LXR DKO mice deficient of both the α and β isoforms exhibited heightened cholestatic sensitivity. LXR-mediated cholestatic resistance was associated with an increased expression of Sult2a9/2a1 and several bile acid transporters, whereas the basal expression of these gene products was reduced in the LXR DKO mice. In the same study, promoter analysis established Sult2a9/2a1 as a LXR target gene [13]. We also showed that activation of LXRs suppressed the expression of the oxysterol 7α-hydroxylase (Cyp7b1), which may lead to increased levels of the LXR-activating oxysterols. On the basis of these results, we propose that LXRs have evolved to have dual function in maintaining cholesterol and bile acid homeostasis by increasing cholesterol catabolism and, at the same time, preventing toxicity from bile acid accumulation [13].

In another independent study, we showed that LXR controls estrogen homeostasis by regulating the basal and inducible hepatic expression of Est/Sult1e1, a designated SULT that catalyzes the sulfonation and deactivation of estrogens [73]. Genetic or pharmacological activation of LXR resulted in Est/Sult1e1 induction, which in turn inhibited estrogen-dependent uterine epithelial cell proliferation and gene expression, as well as estrogen-dependent breast cancer growth in a nude mouse model of tumorigenicity. We further established that Est/Sult1e1 is a transcriptional target of LXR and a deletion of the Est/Sult1e1 gene in mice abolished the LXR effect on estrogen deprivation [14]. In the same study, it was found that Est/Sult1e1 regulation by LXR appeared to be liver specific, further underscoring the role of liver in estrogen metabolism. Activation of LXR failed to induce other major estrogen-metabolizing enzymes, suggesting that the LXR effect on estrogen metabolism is Est/Sult1e1 specific [14]. This study has revealed a novel mechanism controlling estrogen homeostasis *in vivo* and may have implications for drug development in the treatment of breast cancer and other estrogen-related cancerous endocrine disorders.

In addition to its regulation on SULTs, LXR may influence the expression of phase I enzymes and phase II enzymes other than SULTs. It was reported that loss of both LXR isoforms in mice resulted in an increased basal expression of Cyp3a11 and 2b10 [74]. In contrast, activation of LXR resulted in a suppression of Cyp3a11 expression, but had little effect on the expression of Ugt1a1 [13]. The mechanisms by which LXR affects the expression of phase I enzymes remain to be established.

11.5   A FUNCTIONAL CROSS TALK BETWEEN RORα AND LXR IN THE REGULATION OF XENO- AND ENDOBIOTIC GENES

The possibility of RORα–LXR cross talk was initially indicated by the remarkable overlap in the pattern of genes affected in livers from the RORα$^{sg/sg}$ and LXR-activated mice. As mentioned earlier, activation of LXR in mice induced the expression of Est/Sult1e1 and Sult2a9/2a1. In the same LXR-activated mice, the expression of Cyp7b1 was suppressed [13], whereas the expression of CD36, a fatty acid uptake transporter, was induced [75]. Remarkably, the same pattern of gene regulation was observed in the RORα$^{sg/sg}$ mice. These results suggest that RORα and LXR may be mutually suppressive *in vivo*.

To obtain support for this hypothesis, we examined the mutual suppression between RORα and LXR using the regulation of Cyp7b1 as a model system. Having established RORα as a positive Cyp7b1 regulator and knowing LXR suppresses the expression of Cyp7b1, we hypothesized that LXR may suppress Cyp7b1 gene expression by inhibiting the RORα activity. Indeed, we showed that the activation of Cyp7b1 promoter by RORα was suppressed by cotransfection of LXRα, even in the absence of LXR agonists. The inhibitory effect of LXRα was enhanced by the LXR agonist TO1317. The inhibitory effect of LXRα was largely abolished when the RORE was mutated, suggesting that the inhibition was mediated by RORα [52]. The inhibitory effect of LXR on ROR was also seen when the RORE-containing synthetic reporter genes were used [52]. The LXRα activity was reciprocally suppressed by RORα. tk-MTV is a LXR-responsive reporter gene [76]. We showed that the activation of tk-MTV by LXRα was inhibited by cotransfection of RORα in a dose-dependent manner.

Our further studies suggest that, at least in cultured cells, the mutual suppression between RORα and LXR may be due to their competition for the common nuclear receptor coactivators. RORα is known to interact with nuclear receptor coactivators without an exogenously added ligand [77, 78]. We showed that LXRα also exhibited ligand-independent interaction with the nuclear receptor coactivator SRC-1 as confirmed by both mammalian two-hybrid assay and chromatin immunoprecipitation (ChIP) analysis. We hypothesize that the ligand-independent recruitment of coactivator accounts for the constitutive activities of both RORα and LXR, and coactivator competition may represent a plausible mechanism for the mutual suppression of transcriptional activity between these two receptors. However, we cannot exclude the possibility that the "constitutive" activity of RORα may have resulted from the binding of an endogenous ligand to this "orphan receptor."

The potential functional cross talk between RORα and LXR was further investigated *in vivo*. For this purpose, we measured the expression of LXR target genes and ROR target genes in the RORα$^{sg/sg}$ and LXR DKO mice, respectively. As summarized in Table 11.3, in the female RORα$^{sg/sg}$ mice, in addition to the activation of Est/Sult1e1, Sult2a9/2a1, Cd36, and Cyp7b1, the expression of other LXR target genes, such as lipoprotein lipase (Lpl) [79], aldo-ketoreductase 1d1 (Akr1d1) [80],

TABLE 11.3 Regulation of LXR Target Genes in Female RORα Null Mice

LXR Target Genes	Regulation^a
Est/Sult1e1	↑
Sult2a9/2a1	↑
Cd36	↑
Lpl	↑
Fas	↑
Akr1d1	↑
SR-B1	↑
Cyp7a1	→
Acc-1	↑
ApoE	→
Scd1	→
Srebp-1c	↓

^{a}↑, Up-regulated; →, no change; ↓, down-regulated.

TABLE 11.4   Regulation of ROR Target Genes in Female LXR DKO Mice

ROR Target Genes	Regulation[a]
Bmal1	↑
ApoAI	↑
p21	↑
ApoCIII	→
Ikkβ	↑
Rev-erbα	→

[a] ↑, Up-regulated; →, no change.

scavenger receptor BI (SR-BI) [81], and acetyl CoA carboxylase 1 (Acc-1), was also significantly induced. However, the expression of Srebp-1c was significantly suppressed, whereas the expression of ApoE, ApoAI, and Sed-1 was not affected. When the expression of RORα target genes was measured in the LXR DKO mice, we found that the expression of Bmal1 [82], ApoAI [49], p21 [83], and Ikkβ [46] was induced in LXR DKO female mice, but the expression of ApoCIII [50], and Rev-erbα [77] was not significantly altered (Table 11.4). It is interesting to note that the mutual activation of target gene expression in the RORα$^{sg/sg}$ and LXR DKO mice is gene specific. The mechanism for this selective gene regulation remains to be determined. Cross talk between RORα and LXR can involved several mechanisms, including competition for coactivators and DNA binding sites. In addition, the promoter context might be a determining factor. The inhibition of RORα-mediated Cyp7b1 activation by LXR appears to involve competition for common coactivators. Repression of LXR target genes by RORα may also be mediated through cross talk involving adjacent or distant ROREs.

The activation of LXR target genes in the RORα$^{sg/sg}$ mice has its physiological outcome. We showed the RORα$^{sg/sg}$ mice had increased hepatic triglyceride accumulation. The expression of Srebp-1c, a LXR target gene, however, was not induced in the RORα$^{sg/sg}$ mice [52]. We reason that the hepatic steatotic phenotype in RORα$^{sg/sg}$ mice is most likely accounted by the activation of Cd36, another LXR target gene. Cd36, a fatty acid transporter, facilitates the uptake of free fatty acids from the circulation and their subsequent convertion into triglycerides. We have recently shown that the steatotic effect of both PXR [84] and LXR [75] was associated with the activation of Cd36. Moreover, the steatotic effect of LXR agonists was largely abolished in mice deficient of Cd36 [75]. The LXR–ROR cross talk and its potential implications in physiology are summarized in Figure 11.3.

11.6   CLOSING REMARKS

Our recent findings have clearly suggested that RORs can positively or negatively regulate the expression of drug-metabolizing enzymes in a gene-specific manner.

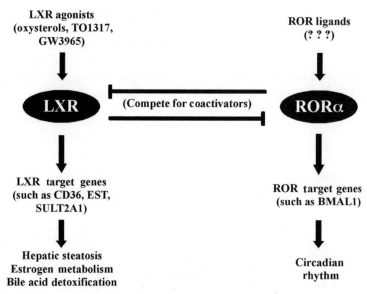

FIGURE 11.3   Proposed model of ROR–LXR cross talk and its potential implications in physiology.

Although RORα regulates the expression of certain drug-metabolizing enzymes, it is questionable that RORα is a traditional "xenobiotic receptor." Unlike PXR and CAR that are mostly associated with positive xenobiotic gene regulation, RORα can exert both positive and negative regulation in a gene-specific manner. Another hallmark of xenobiotic receptors is their wide spectrum of xenobiotic ligands. Unlike their xenobiotic receptor counterparts, no physiological ROR agonists have been reported despite efforts from several laboratories [29, 32].

The cross talk between RORα and LXR is of particular interest. The cross talk between xenobiotic receptors PXR and CAR has been reported, in which the mechanism appears to be the share of DNA response elements and thus share of target genes [18]. The RORα–LXR cross talk appears to involve the competition of common coactivators, adding another dimension of complexity to the nuclear receptor-mediated metabolic safety net.

Although the role of RORs in xeno- and endobiotic gene regulation has emerged, there are several outstanding challenges. First, the physiological relevance of ROR-mediated xeno- and endobiotic gene regulation remains to be further defined. For example, the oxysterol levels were increased in mice deficient of Cyp7b1, presumably due to a defect in the conversion of oxysterols to bile acids [85]. It would be interesting to know whether the decreased basal expression of Cyp7b1 in the RORα$^{sg/sg}$ mice is associated with accumulation of oxysterols, the endogenous LXR agonists. It is also unknown whether the activation of Sult2a9/2a1 and Est/Sult1e1 in the RORα$^{sg/sg}$ mice will be associated with decreased sensitivity to bile acid toxicity and compromised estrogen responses, respectively. Second, having established Cyp7b1 as a RORα target gene, the molecular mechanism for the activation of Sult2a9/2a1 and Est/Sult1e1 in the RORα$^{sg/sg}$ mice remains to be established. It would be interesting to determine whether RORs exert their effect on Sult2a9/2a1 and Est/Sult1e1 gene expression via cross talk with LXR. Finally, continued effort should be made to identify or develop functional ROR ligands, which will provide valuable pharmacological tools to dissect the function of RORs.

REFERENCES

1. Sonoda, J., Rosenfeld, J. M., Xu, L., Evans, R. M., and Xie, W. (2003) A nuclear receptor-mediated xenobiotic response and its implication in drug metabolism and host protection. *Curr. Drug Metab.* **4**, 59–72.

2. Xie, W., Uppal, H., Saini, S. P., Mu, Y., Little, J. M., Radominska-Pandya, A., and Zemaitis, M. A. (2004) Orphan nuclear receptor-mediated xenobiotic regulation in drug metabolism. *Drug Discov. Today* **9**, 442–449.

3. Kliewer, S. A., Moore, J. T., Wade, L., Staudinger, J. L., Watson, M. A., Jones, S. A., McKee, D. D., Oliver, B. B., Willson, T. M., Zetterstrom, R. H., et al. (1998) An orphan nuclear receptor activated by pregnanes defines a novel steroid signaling pathway. *Cell* **92**, 73–82.

4. Blumberg, B., Sabbagh, W., Jr., Juguilon, H., Bolado, J., Jr., van Meter, C. M., Ong, E. S., and Evans, R. M. (1998) SXR, a novel steroid and xenobiotic-sensing nuclear receptor. *Genes Dev.* **12**, 3195–3205.

5. Xie, W., Barwick, J. L., Downes, M., Blumberg, B., Simon, C. M., Nelson, M. C., Neuschwander-Tetri, B. A., Brunt, E. M., Guzelian, P. S., and Evans, R. M. (2000) Humanized xenobiotic response in mice expressing nuclear receptor SXR. *Nature* **406**, 435–439.

6. Staudinger, J. L., Goodwin, B., Jones, S. A., Hawkins-Brown, D., MacKenzie, K. I., LaTour, A., Liu, Y., Klaassen, C. D., Brown, K. K., Reinhard, J., et al. (2001) The nuclear receptor PXR is a lithocholic acid sensor that protects against liver toxicity. *Proc. Natl Acad. Sci. USA* **98**, 3369–3374.

7. Honkakoski, P., Zelko, I., Sueyoshi, T., and Negishi, M. (1998) The nuclear orphan receptor CAR-retinoid X receptor heterodimer activates the phenobarbital-responsive enhancer module of the CYP2B gene. *Mol. Cell. Biol.* **18**, 5652–5658.

8. Wei, P., Zhang, J., Egan-Hafley, M., Liang, S., and Moore, D. D. (2000) The nuclear receptor CAR mediates specific xenobiotic induction of drug metabolism. *Nature* **407**, 920–923.

9. Makishima, M., Lu, T. T., Xie, W., Whitfield, G. K., Domoto, H., Evans, R. M., Haussler, M. R., and Mangelsdorf, D. J. (2002) Vitamin D receptor as an intestinal bile acid sensor. *Science* **296**, 1313–1316.

10. Hayhurst, G. P., Lee, Y. H., Lambert, G., Ward, J. M., and Gonzalez, F. J. (2001) Hepatocyte nuclear factor 4alpha (nuclear receptor 2A1) is essential for maintenance of hepatic gene expression and lipid homeostasis. *Mol. Cell. Biol.* **21**, 1393–1403.

11. Ito, O., Yasuhiro-Nakamura, Y., Tan, L., Ishizuka, T., Sasaki, Y., Minami, N., Kanazawa, M., Ito, S., Sasano, H., and Kohzuki, M. (2006) Expression of cytochrome P-450 4 enzymes in the kidney and liver: regulation by PPAR and species-difference between rat and human. *Mol. Cell. Biochem.* **10**, 141–148.

12. Marschall, H. U., Wagner, M., Bodin, K., Zollner, G., Fickert, P., Gumhold, J., Silbert, D., Fuchsbichler, A., Sjövall, J., and Trauner, M. (2006) Fxr(−/−) mice adapt to biliary obstruction by enhanced phase I detoxification and renal elimination of bile acids. *J. Lipid Res.* **47**, 582–592.

13. Uppal, H., Saini, S. P., Moschetta, A., Mu, Y., Zhou, J., Gong, H., Zhai, Y., Ren, S., Michalopoulos, G. K., Mangelsdorf, D. J., et al. (2007) Activation of LXRs prevents bile acid toxicity and cholestasis in female mice. *Hepatology* **45**, 422–432.

14. Gong, H., Guo, P., Zhai, Y., Zhou, J., Uppal, H., Jarzynka, M. J., Song, W., Cheng, S., and Xie, W. (2007) Estrogen deprivation and inhibition of breast cancer growth in vivo through activation of the orphan nuclear receptor LXR. *Mol. Endocrinol.* **21**, 1781–1790.

15. Gonzalez, F. J. (1992) Human cytochromes P450: problems and prospects. *Trends Pharmacol. Sci.* **13**, 346–352.

16. Huang, W., Zhang, J., Chua, S. S., Qatanani, M., Han, Y., Granata, R., and Moore, D. D. (2003) Induction of bilirubin clearance by the constitutive androstane receptor (CAR). *Proc. Natl Acad. Sci. USA* **100**, 4156–4161.

17. Xie, W., Yeuh, M. F., Radominska-Pandya, A., Saini, S. P., Negishi, Y., Bottroff, B. S., Cabrera, G. Y., Tukey, R. H., and Evans, R. M. (2003) Control of steroid, heme, and carcinogen metabolism by nuclear pregnane X receptor and constitutive androstane receptor. *Proc. Natl Acad. Sci. USA* **100**, 4150–4155.

18. Saini, S. P., Mu, Y., Gong, H., Toma, D., Uppal, H., Ren, S., Li, S., Poloyac, S. M., and Xie, W. (2005) Dual role of orphan nuclear receptor pregnane X receptor in bilirubin detoxification in mice. *Hepatology* **41**, 497–505.

19. Xie, W., Radominska-Pandya, A., Shi, Y., Simon, C. M., Nelson, M. C., Ong, E. S., Waxman, D. J., and Evans, R. M. (2001) An essential role for nuclear receptors SXR/PXR in detoxification of cholestatic bile acids. *Proc. Natl Acad. Sci. USA* **100**, 3375–3380.

20. Uppal, H., Toma, D., Saini, S. P., Ren, S., Jones, T. J., and Xie, W. (2005) Combined loss of orphan receptors PXR and CAR heightens sensitivity to toxic bile acids in mice. *Hepatology* **41**, 168–176.

21. Zhang, J., Huang, W., Qatanani, M., Evans, R. M., and Moore, D. D. (2004) The constitutive androstane receptor and pregnane X receptor function coordinately to prevent bile acid-induced hepatotoxicity. *J. Biol. Chem.* **279**, 49517–49522.

22. Stedman, C. A., Liddle, C., Coulter, S. A., Sonoda, J., Alvarez, J. G., Moore, D. D., Evans, R. M., and Downes, M. (2005) Nuclear receptors constitutive androstane receptor and pregnane X receptor ameliorate cholestatic liver injury. *Proc. Natl Acad. Sci. USA* **102**, 2063–2068.

23. Jetten, A. M., Kurebayashi, S., and Ueda, E. (2001) The ROR nuclear orphan receptor subfamily: critical regulators of multiple biological processes. *Prog. Nucleic Acid Res. Mol. Biol.* **69**, 205–247.

24. Carlberg, C., Hooft van Huijsduijnen, R., Staple, J. K., DeLamarter, J. F., and Becker-Andre, M. (1994) RZRs, a new family of retinoid-related orphan receptors that function as both monomers and homodimers. *Mol. Endocrinol.* **8**, 757–770.

25. Hamilton, B. A., Frankel, W. N., Kerrebrock, A. W., Hawkins, T. L., FitzHugh, W., Kusumi, K., Russell, L. B., Mueller, K. L., van Berkel, V., Birren, B. W., et al. (1996) Disruption of the nuclear hormone receptor RORalpha in staggerer mice. *Nature* **379**, 736–739.

26. Steinmayr, M., Andre, E., Conquet, F., Rondi-Reig, L., Delhaye-Bouchaud, N., Auclair, N., Daniel, H., Crepel, F., Mariani, J., Sotelo, C., et al. (1998) Staggerer phenotype

in retinoid-related orphan receptor alpha-deficient mice. *Proc. Natl Acad. Sci. USA* **95**, 3960–3965.

27. Andre, E., Gawlas, K., Steinmayr, M., and Becker-Andre, M. (1998) A novel isoform of the orphan nuclear receptor RORbeta is specifically expressed in pineal gland and retina. *Gene* **216**, 277–283.

28. Medvedev, A., Yan, Z. H., Hirose, T., Giquere, V., and Jetten, A. M. (1996) Cloning of a cDNA encoding the murine orphan receptor RZR/ROR gamma and characterization of its response element. *Gene* **181**, 199–206.

29. Kallen, J. A., Schlaeppi, J. M., Bitsch, F., Geisse, S., Geiser, M., Delhon, I., and Fournier, B. (2002) X-ray structure of the hRORalpha LBD at 1.63 Å:structural and functional data that cholesterol or a cholesterol derivative is the natural ligand of RORalpha. *Structure (Camb.)* **10**, 1697–1707.

30. Missbach, M., Jagher, B., Sigg, I., Nayeri, S., Carlberg, C., and Wiesenberg, I. (1998) Thiazolidine diones, specific ligands of the nuclear receptor retinoid Z receptor/retinoid acid receptor-related orphan receptor a with potent antiarthritic activity. *J. Biol. Chem.* **271**, 13515–13522.

31. Wiesenberg, I., Missbach, M., Kahlen, J. P., Schräder, M., and Carlberg, C. (1995) Transcriptional activation of the nuclear receptor RZR alpha by the pineal gland hormone melatonin and identification of CGP 52608 as a synthetic ligand. *Nucleic Acids Res.* **23**, 327–333.

32. Stehlin-Gaon, C., Willmann, D., Zeyer, D., Sanglier, S., Van Dorsselaer, A., Renaud, J. P., Moras, D., and Schule, R. (2003) All-trans retinoic acid is a ligand for the orphan nuclear receptor ROR beta. *Nat. Struct. Biol.* **10**, 820–825.

33. Giguere, V., Tini, M., Flock, G., Ong, E., Evans, R. M., and Otulakowski, G. (1994) Isoform-specific amino-terminal domains dictate DNA-binding properties of ROR alpha, a novel family of orphan hormone nuclear receptors. *Genes Dev.* **8**, 538–553.

34. Harding, H. P., Atkins, G. B., Jaffe, A. B., Seo, W. J., and Lazar, M. A. (1997) Transcriptional activation and repression by RORalpha, an orphan nuclear receptor required for cerebellar development. *Mol. Endocrinol.* **11**, 1737–1746.

35. Atkins, G. B., Hu, X., Guenther, M. G., Rachez, C., Freedman, L. P., and Lazar, M. A. (1999) Coactivators for the orphan nuclear receptor RORalpha. *Mol. Endocrinol.* **13**, 1550–1557.

36. Gold, D. A., Baek, S. H., Schork, N. J., Rose, D. W., Larsen, D. D., Sachs, B. D., Rosenfeld, M. G., and Hamilton, B. A. (2003) RORalpha coordinates reciprocal signaling in cerebellar development through sonic hedgehog and calcium-dependent pathways. *Neuron* **40**, 1119–1131.

37. Jetten, A. M. and Joo, J. H. (2006) Retinoid-related orphan receptors (RORs): roles in cellular differentiation and development. *Adv. Dev. Biol.* **16**, 314–354.

38. Liu, C., Li, S., Liu, T., Borjigin, J., and Lin, J. D. (2007) Transcriptional coactivator PGC-1alpha integrates the mammalian clock and energy metabolism. *Nature* **447**, 447–481.

39. Jarvis, C. I., Staels, B., Brugg, B., Lemaigre-Dubreuil, Y., Tedgui, A., and Mariani, J. (2002) Age-related phenotypes in the staggerer mouse expand the RORalpha nuclear receptor's role beyond the cerebellum. *Mol. Cell. Endocrinol.* **186**, 1–5.

40. Jaradat, M., Stapleton, C., Tilley, S. L., Dixon, D., Erikson, C. J., McCaskill, J. G., Kang, H. S., Angers, M., Liao, G., Collins, J., et al. (2006) Modulatory role for retinoid-related

orphan receptor alpha in allergen-induced lung inflammation. *Am. J. Respir. Crit. Care Med.* **174**, 1299–1309.

41. Stapleton, C. M., Jaradat, M., Dixon, D., Kang, H. S., Kim, S. C., Liao, G., Carey, M. A., Cristiano, J., Moorman, M. P., and Jetten, A. M. (2005) Enhanced susceptibility of staggerer (ROR{alpha}sg/sg) mice to lipopolysaccharide-induced lung inflammation. *Am. J. Physiol. Lung Cell. Mol. Physiol.* **280**, 144–152.

42. Akashi, M. and Takumi, T. (2005) The orphan nuclear receptor RORalpha regulates circadian transcription of the mammalian core-clock Bmal1. *Nat. Struct. Mol. Biol.* **12**, 441–448.

43. Meyer, T., Kneissel, M., Mariani, J., and Fournier, B. (2000) In vitro and in vivo evidence for orphan nuclear receptor RORalpha function in bone metabolism. *Proc. Natl Acad. Sci. USA* **97**, 9197–9202.

44. Lau, P., Bailey, P., Dowhan, D. H., and Muscat, G. E. (1999) Exogenous expression of a dominant negative RORalpha1 vector in muscle cells impairs differentiation: RORalpha1 directly interacts with p300 and myoD. *Nucleic Acids Res.* **27**, 411–420.

45. Besnard, S., Bakouche, J., Lemaigre-Dubreuil, Y., Mariani, J., Tedgui, A., and Henrion, D. (2002) Smooth muscle dysfunction in resistance arteries of the sg/sg mouse, a mutant of the nuclear receptor RORalpha. *Circ. Res.* **90**, 820–825.

46. Delerive, P., Monte, D., Dubois, G., Trottein, F., Fruchart-Najib, J., Mariani, J., Fruchart, J. C., and Staels, B. (2001) The orphan nuclear receptor ROR alpha is a negative regulator of the inflammatory response. *EMBO Rep.* **2**, 42–48.

47. Besnard, S., Silvestre, J. S., Duriez, M., Bakouche, J., Lemaigre-Dubreuil, Y., Mariani, J., Levy, B. I., and Tedgui, A. (2001) Increased ischemia-induced angiogenesis in the sg/sg mouse, a mutant of the nuclear receptor RORalpha. *Circ. Res.* **89**, 1209–1215.

48. Besnard, S., Heymes, C., Merval, R., Rodriguez, M., Galizzi, J. P., Boutin, J. A., Mariani, J., and Tedgui, A. (2002) Expression and regulation of the nuclear receptor RORalpha in human vascular cells. *FEBS Lett.* **511**, 36–40.

49. Vu-Dac, N., Gervois, P., Grotzinger, T., De Vos, P., Schoonjans, K., Fruchart, J. C., Auwerx, J., Mariani, J., Tedgui, A., and Staels, B. (1997) Transcriptional regulation of apolipoprotein A-I gene expression by the nuclear receptor RORalpha. *J. Biol. Chem.* **272**, 22401–22404.

50. Raspe, E., Duez, H., Gervois, P., Fievet, C., Fruchart, J. C., Bensnard, S., Mariani, J., Tedgui, A., and Staels, B. (2001) Transcriptional regulation of apolipoprotein C-III gene expression by the orphan nuclear receptor RORalpha. *J. Biol. Chem.* **276**, 2865–2871.

51. Kang, H. S., Angers, M., Beak, J. Y., Wu, X., Gimble, J. M., Wada, T., Xie, W., Collins, J. B., Grissom, S. F., and Jetten, A. M. (2007) Gene expression profiling reveals a regulatory role for ROR alpha and ROR gamma in phase I and phase II metabolism. *Physiol. Genomics* **31**, 281–294.

52. Wada, T., Kang, H. S., Angers, M., Gong, H., Bhatia, S., Khadem, S., Ren, S., Ellis, E., Strom, S., Jetten, A., et al. Identification of oxysterol 7{alpha}-hydroxylase (Cyp7b1) as a novel ROR{alpha} (NR1F1) target gene and a functional crosstalk between ROR{alpha} and LXR (NR1H3). *Mol. Pharmacol.* **73**, 891–899.

53. Andre, E., Conquet, F., Steinmayr, M., Stratton, S. C., and Becker-Andre, M. (1998) Disruption of retinoid-related orphan receptor beta changes circadian behavior, causes retinal degeneration and leads to vacillans phenotype in mice. *EMBO J.* **17**, 3867–3877.

54. Srinivas, M., Ng, L., Liu, H., Jia, L., and Forrest, D. (2006) Activation of the blue opsin gene in cone photoreceptor development by retinoid-related orphan receptor beta. *Mol. Endocrinol.* **20**, 1728–1741.

55. Kurebayashi, S., Ueda, E., Sakaue, M., Dhavalkumar, D., Patel, D. D., Medvedev, A., Zhang, F., and Jetten, A. M. (2000) Retinoid-related orphan receptor gamma (RORgamma) is essential for lymphoid organogenesis and controls apoptosis during thymopoiesis. *Proc. Natl Acad. Sci. USA* **97**, 10132–10137.

56. Sun, Z., Unutmaz, D., Zou, Y. R., Sunshine, M. J., Pierani, A., Brenner-Morton, S., Mebius, R. E., and Littman, D. R. (2000) Requirement for RORgamma in thymocyte survival and lymphoid organ development. *Science* **288**, 2369–2373.

57. Ivanov, I. I., McKenzie, B. S., Zhou, L., Tadokoro, C. E., Lepelley, A., Lafaille, J. J., Cua, D. J., and Littman, D. R. (2006) The orphan nuclear receptor RORgammat directs the differentiation program of proinflammatory IL-17(+) T helper cells. *Cell* **126**, 1121–1133.

58. Yang, X. O., Pappu, B. P., Nurieva, R., Akimzhanov, A., Kang, H. S., Chung, Y., Ma, L., Shah, B., Panopoulos, A. D., Schluns, K. S., et al. (2008) T Helper 17 lineage differentiation is programmed by orphan nuclear receptors RORalpha and RORgamma. *Immunity* **28**, 29–39.

59. Schwarz, M., Lund, E. G., and Rusell, D. W. (1998) Two 7 alpha-hydroxylase enzymes in bile acid biosynthesis. *Curr. Opin. Lipidol.* **9**, 113–118.

60. Chiang, J. Y. (2004) Regulation of bile acid synthesis: pathways, nuclear receptors, and mechanisms. *J. Hepatol.* **40**, 539–551.

61. Schultz, J. R., Tu, H., Luk, A., Repa, J. J., Medina, J. C., Li, L., Schwendner, S., Wang, S., Thoolen, M., Mangelsdorf, D. J., et al. (2000) Role of LXRs in control of lipogenesis. *Genes Dev.* **14**, 2831–2838.

62. Collins, J. L., Fivush, A. M., Watson, M. A., Galardi, C. M., Lewis, M. C., Moore, L. B., Parks, D. J., Wilson, J. G., Tippin, T. K., Binz, J. G., et al. (2002) Identification of a nonsteroidal liver X receptor agonist through parallel array synthesis of tertiary amines. *J. Med. Chem.* **45**, 1963–1966.

63. Repa, J. J. and Mangelsdorf, D. J. (2002) The liver X receptor gene team: potential new players in atherosclerosis. *Nat. Med.* **8**, 1243–1248.

64. Gong, H. and Xie, W. (2004) Orphan nuclear receptors, PXR and LXR: new ligands and therapeutic potential. *Expert Opin. Ther. Targets* **8**, 49–54.

65. Millatt, L. J., Bocher, V., Fruchart, J. C., and Steals, B. (2003) Liver X receptors and the control of cholesterol homeostasis: potential therapeutic targets for the treatment of atherosclerosis. *Biochim. Biophys. Acta* **1631**, 107–118.

66. Chawla, A., Repa, J. J., Evans, R. M., and Mangelsdorf, D. J. (2001) Nuclear receptors and lipid physiology: opening the X-files. *Science* **294**, 1866–1870.

67. Kim, J. B. and Spiegelman, B. M. (1996) ADD1/SREBP1 promotes adipocyte differentiation and gene expression linked to fatty acid metabolism. *Genes Dev.* **10**, 1096–1107.

68. Shimano, H., Yahagi, N., Amemiya-Kudo, M., Hasty, A. H., Osuga, J., Tamura, Y., Shionoiri, F., Iizuka, Y., Ohashi, K., Harada, K., et al. (1999) Sterol regulatory element-binding protein-1 as a key transcription factor for nutritional induction of lipogenic enzyme genes. *J. Biol. Chem.* **274**, 35840–35844.

69. Repa, J. J., Liang, G., Ou, J., Bashmakov, Y., Lobaccaro, J. M., Shimomura, I., Shan, B., Brown, M. S., Goldstein, J. L., and Mangelsdorf, D. J. (2000) Regulation of mouse

sterol regulatory element-binding protein-1c gene (SREBP-1c) by oxysterol receptors, LXRalpha and LXRbeta. *Genes Dev.* **14**, 2819–2830.

70. Repa, J. J., Turley, S. D., Lobaccaro, J. A., Medina, J., Li, L., Lustig, K., Shan, B., Heyman, R. A., Dietschy, J. M., and Mangelsdorf, D. J. (2000) Regulation of absorption and ABC1-mediated efflux of cholesterol by RXR heterodimers. *Science* **289**, 1524–1529.

71. Joseph, S. B., Castrillo, A., Laffitte, B. A., Mangelsdorf, D. J., and Tontonoz, P. (2003) Reciprocal regulation of inflammation and lipid metabolism by liver X receptors. *Nat. Med.* **9**, 213–219.

72. Joseph, S. B., Bradley, M. N., Castrillo, A., Bruhn, K. W., Mak, P. A., Pei, L., Hongenesch, J., O'connell, R. M., Cheng, G., Miller, J. F., et al. (2004) LXR-dependent gene expression is important for macrophage survival and the innate immune response. *Cell* **119**, 299–309.

73. Song, W. C. (2001) Biochemistry and reproductive endocrinology of estrogen sulfotransferase. *Ann. N. Y. Acad. Sci.* **948**, 43–50.

74. Gnerre, C., Schuster, G. U., Roth, A., Handschin, C., Johansson, L., Looser, R., Parini, P., Podvinec, M., Robertsson, K., Gustafsson, J. A., et al. (2005) LXR deficiency and cholesterol feeding affect the expression and phenobarbital-mediated induction of cytochromes P450 in mouse liver. *J. Lipid Res.* **46**, 1633–1642.

75. Zhou, J., Febbraio, M., Wada, T., Zhai, Y., Kuruba, R., He, J., Lee, J. H., Khadem, S., Ren, S., Li, S., et al. (2008) Hepatic fatty acid transporter Cd36 is a common target of LXR, PXR and PPARγ in promoting steatosis. *Gastroenterology* **134**, 556–567.

76. Willy, P. J., Umesono, K., Ong, E. S., Evans, R. M., Heyman, R. A., and Mangelsdorf, D. J. (1995) LXR, a nuclear receptor that defines a distinct retinoid response pathway. *Genes Dev.* **9**, 1033–1045.

77. Delerive, P., Chin, W. W., and Suen, C. S. (2002) Identification of reverb (alpha) as a novel ROR(alpha) target gene. *J. Biol. Chem.* **277**, 35013–35018.

78. Kurebayashi, S., Nakajima, T., Kim, S. C., Chang, C. Y., McDonnell, D. P., Renaud, J. P., and Jetten, A. M. (2004) Selective LXXLL peptides antagonize transcriptional activation by the retinoid-related orphan receptor RORgamma. *Biochem. Biophys. Res. Commun.* **315**, 919–927.

79. Zhang, Y., Repa, J. J., Gauthier, K., and Mangelsdorf, D. J. (2001) Regulation of lipoprotein lipase by the oxysterol receptors, LXRalpha and LXRbeta. *J. Biol. Chem.* **276**, 43018–43024.

80. Volle, D. H., Repa, J. J., Mazur, A., Cummins, C. L., Val, P., Henry-Berger, J., Caira, F., Veyssiere, G., Mangelsdorf, D. J., and Lobaccaro, J. M. (2004) Regulation of the aldoketo reductase gene akr1b7 by the nuclear oxysterol receptor LXRalpha (liver X receptoralpha) in the mouse intestine: putative role of LXRs in lipid detoxification processes. *Mol. Endocrinol.* **1** (8), 888–898.

81. Malerod, L., Juvet, L. K., Hanssen-Bauer, A., Eskild, W., and Berg, T. (2002) Oxysterolactivated LXRalpha/RXR induces hSR-BI-promoter activity in hepatoma cells and preadipocytes. *Biochem. Biophys. Res. Commun.* **299**, 916–923.

82. Sato, T. K., Panda, S., Miraglia, L. J., Reyes, T. M., Rudic, R. D., McNamara, P., Naik, K. A., FitzGerald, G. A., Kay, S. A., and Hogenesch, J. B. (2004) A functional genomics strategy reveals Rora as a component of the mammalian circadian clock. *Neuron* **43**, 527–537.

83. Schrader, M., Danielsson, C., Wiesenberg, I., and Carlberg, C. (1996) Identification of natural monomeric response elements of the nuclear receptor RZR/ROR. They also bind COUP-TF homodimers. *J. Biol. Chem.* **271**, 19732–19736.

84. Zhou, J., Zhai, Y., Mu, Y., Gong, H., Uppal, H., Toma, D., Ren, S., Evans, R. M., and Xie, W. (2006) A novel pregnane X receptor-mediated and sterol regulatory element-binding protein-independent lipogenic pathway. *J. Biol. Chem.* **281**, 15013–15020.

85. Li-Hawkins, J., Lund, E. G., Turley, S. D., and Russell, D. W. (2000) Disruption of the oxysterol 7alpha-hydroxylase gene in mice. *J. Biol. Chem.* **275**, 16536–16542.

INDEX

Nuclear Receptors in Drug matabolism Edited by Wen Xie
Copyright © 2009 John Wiley & Sons, Inc.